Proceedings of the 2002 International Symposium on

Cosmology and Particle Astrophysics

CosPA 2002

Proceedings of the 2002 International Symposium on

Cosmology and Particle Astrophysics

CosPA 2002

X-G He
National Taiwan University

K-W Ng
Academia Sinica, Taiwan

World Scientific
New Jersey • London • Singapore • Hong Kong

Published by

World Scientific Publishing Co Pte Ltd
5 Toh Tuck Link, Singapore 596224
USA office: 27 Warren Street, Suite 401-402, Hackensack, NJ 07601
UK office: 57 Shelton Street, Covent Garden, London WC2H 9HE

British Library Cataloguing-in-Publication Data
A catalogue record for this book is available from the British Library.

PROCEEDINGS OF THE 2002 INTERNATIONAL SYMPOSIUM ON COSMOLOGY AND PARTICLE ASTROPHYSICS (CosPA 2002)

ISBN-13 978-981-238-284-9
ISBN-10 981-238-284-4

Preface

As we venture into the 21st Century, it just so happens that the human race is becoming capable of making use of their hi-tech's to explore the edge of our Universe or, equivalently, what had happened much early on at the very early universe, shortly after the Big Bang. This has stiffened the competitions among astronomers and particle physicists in their vigorous pursuits for the true theory of cosmology, being un-imaginable even a decade ago. Here in Taiwan and especially at the Center for Cosmology and Particle Astrophysics (CosPA Center), we wish to actively join the crusade of the scientists worldwide in this pursuit of the observation-based cosmology, through the so-called Taiwan CosPA Project, funded by the Research Excellence Initiative of the Ministry of Education and the National Science Council in Taiwan. The 2002 International Symposium on Cosmology and Particle Astrophysics (CosPA2002), held from May 31 to June 2, 2002 in Taipei, Taiwan, was part of this effort. It was organized by the CosPA Center and sponsored by the Ministry of Education, the National Science Council and the National Taiwan University.

The CosPA2002 symposium was intended to bring together scientists to engage in serious discussions on cosmology and particle astrophysics. The topics covered during the symposium include among others, the following ones:

(1) CMB Physics: SZ surveys, polarization, large-scale structures, gravitational lensing, and data analysis.
(2) Dark Energy and Dark Matter: dark matter physics, quintessence and the cosmological constant.
(3) Cosmology of Ultra High Energy Cosmic Rays.
(4) Inflation and New Physics: inflation, noncommutative geometry, branes, and extra dimensions.

It is a great pleasure to thank the members of the International Advisory and the Local Organizing Committees for their efforts in preparing the symposium. We especially Thank the CosPA Center Director, Professor W-Y. Pauchy Hwang, for his enthusiasm and full support. We thank the conference secretary Vicky Chen, and the secretary team, Chih-Hsin Huang, Linda Shao, YuanRu Ho, Huei-Ming Yao, Maggie Wang, and Ada Lin, for their excellent work; Dr. Je-An Gu for his work as scientific secretary; and Je-An Gu and YuanRu Ho for their tremendous help in finalizing this proceedings. Finally we thank all the speakers, session chairs and participants for coming to CosPA2002 and making the symposium a success.

Xiao-Gang He
Kin-Wang Ng

International Advisory Committee

John D. Barrow	(Cambridge)
Francois Bouchet	(Paris)
Tzi-Hong Chiueh	(NTU)
John R. Ellis	(CERN)
Ernest M. Henley	(Seattle)
W-Y. Pauchy Hwang	(Chair, NTU)
Andrew H. Jaffe	(Imperial)
Chung Wook Kim	(KIAS)
Andrew R. Liddle	(Sussex)
Bruce H. J. McKellar	(Melbourne)
Sandip Pakvasa	(Hawaii)
Katsuhko Sato	(Tokyo)
Joe Silk	(Oxford)
Frank H. Shu	(NTHU)
George Smoot	(Berkeley)

Local Organizing Committee

Vicky Chen	(Administrative Secretary)
Je-An Gu	(Scientific Secretary, NTU)
Xiao-Gang He	(Co-Chair, NTU)
Pei-Ming Ho	(NTU)
Win-Fun Kao	(NCTU)
Guey-Lin Lin	(NCTU)
Kin-Wang Ng	(Co-Chair, AS)
J.-H. Proty Wu	(NTU)

CONTENTS

Cosmology: An Experimental Science for the New Century

W-Y. Pauchy Hwang

COSMOLOGY: AN EXPERIMENTAL SCIENCE FOR THE NEW CENTURY

W-Y. PAUCHY HWANG
Center for Academic Excellence on Cosmology and Particle Astrophysics
Department of Physics, National Taiwan University
Taipei, Taiwan 106, R.O.C.
E-mail: wyhwang@phys.ntu.edu.tw

In this opening talk, I wish to stress that, at the turn of the century, cosmology has turned itself into an experimental science - the science in which key predictions of basic theoretical ideas can now be tested in details against the observational data. The 11 science questions for the new century, as addressed in the report of the Committee on the Physics of the Universe (CPU), National Research Council of U.S.A., summarizes succinctly possible major directions of the future research in physics and astronomy. Taiwan is joining this red-hot competition by carrying out the project in search of academic excellence on "cosmology and particle astrophysics", dubbed as "Taiwan CosPA Project", which has the science goals as addressed in the CPU report.

1. Introduction

The subject of "cosmology", stemming from the urge for understanding the astro/cosmo environment around us, has fascinated people in all different walks of life, "scientists" or philosophers of all ages, in different countries and in all ancient civilizations. As shall be explained in this opening talk, however, there has never been a moment like today as we venture into the 21st century, the moment when "cosmology" itself is turning into an experimental science, such that many basic aspects in relation to cosmology can be tested against the observational data. Owing to the great leaps in the science and technology of the last century, phenomena which took place near the edge of the Universe and carry important clues regarding the origin of our universe can now be observed with precision, thereby helping transform cosmology into an experimental science. As science is always based upon, or defined by, both the theoretical and experimental aspects, the moment that cosmology is turning into a real science will be remembered as a unique historic moment in the civilization of the human kind. Our last experience of inching toward the "truth" could well be identified as the discovery of the smallest world in late 1960's and its eventual spell as the "standard model of particle physics", but I would be tempting to place the present historic moment regarding cosmology as above all subject sciences to our knowledge.

It would not be easy to associate the present historic moment with some specific events, as this might not be fair to some of the most important developments. Most recently, however, I would be tempted to point to the 1992 COBE/DMR discovery[1] of the anisotropy in the cosmic microwave background radiation (CMBR) as well as to the 1999 discovery[2] of the accelerating expansion of our universe. In the time scale of centuries or much longer, the present historic moment may nevertheless be crowded, or overshadowed, by Einstein's invention of general relativity, Hubble's discovery of the expansion of our universe, Gamow's big bang conjecture, and the Penzias-Wilson discovery of CMBR. Most of us the human being will live through the critical historic moment without even recognizing it, but it would be a pity if some of the most talented young scientists would do so.

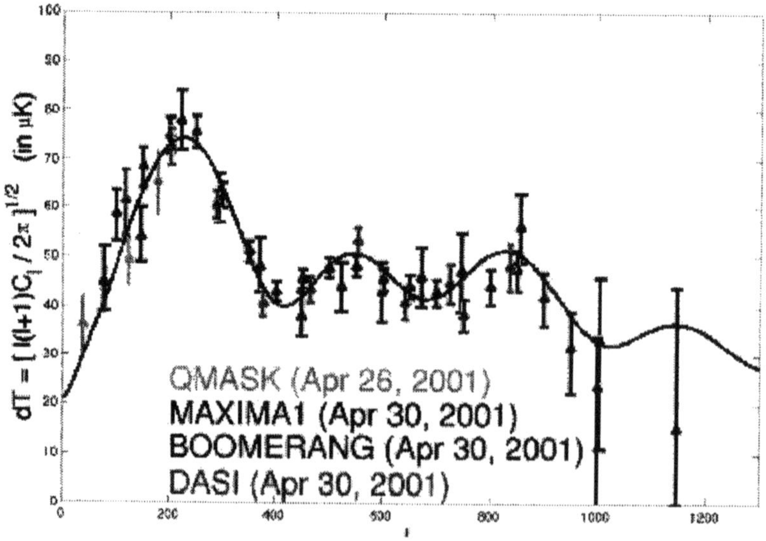

Figure 1. An update of the CMBR data observed up to 30 April 2001.

Indeed, the 1992 discovery[1] of fluctuations or anisotropies, at the level of 10^{-5}, associated with the CMBR has helped transformed the physics of the early universe into a main-stream research area in astronomy and in particle astrophysics, both theoretically and observationally.[3] Dozens of efforts to observe CMB anisotropies and polarizations are now underway and, starting from the Spring of 2000, Boomerang released their first result showing clearly the position of the first peak of the CMBR spectrum near $\ell \approx 200$ corresponding to the flat universe or the overall density being critical. Such discovery gets into the news media easily, often generating unscientific speculations. A sample summary of the CMBR observational data (from Boomerang, MAXIMA, QMASK, and DASI)up to the Spring of 2001 is illustrated in Figure 1 and it is fair to say that the attention which has been generated in the news media and the general public completely overwhelms all the physicists and astronomers. [Another update on the CMBR data just arrived a week ago and would be included in the other talks in the proceedings.]

2. A Bit on the Quantitative Side

A prevailing view regarding our universe is that it originates from the joint making of Einstein's general relativity and the cosmological principle while the observed anisotropies associated with the cosmic microwave background (CMBR), at the level of about one part in 100,000, may stem from quantum fluctuations in the inflation era. In what follows, we wish to first outline very briefly a few key points in the standard scenario of cosmology, a framework which we may employ to address different questions in our quest for understanding the origin of our Universe.

Based upon the cosmological principle which state that our universe is homogeneous and isotropic, we use the Robertson-Walker metric to describe our universe.[4]

$$ds^2 = dt^2 - R^2(t)\{\frac{dr^2}{1-kr^2} + r^2 d\theta^2 + r^2 sin^2\theta d\phi^2\}. \tag{1}$$

Here the parameter k describes the spatial curvature with $k = +1, -1$, and 0 referring to an open, closed, and flat universe, respectively. The scale factor $R(t)$ describes the size of the universe at time t.

To a reasonable first approximation, the universe can be described by a perfect fluid, i.e., a fluid with the energy-momentum tensor $T^\mu{}_\nu = diag\,(\rho, \;, -p, -p, -p)$ where ρ is the energy density and p the pressure. Thus, the Einstein equation, $G^\mu{}_\nu = 8\pi G_N T^\mu{}_\nu + \Lambda g^\mu{}_\nu$, gives rise to only two independent equations, i.e., from $(\mu, \nu) = (0, 0)$ and (i, i) components,

$$\frac{\dot{R}^2}{R^2} + \frac{k}{R^2} = \frac{8\pi G_N}{3}\rho + \frac{\Lambda}{3}. \tag{2}$$

$$2\frac{\ddot{R}}{R}+\frac{\dot{R}^2}{R^2}+\frac{k}{R^2}=-8\pi G_N\,p+\Lambda. \tag{3}$$

Combining with the equation of state (EOS), i.e. the relation between the pressure p and the energy density ρ, we can solve the three functions $R(t)$, $\rho(t)$, and $p(t)$ from the three equations. Further, the above two equations yields

$$\frac{\ddot{R}}{R}=-\frac{4\pi G_N}{3}(\rho+3p)+\frac{\Lambda}{3}, \tag{4}$$

showing either that there is a positive cosmological constant or that $\rho+3p$ must be somehow negative, if the major conclusion of the Supernova Cosmology Project is correct [2], i.e. the expansion of our universe still accelerating ($\frac{\ddot{R}}{R}>0$).

Assuming a simple equation of state, $p=w\rho$, we obtain, from Eqs. (2) and (3),

$$2\frac{\ddot{R}}{R}+(1+3w)(\frac{\dot{R}^2}{R^2}+\frac{k}{R^2})-(1+w)\Lambda=0, \tag{5}$$

so that, with $p=-\rho$ and $k=0$, we find

$$\ddot{R}-\frac{\dot{R}^2}{R}=0, \tag{6}$$

which has an exponentially growing, or decaying, solution $R\propto e^{\pm\alpha t}$, compatible with the so-called "inflation" or "big inflation". In fact, considering the simplest case of a real scalar field $\phi(t)$, we have

$$\rho=\frac{1}{2}\dot{\phi}^2+V(\phi), \qquad p=\frac{1}{2}\dot{\phi}^2-V(\phi), \tag{7}$$

so that, when the "kinetic" term $\frac{1}{2}\dot{\phi}^2$ is negligible, we have an equation of state, $p\sim-\rho$. In addition to its possible role as the "inflaton" responsible for inflation, such field has also been invoked to explain the accelerating expansion of the present universe, as dubbed as "quintessence" [5]. In the case of "quintessence", the kinetic term is not required to be negligible compared to the potential.

Another simple consequence of the homogeneous model is to derive the continuity equation from Eqs. (2) and (3):

$$d(\rho R^3)+pd(R^3)=0. \tag{8}$$

Accordingly, we have $\rho\propto R^{-4}$ for a radiation-dominated universe ($p=\rho/3$) while $\rho\propto R^{-3}$ for a matter-dominated universe ($p<<\rho$). The present universe has a matter content of about $3\times10^{-30}g/cm^3$ (including the majority in dark matter), much bigger than its radiation content $5\times10^{-35}g/cm^3$, as estimated from the 3° black-body radiation. However, as $t \to 0$, we anticipate $R \to 0$, extrapolated back to a very small universe as compared to the present one. Therefore, the universe is necessarily dominated by the radiation during its early enough epochs.

For the radiation-dominated early epochs of the universe with $k = 0$ and $\Lambda = 0$ (for the sake of simple arguments), we could deduce, also from Eqs. (2) and (3),

$$\rho = \frac{3}{32\pi G_N} t^{-2}, \qquad T = \{\frac{3c^2}{32\pi G_N a}\}^{\frac{1}{4}} t^{-\frac{1}{2}} \cong 10^{10} t^{-1/2} ({}^\circ K). \qquad (9)$$

These equations tell us a few important times in the early universe, such as $10^{-11} sec$ when the temperature T is around 300 GeV during which the electroweak (EW) phase transition is expected to occur, or somewhere between $10^{-5} sec$ ($\cong$ 300 MeV) and $10^{-4} sec$ ($\cong$ 100 MeV) during which quarks and gluons undergo the QCD confinement phase transition.

So, what have we learned up to now? The CMBR data shown in Figure 1 demonstrate clearly that the first peak of the power spectrum occurs at $l \approx 200$, indicating that our universe is flat and that the present energy density is of the critical value. The occurrence of the second and further peaks suggest that inhomogeneities seen with the COBE/DMR data may have little to do with topological defects possibly produced in the early universe. On the other hand, the supernova type-Ia measurements, when combined with the current CMBR, suggest that the present universe has the dominant component in the form of "dark energy", either a positive cosmological constant or quintessence, at the level of about 65 % of the total energy density. The various high$-z$ surveys in astronomy have also pointed to a baryon content of about 5%, leaving the remaining 30 % in the mysterious form of "dark matter".

3. Connecting Quarks with the Cosmos

The current status regarding our understanding of the Universe has been summarized very well in a report just released by the Committee on the Physics of the Universe (CPU), Board on Physics and Astronomy (BPA), National Research Council (NRC) of U.S.A. The report summarizes the opportunities in the joint disciplines of physics and astronomy into "Eleven Science Questions for the New Century" while making seven key recommendations on how to realize these opportunities.

3.1. *Eleven Science Questions for the New Century*

The "Eleven Science Questions for the New Century", as posed in the CPU/BPA/NRC report, are listed below, without detailed qualifying statements accompanying each question in the original report.

- **Question No. 1:** What is the dark matter?
- **Question No. 2:** What is the nature of the dark energy?
- **Question No. 3:** How did the universe begin?
- **Question No. 4:** Did Einstein have the last word on gravity?
- **Question No. 5:** What are the masses of the neutrinos, and how have they shaped the evolution of the universe?
- **Question No. 6:** How do cosmic accelerators work and what are they accelerating?
- **Question No. 7:** Are protons unstable?
- **Question No. 8:** Are there new states of matter at exceedingly high density and temperature?
- **Question No. 9:** Are there additional spacetime dimensions?
- **Question No. 10:** How were the elements from iron to uranium made?
- **Question No. 11:** Is a new theory of matter and light needed at the highest energies?

It is essential to stress that these questions are "experimental" in nature. That is, we must seek the answers to these questions through some suitable experimental means and/or innovative observational projects. One should not anticipate that such answers could come in any way "cheap" and "easy".

3.2. *Seven Recommendations*

In order to answer the 11 science questions for the new century and to realize the opportunities offered by the experimental or observational efforts to bring out the answers, the Committee has made seven specific recommendations which are stated below:

- **Recommendation No. 1:** Measure the polarization of the cosmic microwave background with the goal of detecting the signature of inflation. The Committee recommends that NASA, NSF, and DOE undertake research and development to bring the needed experiments to fruition.
- **Recommendation No. 2:** Determine the properties of the dark energy. The Committee supports the Large Synoptic Survey Telescope project, which has significant promise for shedding light on

the dark energy. The Committee further recommends that NASA and DOE work together to construct a wide-field telescope in space to determine the expansion history of the universe and fully probe the nature of the dark energy.

- **Recommendation No. 3:** Determine the neutrino masses, the constituents of the dark matter and the lifetime of the proton. The Committee recommends that DOE and NSF work together to plan for and to fund a new generation of experiments to achieve these goals. We further recommend that an underground laboratory with sufficient infrastructure and depth be built to house and operate the needed experiments.
- **Recommendation No. 4:** Use space to probe the basic laws of physics. The Committee supports the Constellation-X and Laser Interferometer Space Antenna missions, which have high promise for studying black holes and for testing Einstein's theory in new regimes. The Committee further recommends that the agencies proceed with an advanced technology program to develop instruments capable of detecting gravitational waves from the early universe.
- **Recommendation No. 5:** Determine the origin of the highest energy gamma rays, neutrinos and cosmic rays. The Committee supports the broad approach already in place, and recommends that the United States ensure the timely completion and operation of the Southern Auger array.
- **Recommendation No. 6:** Discern the physical principles that govern extreme astrophysical environments through the laboratory study of high-energy-density physics. The Committee recommends that the agencies cooperate in bringing together the different scientific communities that can foster this rapidly developing field.
- **Recommendation No. 7:** Realize the scientific opportunities at the intersection of physics and astronomy. The Committee recommends establishment of an Interagency Initiative on the Physics of the Universe, with the participation of DOE, NASA, and NSF. This initiative should provide structures for joint planning and mechanisms for joint implementation of cross-agency projects.

4. Taiwan CosPA Project

The Project in Search of Academic Excellence on "Cosmology and Particle Astrophysics (CosPA)", a multi-institutional research project funded for a period of four years beginning from January 2000 by the Ministry of Education of R.O.C. (Taiwan) and dubbed as "Taiwan CosPA Project", consists of five subprojects and an overseeing project and aims at building up Taiwan's astronomy through research efforts in the hotly-contested areas of cosmology and particle astrophysics. The five subprojects and their missions or science goals are described very briefly as follows:

- **Subproject No. 1:** Array for Microwave Background Radiation (AMiBA): From Construction and Operation to Data Acquisition and Analysis (P.I.: K. Y. Lo). [As Fred K.Y. Lo has assumed the directorship of the National Radio Astronomy Observatory of the U.S.A. starting from 1 September 2002, the role of the principal investigator has been assumed by Paul Ho.]
- **Subproject No. 2:** Experimental Particle Physics Studies on Issues related to "Early Universe, Dark Matter, and Inflation" (P.I.: W. S. Hou).
- **Subproject No. 3:** Theoretical Studies of Cosmology and Particle Astrophysics (P.I.: W-Y. Pauchy Hwang).
- **Subproject No. 4:** Frontier Observation in Optical and Infrared Astronomy (P.I.: Typhoon Lee).
- **Subproject No. 5:** National Infrastructure (P.I.: Wing Ip).

The total budget of the project is at the level of about 15 million U.S. dollars over 4 Years. What do we attempt to accomplish through these projects?

On Subproject No. 1 for radio astronomy, we wish to (1) build the AMiBA telescope, (2) attempt to measure the CMB polarizations, and (3) conduct SZ survey of high-z clusters.

On Subproject No. 2 for particle astrophysics, the experimental high energy physics team based at National Taiwan University is the basic manpower infrastructure to carry out the experiments related to dark matter, the early universe, and/or inflation. The group has been an active player in the KEK/Belle Collaboration on the B & CP studies - they were in the news of major discovery, both in July 2001 and in February 2002. During the first year of this project, the team has completed the feasibility study of a dark matter search. Lately, their interests have switched to neutrino astrophysics and now is attempting to build a prototype neutrino telescope for the detection of very high energy cosmic neutrinos. In addition, the group attempts to phase-in a meaningful participation of the GLAST project (NASA / DOE) through the Research University Initiative (R.U.I.).

On Subproject No. 3 for theoretical studies of cosmology and particle astrophysics, the science goal is to make significant progresses, hopefully some breakthroughs, in the prime area of cosmology, i.e. the physics of the early universe. Subjects under intensive studies include CMB polarization and anisotropy, dark energy and the accelerating universe, noncommutative spacetime and cosmology, roles of phase transitions in the early universe, and physics of ultra high energy cosmic rays (UHECR's).

On Subproject No. 4 for optical and infrared astronomy, an agreement between National Taiwan University and Canada-French-Hawaii Telescope (CFHT) Corporation was ironed out in July 2001 in order that the CosPA team will participate the construction efforts of the Wide-Field Infrared Camera (WIRCAM) and that the CosPA team will have 68 nights to use CFHT/OIR to conduct the large-scale-structure (LSS) survey to complement the SZ survey of the AMiBA.

On Subproject No. 5 for national infrastructure, we are trying to make the links between education and research so that there will be adequate young manpower to sustain the growth of the astronomy as a field. Through this sub-project, we establish, on top of the Lu-lin Mountain an observatory which house small research telescopes such as one of the TAOS survey network telescopes.

It is of some interest to note that CosPA's overall scientific objectives sit amazingly well on top of the CPU/BPA/NRC report. However, there are currently many other projects around the world, with a scale similar to our CosPA project, such as Center for Cosmology at University of Chicago, several distinguished projects going on at California Institute of Technology, Chen Institute for Particle Astrophysics and Cosmology at Stanford University, and Research Center for the Early Universe (RESCEU) at University of Tokyo. Cosmology has indeed become a hotly-contested area of forefront research in physics and astronomy.

5. Prospects

At the turn of the century, cosmology is transforming itself into an experimental science. It has become the main-stream research in astronomy, as well as in particle astrophysics. It may take 20 or 30 years before the competition is over.

Taiwan is joining this red-hot race through the Taiwan CosPA Project. The project, if successful, should help us to build a world-class, research-based, respectable astronomy in Taiwan.

Acknowledgement

This Conference, together with the present work, is supported primarily through the Taiwan CosPA Project, as funded by the Ministry of Education (89-N-FA01-1-0 up to 89-N-FA01-1-5). The present work is also supported in part by National Science Council of R.O.C. (NSC 90-2112-M002-028 & NSC 91-2112-M002-041).

References

1. G. Smoot *et al.*, *Astrophys. J.* **396**, L1 (1992); C. Bennett *et al.*, *Astrophys. J.* **396**, L7 (1992); E. Wright *et al.*, *Astrophys. J.* **396**, L11 (1992).
2. S. Perlmutter *et al.* [Supernova Cosmology Project], *Astrophys. J.* **517**, 565 (1999); A. G. Riess *et al.* [Supernova Search Team], *Astron. J.* **116**, 1009 (1998).
3. C. L. Bennett, M. S. Turner, and M. White, *Physics Today*, November 1997, p. 32, for an early general review.
4. E.W. Kolb and M.S. Turner, *The Early universe*, *Addison-Wesley Publishing Co.* (1994).
5. R.R. Caldwell, R. Dave, and P.J. Steinhardt, *Phys. Rev. Lett.* **80**, 1582 (1998).

Cosmic Microwave Fluctuations, Present and Future

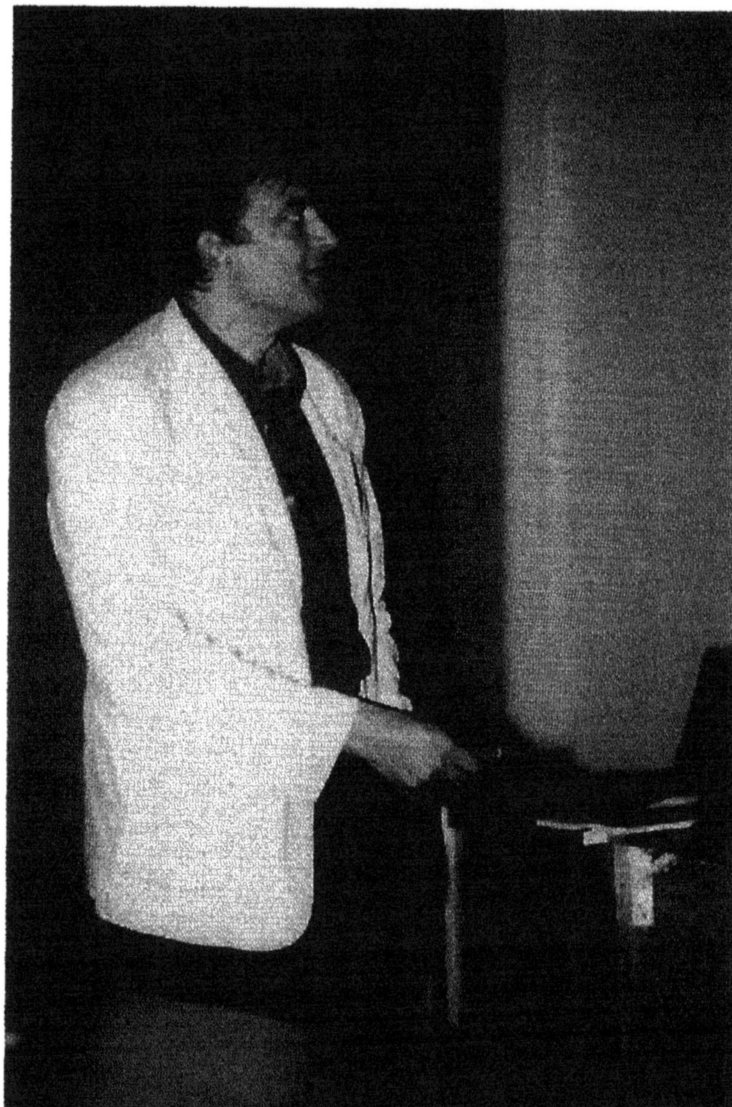

Francois Bouchet

COSMIC MICROWAVE FLUCTUATIONS, PRESENT AND FUTURE

F. R. BOUCHET
Institut d'Astrophysique de Paris, CNRS,
98 bis Boulevard Arago, F-75014, Paris, France
E-mail: bouchet@iap.fr

& THE ARCHEOPS COLLABORATION (FOR §2)
http://www.archeops.org

The knowledge of the characteristics of the cosmic Microwave Background (CMB) anisotropies has been fast and steadily increasing over the last few years. After a brief historical overview, I show that a consistent picture has emerged, from the analysis of CMB data alone, from the analysis of Large Scale Structures (LSS) data, and from their combination. I then review the status of the ARCHEOPS balloon borne experiment whose first results should be available within the same time frame than those of MAP. I will then present what to expect from the two satellite projects in this area, the MAP experiment which is currently operating, and the Planck experiment which should be launched mid-2007.

1. Introduction & CMB experiments status

In the standard cosmological model, the origin of structures in the universe is taking roots in the early universe. Inflation is a mechanism which can bring quantum fluctuations to macroscopic scales. Phase transitions in the early universe can also generate topological defects that can contribute to structure formation. Nevertheless, the present CMB anisotropies data do not allow these to be the dominant source for structure formation (but hybrid models are only weakly constrained by the data).

Whatever is the source of these primordial fluctuations, one expects that their statistical properties can be described with a handful of numbers like the amplitude and the logarithmic slope of the density power spectrum. For a given set of cosmological parameters the linear development of perturbations can be fully computed for the different components: radiation, neutrinos, baryons, cold dark matter...

At z= 1000 the universe becomes transparent and the radiation propagates with little interactions with the matter, which can freely collapse under it's own gravity. The matter fluctuations at that time can then be seen as the initial conditions for the gravitational formation of Large Scale Structures, while their imprint on the last scattering surface leads to the CMB anisotropies we detect today. The matter and radiation statistical properties at recombination should thus only depend of the few numbers describing the characteristics i) of the primordial fluctuations and ii) of the Universe which affect their development.

The large scale distribution of galaxies which we measure today reflects the properties of the matter fluctuations at recombination, but one has to further take into account that the evolution since then has been non-linear on most of the (relatively small) scales probed by galaxy catalogs, that this evolution depends on the Cosmological parameters, and that galaxies may be biased tracers of the matter distribution. The connection with the few numbers that cosmologists seek to determine is thus less direct than with CMB anisotropy observations.

Indeed the essentially free streaming of photons till recombination implies a straight relation between the angular power spectrum of radiation today and the recombination properties, and thus a safer access to these numbers. Still, it is a crucial verification of the global paradigm that both approaches lead to consistent answers. Furthermore, the effect of different parameters can be partially degenerate, and using different probes may help lifting these degeneracies. Additionally, one should note that CMB intensity and polarized data should be correlated in a specific fashion. Polarization measurements of the CMB will thus also check the consistency of the paradigm and help removing degeneracies.

1.1. *CMB observational requirements*

The observational requirements for CMB measurements are closely related to the nature of the physical processes that we want to unveil:

(1) We want to measure quite tiny relative temperature fluctuations (the needed $\Delta T/T$ accuracy is close to 10^{-6}). As a consequence, we aim at getting maps of the sky which are photon noise limited at the frequencies where the CMB dominates (100 to 200 GHz).
(2) We are interested in the statistics of these fluctuations, the simplest of which being, the (angular) Power Spectrum, $C(\ell)$. If the observed fraction of the sky, f_{sky}, is too small, the statistics of the observation will not be well determined (ΔC is approximatively $\propto f_{sky}^{-1/2}$). In order to retrieve precise physics from the measurement, they must

include a large number of independent samples of the fluctuations, i.e. most of the sky.

(3) The information we need is mostly contained in the first 3000 ℓ-modes of the decomposition into spherical harmonics.This corresponds to angular resolutions of about 4 arc minutes. Higher angular frequencies should be completely dominated by foreground point sources (& secondary effects).

(4) At the same time, strong sources (such as the Galaxy) far from the optical axis will contaminate the measurement. Very low optical side lobes are needed up to large angular distance from the main beam to keep the level of contamination comparable to or smaller than the instrumental noise. Such very low side lobes (rejection of 10^{12} or better at large angles for frequencies larger than 300 GHz) cannot be fully measured on the ground and upper limits or measurements must be extracted from the data themselves with the required accuracy.

(5) Unidentified sources of noise can be confused with the cosmological signal. The best way to fight these noises is to get data with a high level of redundancy. This high level of redundancy is also used to identify any systematic effect including the side lobes signal.

Requirements 1 to 3 are difficult to meet altogether. Small beams and large sky coverage imply a large number of pixels (4×10^7 measurements for the full sky with 5 arc minute beams sampled $\sim$ 2.5 times per beam). For a given duration of the observation, this limits the possible integration time per pixel, and thus the map sensitivity. In the same time, the RMS value of the expected signal falls rapidly at high ℓ numbers.

It can be shown that for a given instrument and a given duration of the observation, there are an optimal sky coverage and an optimal angular resolution.[6] They correspond roughly to the case when the contributions to the RMS which come from the signal and the detector noise are about equal at the scale of the elementary map pixel.[a] A higher signal to noise, while not statistically optimal, might nevertheless help with the identification and removal of weak systematic effects.

Requirement 5 implies a need for redundancy at many different time scales, which means that each measurement has to be made many times, in as varied conditions as possible.

[a]This is indeed the case for the 5 arc minute channel at 217 GHz of PLANCK-HFI(signal and noise C_ℓ cross at $\ell \sim 2000$).

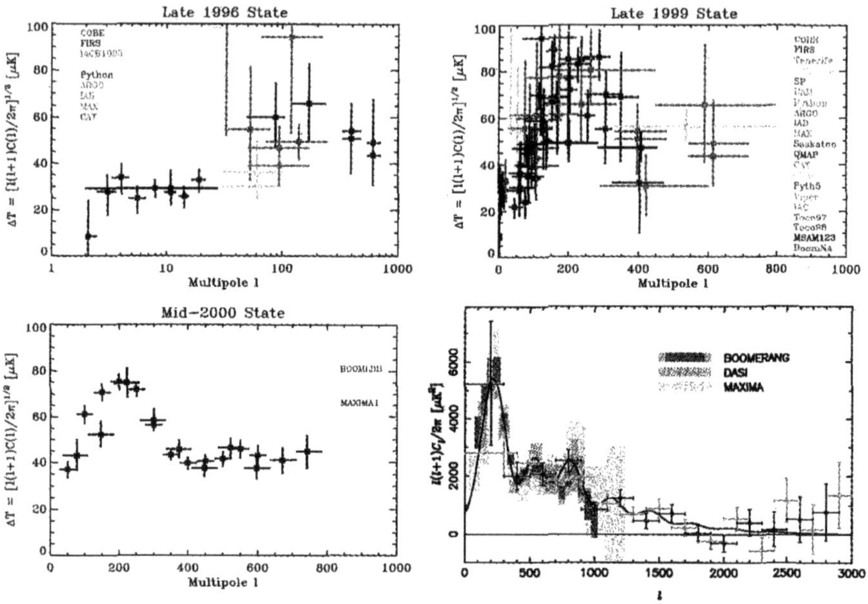

Figure 1. Measurements of the power spectrum. Top row: the first plot shows all published detection at the end of 1996, while the second plot is an update at the end of 1999. Bottom row: the left plot shows the results published in may 2000 by the BOOMERanG and MAXIMA teams (with each curve moved by + or - 1 σ of their respective calibration, see Ref. 3, and the right plot (reprinted from Sievers et al.[5] compares the CBI results to previous measurements.

In the following, we shall see how past, present, and future CMB experiments have come close to the ideal set forth above.

1.2. *Present status of observations*

CMB anisotropy experiments can be classified in three generations. The DMR experiment on COBE got the first detection of the CMB anisotropies with a 7 deg beam and a signal to noise per pixel around 1.

The second generation experiments are ongoing or soon to be operated. They have an angular resolution $\sim$ 10 arc minutes. The sensitivity is limited by uncooled detectors for most of them and by ambient temperature telescopes for the most recent balloon borne bolometer experiments: MAXIMA, BOOMERANG, ARCHEOPS. These experiments are well within a factor of 2 of the photon noise of their ambient temperature telescope.

The third generation will be the PLANCK mission which will have a low

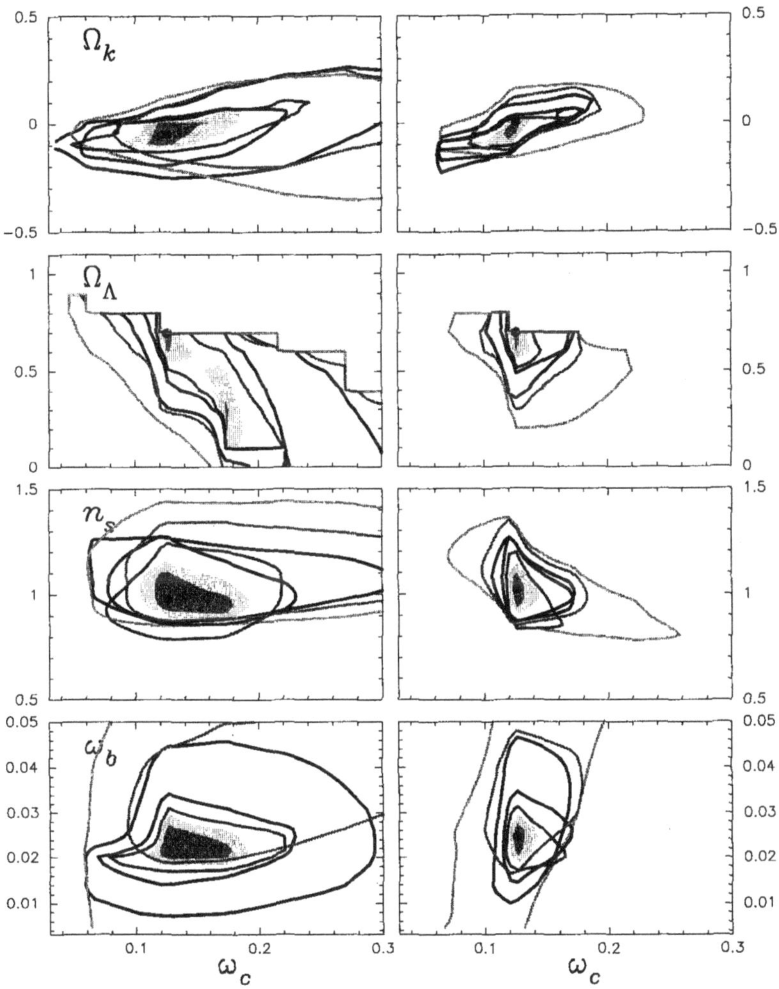

Figure 2. Two-σ contour plots for some cosmological parameters, obtained by combining DMR with various combination of experiments (with respectively black, purple, dark blue, green and turquoise corresponding to CBI, BOOMERANG, DASI, MAXIMA, and a combination of Boomerang North America + TOCO & all 17 others prior to April 1999). The filled contours correspond to all experiments taken together at the one and two σ level. All panels were made for the "weak-h" prior (i.e. $0.45 < h < 0.9$ & $t_0 >$ 10 GYr & $\Omega_m > 0.1$), while the n_s & Ω_b panels add Ω_k=0. Note that the hatched regions were not searched. Finally, the left panels additionally include a constraint on σ_8 and Γ_{eff} which comes form Large Scale Structures studies. These plots reprinted from Sievers et al.[5] illustrate the great consistency achieved between various probes, and the current level of accuracy.

background provided by its 50 K passively cooled telescope large enough to provide 5 arc minute resolution and 0.1 K bolometers which will be photon noise limited.

Figure 1 summarizes progresses made till now. COBE/DMR revealed the normalization of the spectrum and the consistency of the low-ℓ part of the spectrum with a near scale invariance of the primordial fluctuations. Till the end of 1996 we had only a hint of the presence of the expected first peak. Three years later, many more experiments suggested the shape and position of that peak (fig.1.b), although it was not clear whether individual experiments (unaccounted for) systematics, and cross-experiments inter-calibration errors could or not change the peak characteristics.

In mid 2000, the Boomerang[2,4] and Maxima collaborations[3] each announced independently and at about the same time a full mapping of the first peak and some constraints on the second peak (fig.1.c), both using cooled spider-web bolometers in total power mode. Each could by itself tell that the Universe is spatially flat under the same priors (adiabatic initial conditions). Further results since then remained in agreement, and in particular those announced the week prior the meeting by two interferometric measurements, VSA and CBI. CBI has further mapped the smaller scales and confirmed the exponential damping of the fluctuations at small scale due to diffusion in the imperfect fluid during recombination (fig.1.d).

Figure 2 shows the constraints on some combination of parameters which can be derived from some or all these experiments, in conjunction with further constraints derived form LSS studies. It is quite comforting to see that different datasets lead to the same picture and that the preferred values (points, derived from nucleosynthesis argument, Supernovae analysis, and generic inflation prediction) lie well inside the tightest contours. The analysis by the VSA team also concludes to a quantitative agreement between it's determination of the cosmological parameters and the others (for similar priors).

2. The ARCHEOPS experiment

The balloon borne ARCHEOPS experiment uses cooled bolometers as the Boomerang and Maxima ones. The ARCHEOPS consortium is headed by Alain Benoit from CRTBT, and is constituted of members from laboratories in France (CESR, CRTBT, CSNSM, IAP, IAS, ISN, LAL, LAOG, PCC/CdF, OMP, SPP/CEA), Italy (Univ. La Sapienza of Rome, IROE CNR), Russia (Landau Ins. of Theoretical Physics), UK (QMW now in Cardiff), and the USA (CALTECH, JPL, Univ. Of Minnesota). Detailed information on the ARCHEOPS experiment may be obtained from the web site *http://www.archeops.org*.

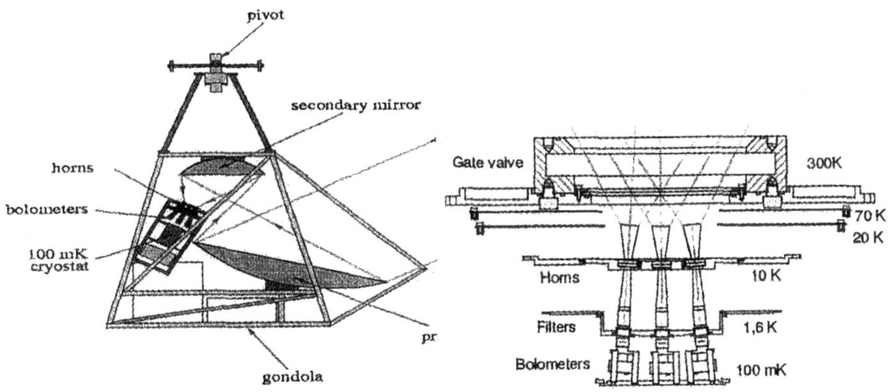

Figure 3. Left: gondola and experimental arrangement. Note that for KS3 the pivot was deported away to the other end of the flight chain to minimize electromagnetic and mechanical interferences (an a posteriori key improvement for the KS3 flight). Right: bolometers, horns & filters arrangement together with the various cryogenic stages. The cryostat is closed by a gate valve which opens automatically at low pressure.

ARCHEOPS is in many ways a prototype for the HFI instrument aboard the Planck satellite (see §3 below). It uses:

- Planck HFI spider web bolometers operating at 100 mK (JPL/CALTECH, Mauskopf et al., Appl. Opt., 36, 1997),
- Cold optics with back-to-back corrugated horns (QMW) at 10K, with filters at 1.6K (QMW),
- Planck like total power read out electronics with sampling frequency 180Hz and no 1/f noise down to the spin frequency,
- a helium cryostat at 4K (whereas Planck has Passive cooling to 55K, 20K hydrogen Sorption cooler and a 4K helium compressor cryocooler),
- a Planck-like dilution cooler system to reach 1.6K and 0.1K)
- an off-axis Gregorian telescope (with a 1.5 m primary) made of aluminum (U. of Minnesota) and based on an early design of the Planck telescope (which is nevertheless very close to the final one), allowing a high angular resolution of 8 arc minute at 143 GHz,
- the gondola has a rotation speed of 2 to 3 rpm (2rpm correspond to 3 arc-minutes in 5 milliseconds, which requires fast bolometers) and stellar sensors (provided by the U. La Sapienza from Rome) are used to get a very good a posteriori attitude reconstruction.

The instantaneous sensitivity is limited by the photon noise. The background due to the warm telescope and the atmosphere is larger than the background expected in Planck by a factor 10 to 15, which reduces the achievable instantaneous sensitivity by a factor 3 to 4 (it is nevertheless rather close to the Planck instantaneous sensitivity). ARCHEOPS is thus an excellent test bed of the concept of the HFI focal plane unit.

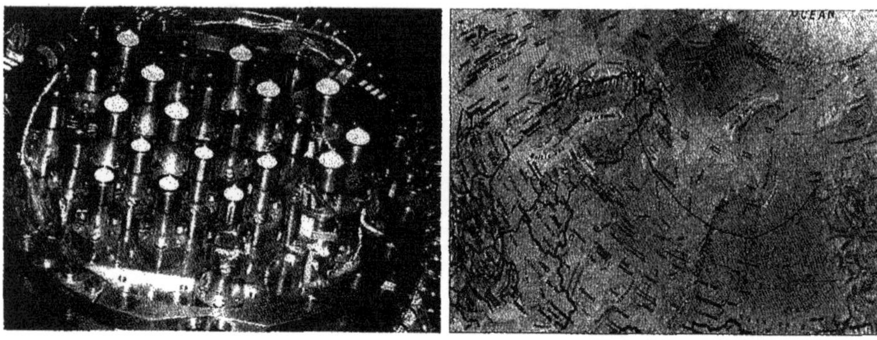

Figure 4. Left: 100 mK focal plane with the encasing of the spider Web bolometers (8 @ 143GHz, 6 @ 217GHz, 6 polarized @ 353GHz (3 OMT pairs), 2@ 545 GHz, 1 blind). Note the winding pipe on the right hand side of the image which is the heat exchanger where 3He & 4He flow. This 100 mK stage is supported by kevlar chords. Right: KS3 ground trajectory from Kiruna (Sweden) to Norilsk (Siberia).

ARCHEOPS was also conceived as a complementary experiment to the BOOMERANG and MAXIMA balloon borne one, both scientifically (coverage of large fraction of the sky of $\sim$ 30% which enables measuring the low-ℓ part of the CMB angular power spectrum) and technically with a scanning strategy similar to that of Planck (large circles on the sky at 41 degrees elevation) testing the use of a Planck like redundancy.

2.1. *The ARCHEOPS flights:*

ARCHEOPS flew for the first time in a cross-Mediterranean technical flight from Trapani (Sicily) in 1999. The focal plane did reach 100mK, the gondola was fully tested, but only limited night data was obtained, and a rather strong parasitic signal due to reflection on one element of the baffle was detected.

The first scientific flight was from Kiruna (KS1) last winter, but was short due to very strong high altitude winds.

During the last campaign, this winter, there were two flights (KS2 and KS3), also from Kiruna: one aborted flight (KS2) due to balloon problem, one successful flight on the 7th of February 2002. The balloon was launched at 12h44 and landed at 10h20 UT on the 8th of February, close to Noril'sk (Siberia), see fig.4.b; the ceiling altitude was 34 km. We obtained 12.5 hours of night data covering 30% of the sky (plus 6.5 hours of daytime), with 7 million night samples per bolometer which gives 110 000 pixels of 20 arc minute with 0.40 seconds/pixel/bolometer.

2.2. *Pointing reconstruction*

This reconstruction is made with the data on visible stars from a small telescope (0.40 m diameter) equipped with a photodiode linear array of 46 detectors spanning a 1.5° elevation (1.9 arc min per diode). One has typically 1 to 5 (magnitude 6) stars per second allowing a resolution of about 1 arc min (with at least 150 stars per revolution). The attitude reconstruction is helped by using further informations from 1) a GPS which gives the position of the gondola and the UT time 2) Gyroscopes measuring the rotation speed and pendulation 3) a magnetometer providing the magnetic North 4) a turn Counter on the pivot yielding the balloon rotation.

2.3. *Calibration and sensitivity:*

The beam shape measurements with Jupiter confirmed that beams have a FWHM close to 10'. Using the calibration on Jupiter (photometric and time response), we obtain the sensitivity of ARCHEOPS during the KS3 flight. The photometric point source calibration agrees at the 10% level with that obtained with extended sources (the cosmological dipole and the Galaxy).

The best 8 detectors in the 143 and 217 GHz channels have an average sensitivity of 180 μK_{CMB} in one second of integration. One can thus expect 1 sensitivities of about 100 and $\mu 150$ K_{CMB} at respectively 143 and 217 GHz per 0.3 degrees pixel. This sensitivity level reached in 12 hours compares well with the MAP nominal sensitivity of 20 μK_{CMB} for one year of integration (*http://map.gsfc.nasa.gov/m_mm/ob_tech1.html*).

One can further deduce from the above that the end-to-end optical efficiency, as measured on the sky, reaches 50% for the best channels. This is excellent news since this global measurement is very hard to obtain from ground measurements and it confirms that the 50% overall efficiency goal for HFI can indeed by reached (NB: the HFI requirement is 25%, and the best estimate prior to this measurement was 35%).

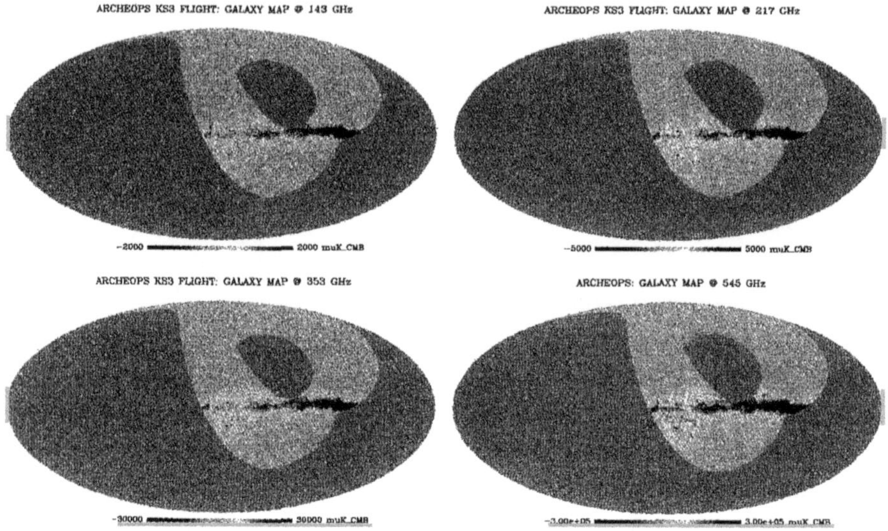

Figure 5. KS3 preliminary maps. Although not usable yet for scientific analysis, they clearly show the large sky coverage (which is likely to remain unique at these frequency, till Planck flies), and provide an idea of the sensitivity achieved.

2.4. *Scientific Prospect:*

Figure 5 shows very preliminary maps at the 4 frequencies of the experiment as deduced from the KS3 flight. These provide the first large scale view of the Galaxy at these frequencies.

Since Archeops scans large circles which are progressively shifted on the sky due to the earth rotation, the area of sky covered is a quasi circular band (not complete in a 12 hour flight), with increased redundancy when the scans tangent each other, both on the inside and the outside edge of that band.

The prospects for CMB measurements in the high redundancy regions where there are as many as 150 to 300 observations per HEALPIX pixel (at nside =128), are quite good. Indeed sum and difference maps in these regions for the 132 and 217 GHz channel, when both are expressed in thermodynamic temperature, directly demonstrate the good (CMB) signal to noise ratio achieved.

The Archeops data should thus allow reducing substantially the current error bars in the region of the CMB power spectrum between the COBE/DMR data and the Boomerang / Maxima / DASI / VSA/ CBI data, as is illustrated in figure 6 which shows from a simulation the ex-

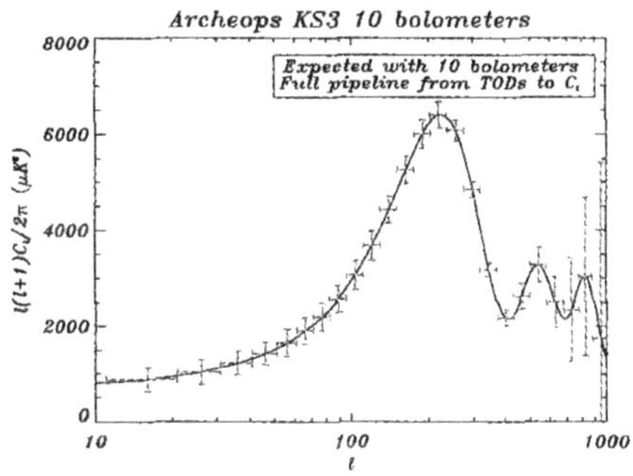

Figure 6. Expected precision for the KS3 flight, using 10 bolometers with 200 μK/$\sqrt{\text{Hz}}$, assuming a 15' resolution for 5 bolometers,, and 11' for the others.

pected error bars on the CMB power spectrum. We hope to release our analyses in time for a useful comparison with the first results from MAP, which should appear early next year.

3. The MAP and Planck sattelites

Two satellites, MAP and PLANCK, are devoted to CMB observations. MAP is American, and was launched by NASA in June 2001. PLANCK is mainly European and will be launched in 2007. Both will map the full sky, from an orbit around the Lagrangian point L2 of the Sun-Earth system, to minimise parasitic radiation from Earth. Both are based on the use of off-axis gregorian telescopes in the 1.5m class.

MAP has been designed for rapid implementation, and is based on fully demonstrated solutions. It's observational strategy uses a differential scheme. Two telescopes are put back to back and feed differential radiometers. These radiometers use High Electronic Mobility Transistors (HEMTs) for direct amplification of the radio-frequency (RF) signal. Angular resolutions are not better than 10 minutes of arc.

PLANCK is a more ambitious and complex project, which is designed to be the ultimate experiment in several aspects. In particular, several channels of the High Frequency Instrument (HFI) will reach the ultimate possible sensitivity, limited by the photon noise of the CMB itself. Bolometers cooled at 0.1 K will allow reaching this sensitivity and, at the same time, reach an angular resolution of 5 minutes of arc. The Low Frequency Instrument (LFI) limited at frequencies less than 100GHz, will use HEMT amplifiers cooled at 20 Kelvin to increase their sensitivity. The scan strategy is of the total power type. Both instruments use internal references to obtain this total power measurement. This is a 0.1 K heat sink for the bolometers, and a 4 K radiative load for the LFI. The combination of these two instruments on Planck is motivated by the necessity to map the foregrounds in a very broad frequency range: 30 to 850 GHz.

Table 1. Summary of (only indicative) experimental characteristics used for comparing experiments. Central band frequencies, ν, are in Gigahertz, the FWHM angular sizes, are in arcminute, and ΔT sensitivities are in μK per $\theta_{FWHM} \times \theta_{FWHM}$ square pixels; the implied noise spectrum normalisation $c_{noise} = \Delta T(\Omega_{FWHM})^{1/2}$, is expressed in μK.deg.

	MAP (as of January 1998)					BoloBall				
ν	22	30	40	60	90	143	217.	353	-	-
FWHM	55.8	40.8	28.2	21.0	12.6	8.0	5.5	5	-	-
ΔT	8.4	14.1	17.2	30.0	50.0	35	78	193	-	-
c_{noise}	8.8	10.8	9.1	11.8	11.8	11.3	17.4	39.0	-	-
	Proposed LFI				Proposed HFI					
ν	30	44	70	100	100	143	217.	353	545	857
FWHM	33	23	14	10	10.7	8.0	5.5	5.0	5.0	5.0
ΔT	4.0	7.0	10.0	12.0	4.6	5.5	11.7	39.3	401	-
c_{noise}	2.5	3.0	2.6	2.3	0.9	0.8	1.2	3.7	38	1711

In summary, MAP & PLANCK both map the full sky, from L2, with polarization capability, making highly redundant measurements. Details are given in table 1, but he differences most affecting their potential for constraining parameters are:

- **Resolution**, σ_{Beam} (assuming a Gaussian, symmetrical lobe), which is $\sim 10'$ for MAP (at 90 GHz) and improves to $\sim 5'$ for PLANCK (at 217 GHz), *i.e.* $\sigma_{MAP}/\sigma_{PLANCK} \sim 2$
- **Sensitivity**, $S = \sigma_{pix}\Omega_{pix}^{1/2}$ (the detector noise in a pixel times the pixel area gives the intrinsic sensitivity[b]), which is 11.8 μK.deg for MAP and improves to 0.8 μK.deg for PLANCK, yielding $S_{MAP}/S_{PLANCK} >\sim 10$; (this ratio depends on the duration of each mission, since $S \propto t^{-1/2}$, and both were assumed nominal)
- **Frequency coverage:** MAP will provide measurements at 30, 44, 70, and 90 GHz, in the Rayleigh-Jeans side of the back-body frequency spectrum. For PLANCK, the LFI instrument also covers the range probed by MAP (& COBE) (30, 44, 70, and 100 GHz) while the HFI will for the first time in space provide measurements on the Wien side of the spectrum at 100, 143, 217, 354, 550, and 857 GHz (and constrain internally the sub-mm emissions from our and other galaxies).

[b]This is independent of the assumed pixel size since $\sigma_{pix} \propto \Omega_{pix}^{-1/2}$

Further MAP and PLANCK References may be obtained from the web at:
`http://map.gsfc.nasa.gov/m_mm/ob\_tech1.html`
`http://tonno.tesre.bo.cnr.it/Research/PLANCK/Redbook`
`http://tonno.tesre.bo.cnr.it/Research/PLANCK/ONLY_SOMEONE/AODOCS/intro.html`
`http://astro.estec.esa.nl/SA-general/Projects/Planck/`

which give access (respectively) to the MAP home page, and for PLANCK, the "red book" (the report produced at the end of the phase A study, basis of ESA's selection in 1996), the LFI answer to ESA's announcement of opportunity for building Planck instruments, and the Planck Science Team pages. The PLANCK "blue book" should soon appear and it will present "The scientific program of Planck", as of 2002.

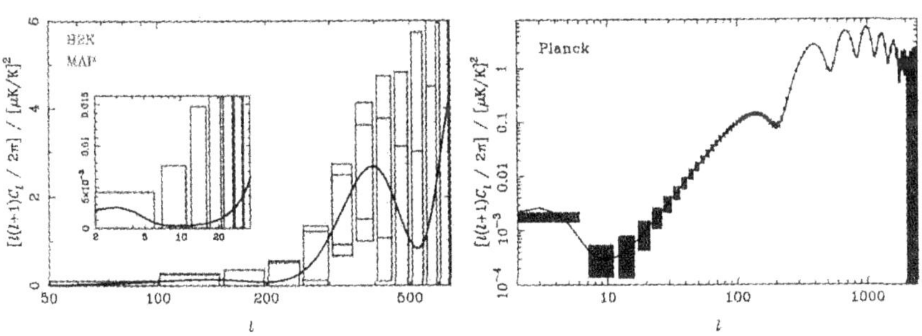

Figure 7. Expected errors on the amplitudes of the E type polarisation for the future Boomerang flight at the end the year and for MAP on the left, and for Planck on the right.

Figure 7 from the "Blue Book" shows the gain of sensitivity to expect between second (left) and third generation (right) experiments for the measurement of the E-type polarisation. One computes similar improvement for the cross-correlation spectrum (between the temperature and E type fluctuations). This is only illustrative though, since the actual precision reached will depend on how precisely the effect of astrophysical foregrounds fluctuations can be removed.

The polarisation signal is expected to be quite weaker than the temperature signal, by at least a factor of ten, and the polarisation properties of the foregrounds are barely known at all. It is therefore quite uneasy if at all possible to assess realistically to what extent foregrounds will decrease our ability at mapping the polarisation of fluctuations at recombination.

For the temperature signal, much more is already known and one can attempt more confidently at forecasting the outcome of the separation of components, by assuming a particular sky model and by specifying which separation method might be used. Bouchet and Gispert[1] gathered the output of many to conduct such a study for PLANCK's red book, and it was further refined for HFI's answer to ESA's announcement of opportunity for building Planck instruments.

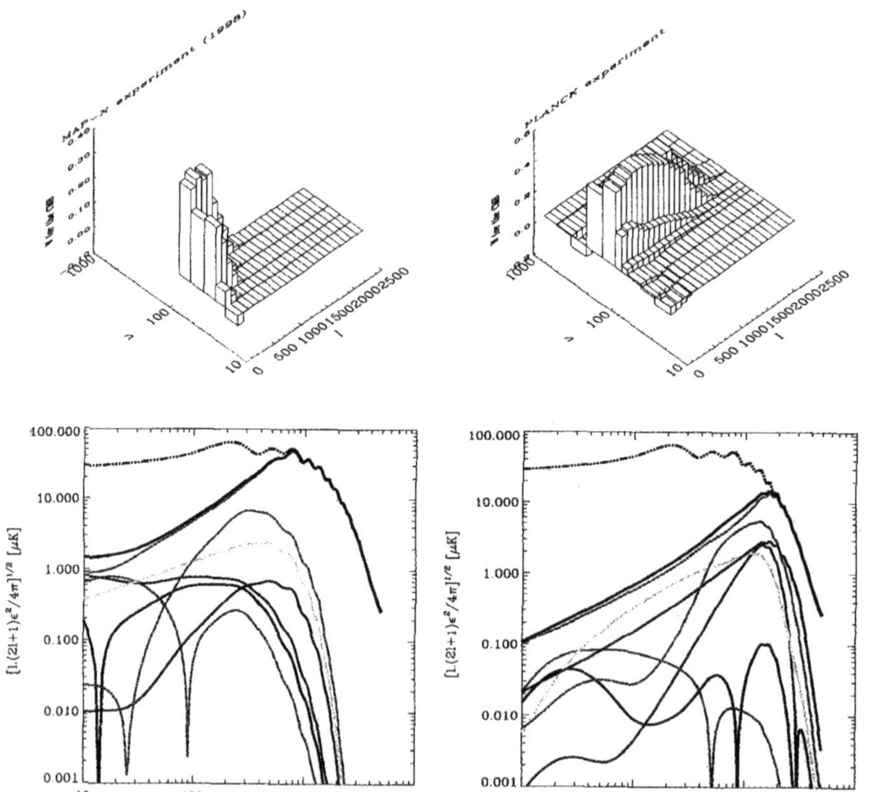

Figure 8. Top row: CMB Wiener matrix elements at $\ell > 200$ for MAP (left) and PLANCK (right). Each bin in ℓ is the average over one hundred contiguous values. The frequency bins are centred at the average frequency of each channel, but their width in not representative of their spectral width. Bottom row: CMB reconstruction error contributed by each component (orange is for the noise, red, blue, and green are for the Galactic components residuals, yellow is for the residual SZ contribution, the 2 others corresponding to unresolved sources). The total error in black can then be compared with the primary $\ell C(\ell)^{1/2}$ in dashed-dotted line. The integral of $\ell(\ell+1)\varepsilon^2/2\pi$ would give the total reconstruction error of the map. Note that for PLANCK, $\varepsilon(\ell) \lesssim 0.025$ μK, a factor more than six below the MAP case. Reprinted from Ref. 1.

If one assumes that Wiener filtering of the frequency maps will be used to produce linearly optimal component maps, one can already compute the characteristing of the Wiener filters corresponding to a particular experimental arrangement. Figure 8.a offers a graphical presentation of the resulting Wiener matrix coefficients (for extracting the CMB component) when they use their sky model. It shows how the different frequency channels are weighted at different angular scales and thus how the "$\nu - \ell$" information gathered by the experiment is used.

In the MAP case, most weight is given to the 90 GHz channel. It is the only one to gather CMB information at $\ell \gtrsim 600$, and it's own weight becomes negligible at $\ell \gtrsim 1000$. In the PLANCK case, the 143 GHz channel is dominant till $\ell \sim 1400$ and useful till $\ell \sim 2000$, while the 217 GHz channel becomes dominant at $\ell \gtrsim 1400$ and gather CMB information till $\ell \sim 2300$. The 100 GHz channel contribution (split in two on the graphics to show the relative HFI and LFI contributions) to the CMB determination is only modest, and peaks at $\ell \sim 800$.

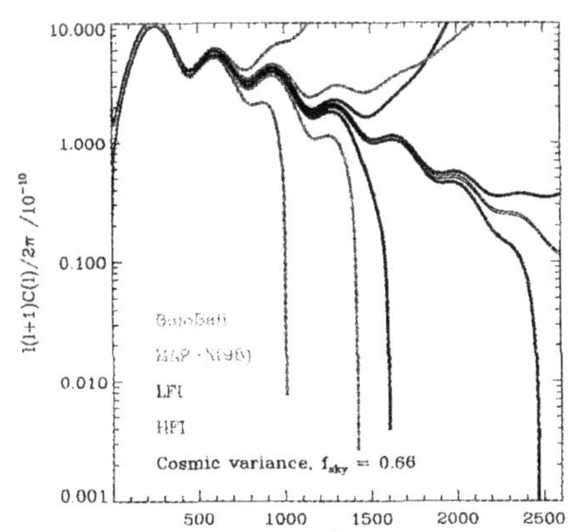

Figure 9. Expected errors on the amplitudes of each mode individually (no band averaging) for different experiments (the full PLANCK is indistinguishable from the HFI case). The thin central lines gives the target theory plus or minus the cosmic variance, for a coverage of 2/3 of the sky. The taget theory used here is Lambda CDM (with $\Omega_b = 0.05$, $\Omega_{CDM} = 0.25$, $\Omega_\Lambda = 0.7$, $h = 0.5$). Reprinted from Ref. 1.

One can then deduce the uncertainty added by the noise and foreground removal to the CMB power spectrum. Figure 9 shows the envelope of the 1-σ error expected from the the current design of MAP(green), the LFI (blue), and the HFI or the full PLANCK (red), as well as a "boloBall" experiment meant to show what bolometers on balloon might achieve soon; the experimental characteristics used in this comparison may be found in table 1. *Note that there has been no "band averaging" in this plot,*[c] which means that there is still cosmological signal to be extracted from the HFI at $\ell \sim 2500$ (a 10% band width around $\ell = 1000$ spans 100 modes...).

Bouchet and Gispert[1] also looked at the impact of changing the foreground model and found very little impact for rather large variations of the assumed sky model, a result confirmed by an alternative Fisher matrix analysis.[7] The message from these plot is thus excellent news since it tells us that even accounting for foregrounds, PLANCK will be able to probe the very weak tail of the power spectrum ($\ell \geq 2000$) and allow breaking the near degeneracy between most of the cosmological parameters.

The ability of Planck to efficiently removed foregrounds thanks to its wide frequency coverage also implies that the foregrounds themselves will be well determined. It is anticipated that Planck/HFI will detect a few tens of thousand clusters thanks to their Sunyaev-Zeldovich effect. Most of those located in the part of the sky covered by the Sloan Digital Sky survey (~ 10000), will have Sloan couterparts, and will already have redshift estimates. Those with no couterparts should be at high z and will be quite interesting targets for follow-up observations (see Bartelman & White 2002).

The determination of a cluster velocity thanks to the kinetic SZ effect is noisy due to the confusion with other parasitic contributions like the primary CMB fluctuations, the contributions from other clusters, unremoved foregrounds and detector noise. But averaging over large volume of space should allow rather precise determination of the bulk velocity (since the noises are uncorrelated at different cluster positions and should average out) at $z \sim 1$ (see Aghanim, Gorski, Puget 2001 for details). This is unlikely to be accessible by any other means till then and it should be one of the most interesting by-product of PLANCK.

Planck should also unveil a number of rare, cold and luminous sub-mm sources thanks to it's nearly full sky coverage. Making quantitative predictions remains a challenge given the sparceness of sub-millimetric observations, although SIRTF (to be launched next year) should much help in that respect.

Finally, let me remind you that many analysis challenges remain to be solved in order to be able to optimally use the wealth of data that PLANCK will provide. Indeed, most analyses methods of today (for making map,

[c]The band averaging would reduce the error bars on the smoothed $C(\ell)$ approximately by the square root of the number of multipoles in each band, if the modes are indeed independent, which is a reasonable approximation for full sky experiments.

separating components, or extracting the power spectrum...) cannot be extrapolated since they would require ridiculously large computer resources. But the improvement rate is so fast, driven by the fast pace of new experimental data, that one is confident that these methods will be in place when PLANCK is finally launched.

References

1. F. R. Bouchet and R. Gispert. Foregrounds and cmb experiments i. semi-analytical estimates of contamination. *New Astronomy*, 4:443–479, Nov. 1999.
2. P. de Bernardis, P. A. R. Ade, J. J. Bock, J. R. Bond, J. Borrill, A. Boscaleri, K. Coble, B. P. Crill, G. De Gasperis, P. C. Farese, P. G. Ferreira, K. Ganga, M. Giacometti, E. Hivon, V. V. Hristov, A. Iacoangeli, A. H. Jaffe, A. E. Lange, L. Martinis, S. Masi, P. V. Mason, P. D. Mauskopf, A. Melchiorri, L. Miglio, T. Montroy, C. B. Netterfield, E. Pascale, F. Piacentini, D. Pogosyan, S. Prunet, S. Rao, G. Romeo, J. E. Ruhl, F. Scaramuzzi, D. Sforna, and N. Vittorio. A flat Universe from high-resolution maps of the cosmic microwave background radiation. *Nature*, 404:955–959, Apr. 2000.
3. S. Hanany, P. Ade, A. Balbi, J. Bock, J. Borrill, A. Boscaleri, P. de Bernardis, P. G. Ferreira, V. V. Hristov, A. H. Jaffe, A. E. Lange, A. T. Lee, P. D. Mauskopf, C. B. Netterfield, S. Oh, E. Pascale, B. Rabii, P. L. Richards, G. F. Smoot, R. Stompor, C. D. Winant, and J. H. P. Wu. MAXIMA-1: A Measurement of the Cosmic Microwave Background Anisotropy on Angular Scales of 10^{-5} degree. *Ap. J. Lett.*, 545:L5–LL9, Dec. 2000.
4. A. E. Lange, P. A. Ade, J. J. Bock, J. R. Bond, J. Borrill, A. Boscaleri, K. Coble, B. P. Crill, P. de Bernardis, P. Farese, P. Ferreira, K. Ganga, M. Giacometti, E. Hivon, V. V. Hristov, A. Iacoangeli, A. H. Jaffe, L. Martinis, S. Masi, P. D. Mauskopf, A. Melchiorri, T. Montroy, C. B. Netterfield, E. Pascale, F. Piacentini, D. Pogosyan, S. Prunet, S. Rao, G. Romeo, J. E. Ruhl, F. Scaramuzzi, and D. Sforna. Cosmological parameters from the first results of Boomerang. *Phys. Rev. D*, 63:42001–+, Feb. 2001.
5. J. L. Sievers, J. R. Bond, J. K. Cartwright, C. R. Contaldi, B. S. Mason, S. T. Myers, S. Padin, T. J. Pearson, U.-L. Pen, D. Pogosyan, S. Prunet, A. C. S. Readhead, M. C. Shepherd, P. S. Udomprasert, L. Bronfman, W. L. Holzapfel, and J. May. Cosmological parameters from cosmic background imager observations and comparisons with boomerang, dasi, and maxima. *Ap. J., submitted; astro-ph/0205387*, 2002.
6. M. Tegmark. Cmb mapping experiments: a designer's guide. *Phys. Rev. D*, 56:4514–4529, Oct. 1997.
7. M. Tegmark, D. J. Eisenstein, W. Hu, and A. de Oliveira-Costa. "foregrounds and forecasts for the cosmic microwave background". *Ap. J.*, 530:133–165, Feb. 2000.

Cosmology and Astrophysics with the CMB in 2002

Andrew Jaffe

COSMOLOGY & ASTROPHYSICS WITH THE CMB IN 2002

A. H. JAFFE

Astrophysics Group, Blackett Lab, Imperial College, London SW7 2BW UK
E-mail: a.jaffe@ic.ac.uk

Since 2000, the glimpse of the early Universe afforded us by observations of the Cosmic Microwave Background (CMB) has confirmed our pictures of the global cosmological model and the formation of structure in the Universe. I review the physics of the CMB, its measurement, and what these, together, tell us about the cosmological model.

1. Introduction

The Cosmic Microwave Background gives us our earliest direct glimpse of the Universe. It consists of photons that last scattered off of nuclei and electrons about 300,000 years after the Big Bang. At this early time, the Universe was only slightly perturbed from homogeneity, making it easy to relate our observations to the state of matter at that time, and thus to its evolution before and since. Observations of CMB perturbations can then be decoded to give precise measurements of the cosmological parameters describing the constituents and evolution of the Universe.

We concentrate here on the power spectrum of temperature fluctuations. In the year 2000, the CMB power spectrum looked roughly as in Figure 1, left. By 2002, the picture had changed dramatically, with the results from BOOMERANG[1], MAXIMA[2], DASI[3], CBI[4] and VSA[5], as shown in the right panel. Indeed, the initial results from MAXIMA[6] and BOOMERANG[7], both in April, 2000, had immediately changed the qualitative picture from one of a cloud of points to an actual curve. More remarkable was the fact that this curve coincided with the predictions of the so-called inflationary Universe. This predicts a flat Universe ($\Omega_{\rm tot} = 1$) and an initial spectrum of adiabatic perturbations with a power spectrum $P(k) \propto k^{n_s}$ and $n_s \simeq 1$.

Combining these data with other results implying a low matter density ($\Omega_m \simeq 0.3$) leads to the not inexorable but nonetheless compelling conclusion that the Universe is dominated by what has come to be known as "Dark Energy," something behaving today very much like the cosmological constant of General Relativity.

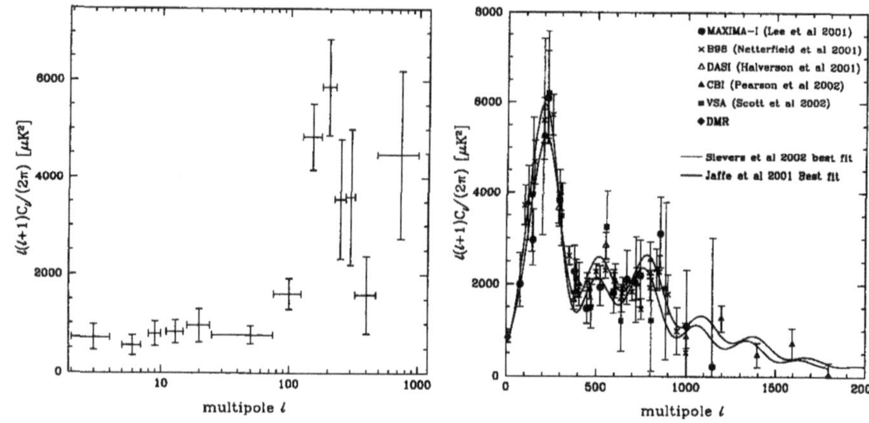

Figure 1. CMB power Spectra, C_ℓ. Left: C_ℓ in 2002, combining the results from COBE/DMR and the degree-scale experiments that reported results in the 1990s. Right: C_ℓ in 2002, showing results from COBE/DMR and other experiments.

2. The CMB and Cosmology

Our goal is to measure the cosmological parameters. The global cosmological model is given by the usual Hot Big Bang Universe and a Friedmann-Robertson-Walker metric. Its evolution is determined by the densities of the various components: relativistic matter and radiation, with density Ω_r; cold matter, (Ω_m), made up of cold dark matter (Ω_c) and baryons (Ω_b), so that $\Omega_m = \Omega_c + \Omega_b$; and a cosmological constant (Ω_Λ). The latter is defined as a component with equation of state $w \equiv p/\rho = -1$ and a constant density. We can instead allow components with $w < 0$, often manifested as a scalar field permeating the Universe, under the general rubric of "Quintessence" (Ω_Q), which may have a time-varying $w(t)$. These densities are defined from the present-day physical density, ρ_i, of some component, as $\Omega_i \equiv 8\pi G\rho_i/(3H_0^2)$ where $H_0 = 100h$km/s/Mpc is the Hubble Constant today. The curvature of the universe is defined by $\Omega_k = 1-\Omega_m-\Omega_r-\Omega_{\Lambda/Q}$; $\Omega_k = 0$ corresponds to a flat Universe.

We also define parameters specifying the distribution of perturbations to the density. The three-dimensional power spectrum of *primordial* density fluctuations is $P(k) = Ak^{n_s}$ where A is a measure of the amplitude of fluctuations (in practice we use a quantity like σ_8^2, the present-day variance of the matter density in $8h^{-1}$Mpc spheres), with the index $n_s = 1$

corresponding to a scale-invariant spectrum (so that the rms amplitude of a mode entering the horizon is constant with time).

These cosmological parameters uniquely determine the statistical properties of CMB fluctuations. Consider the early Universe, after the epoch of Big-Bang Nucleosynthesis, so that the basic material constituents of the Universe — baryons, leptons, photons and dark matter — are in the same form that we observe today. At this time, the Universe was hot and dense, and, by hypothesis, more or less smooth. At high temperatures ($T \gg 1$ eV), the electrons and protons were kept ionized. In this plasma, the photons were tightly coupled to the electrons; the mean free path of photons was small compared to particle horizon. Below a temperature of about 1 eV (considerably lower than the expected 13.6 eV due to the high ratio of the number of baryons to that of photons), the electrons were able to combine with the protons to form neutral hydrogen (the epoch of *recombination*). This neutral gas was suddenly transparent to light, and the photons *decoupled* from the matter. Most have been freely streaming through the Universe ever since; some were fated to be pointing in the direction where, 15 billion years later, we happened to be pointing our microwave telescopes. Thus we observe the $T_0 = 2.73$K CMB.

These photons trace the state of the Universe at *last scattering*, a time when the Universe was hot, dense, and simple. At yet earlier times (in an epoch of Inflation, say), perturbations to the density of the baryons, photons and dark matter were created. The relationships amongst the perturbations to the various components determines whether we are dealing with "adiabatic" or "isocurvature" fluctuations, but we shall concern ourselves only with the former, in which the fractional density perturbation is the same for all species.

We need now recall that the physics of the Universe is governed by causality. Waves in the plasma of photons and baryons in the early Universe could only travel at the speed of sound, c_s. This is $c/\sqrt{3}$ if completely dominated by relativistic species and slightly lower with the admixture of non-relativistic baryons and electrons. Thus, the age of the Universe gives an upper limit to the size of perturbations that can interact and grow — the (sound) Horizon. The size of the sound horizon at some redshift z is given by $\int dt\ c_s(t)/a(t) = \int dz\ c_s(z)E(z)$, where $E(z) = \dot{a}/a = \sqrt{\Omega_m(1+z)^3 + \Omega_\Lambda(1+z)^{3+3w} + \Omega_r(1+z)^4 + \Omega_k(1+z)^2}$ gives the evolution of the scale factor, a. In general, we can trace the evolution of the sound speed with knowledge of the constituents, but the details need not concern us here. Because the Universe had been dominated by radiation for most of its history, the sound horizon is about $1/\sqrt{3}$ of the (classical Big Bang) particle horizon.

But this raises another question: why is the CMB so smooth? CMB photons trace the state of the baryons at Last Scattering, when the Horizon was a few hundred thousand light years, corresponding to roughly one degree on the sky. Yet the CMB—and thus all the matter at Last Scattering—is smooth on scales much larger than this, outside the causal horizon. How is this possible?

Several explanations present themselves. The first unsatisfactorily appeals to initial conditions: the Universe was made that way. A second possibility is to modify the evolution of the Universe so that the causal horizon is much larger than it appears. This is the explanation offered by *inflation.* Inflation posits an epoch of accelerated expansion in the early Universe (so that the standard expressions for the horizon are strictly speaking incorrect). Inflation simultaneously solves this causality problem, sets the curvature of the Universe to zero, and generates the small perturbations that eventually grow into the structure present in the Universe today. There is a large literature on inflation, and we point the reader to some recent reviews.[8,9]

Another, related, possibility is to make the Universe much older than it appears, so that causality has time to work on larger scales. This, in effect, is the solution posited by the Cyclic (a.k.a. ekpyrotic) Universe.[10] In this picture, the Universe evolves toward a rarefied state, eventually sparking another Big Bang, but now with initial conditions set by the previous cycle's evolution. Like Inflation, the Cyclic Universe also flattens the geometry and produces adiabatic fluctuations, so most statements below referring to Inflation apply as well to the Cyclic Universe.

In any event, once the "initial conditions" are set up, the evolution of the particle species and of perturbations to their densities is well understood. We present a very brief review of this evolution and the imprint left on the photons that will become the CMB, recapping material presented at greater length and detail in textbooks[11,12] and review articles[13,14]).

The early Universe was nearly homogeneous with small density perturbations, $\delta\rho/\rho$, in the various components. The simplest inflationary models create perturbations where the fractional perturbations are the same in all species. Since the entropy per particle is therefore spatially constant, such perturbations are referred to as "adiabatic". More complicated models can also produce isocurvature perturbations, in which density fluctuations in some species are initially compensated by fluctuations in others. The data are consistent with pure adiabatic and inconsistent with (at least the simplest) pure isocurvature models. Inflation also generically leads to Gaussian, statistically isotropic perturbations whose properties are described only by a correlation function or, in three-dimensional Fourier space, a power spectrum, $P(k)$.

Just as $P(k)$ quantifies the distribution of matter, we quantify the distribution of temperature fluctuations around the 2.73K mean by the CMB power spectrum, C_ℓ. We define this by first transforming the temperature anisotropy, $\Delta T/T$, into spherical harmonic components,

$$a_{\ell m} = \int d^2\hat{\mathbf{x}} \frac{\Delta T}{T}(\hat{\mathbf{x}}) Y_{\ell m}(\hat{\mathbf{x}}) \tag{1}$$

The power spectrum is then defined by

$$\langle a_{\ell m} a^*_{\ell' m'} \rangle = \delta_{\ell \ell'} \delta_{m m'} C_\ell \,. \tag{2}$$

The Kronecker deltas and the fact that the spectrum is a function only of ℓ enforce isotropy: the average temperature difference between two points only depends on the distance between the points. The angle brackets refer to averages under the distribution of the $a_{\ell m}$ throughout the Universe. For a Gaussian theory, this is simply $P(a_{\ell m}) = (2\pi C_\ell)^{-1/2} \exp[-|a_{\ell m}|^2/(2C_\ell)]$.

When the scale of a density perturbation is greater than that of the Hubble length, the nature of the separate components (dark matter, photons, baryons) is irrelevant. However, because of their different equations of state and inter–species interactions, the small–scale dynamics is different for the various components. When the Universe has aged sufficiently that a wave of some size is of a scale comparable to the Hubble length (we casually say that the scale has "entered the horizon"), pressure and gravitational potential gradients become important and drive sound waves in the plasma.

First, consider waves entering the (sound) horizon around the time of last scattering: these are the largest waves that could have formed a coherent structure by this time. Indeed, by determining the characteristic *angular* scales of the CMB fluctuation pattern, and matching this to the physical scale of the sound horizon at last scattering we can determine the *angular diameter distance* to the last scattering surface, which is mostly dependent on the geometry of the Universe: in a flat universe, angular and physical scales obey the usual Euclidean formulae; in a closed (positively curved) universe, geodesics converge and a given physical scale corresponds to a larger angular scale (and hence smaller multipole ℓ); conversely, in a negatively curved Universe the same physical scale corresponds to a smaller angular scale and larger ℓ.

Consider now a wave that enters the horizon some time considerably before Last Scattering, when the density of the Universe is still dominated by radiation, and the Baryons are tightly-coupled to the photons. Although the dark matter is pressureless, the dominant radiation has pressure $p = \rho/3$, or somewhat less due to the baryons. Although the dark matter can continue to collapse, the radiation rebounds when the pressure and density become sufficiently high. Eventually, gravity may take over and cause the perturbation to collapse again, one or more times. Larger and larger scales, entering the horizon later and later, will experience fewer and fewer collapse and rebound cycles. Moreover, because of the effect of the baryons on the pressure, the strength of the rebound is decreased as we increase the baryon density. It is this cycle of collapse and rebound that we see as peaks in the CMB power spectrum, often called *acoustic peaks* after the waves responsible for them.

We thus use the *heights* of the peaks to measure the relative contributions of baryons and photons to the pressure, and their *angular scale* to determine the geometry, as well as the history of the sound speed in the baryon-photon plasma. Other cosmological parameters affect the spectrum in yet other ways.

Although the photons and baryons are tightly bound to one another via scattering, the coupling is not perfect. Hence, there is a scale (known as the *Silk* damping scale) below which the photons can stream and wash out perturbations. This free-streaming damps perturbations on small scales.

All of these physical effects are included in computer codes CMBFAST[15,16] and CAMB[17,18] which solve the combined Boltzmann and linearized Einstein equations in an expanding Universe, and calculate the CMB power spectrum for any model. A sample of spectra for various input cosmological parameters is shown in Figure 2. We see that a flat Universe generically has a first peak at $\ell \simeq 200$, corresponding to a sound horizon projecting to about 1° on the sky.

The matter transport caused by pressure and gravitational potential gradients means there are *velocity* perturbations as well. Hence, photons scatter off of moving electrons, generating a net linear polarization of the photons.[19] For the sound waves we consider, velocities are greatest when the density contrast is smallest, and vice versa: the velocity is out of phase with the density–and the polarization signal is out of phase with the temperature. Unfortunately, due to the inefficiency of scattering off of the moving electrons, the polarization fraction is only about 10%, and the polarization spectra are correspondingly suppressed. We discuss polarization induced by gravitational waves in § 6.

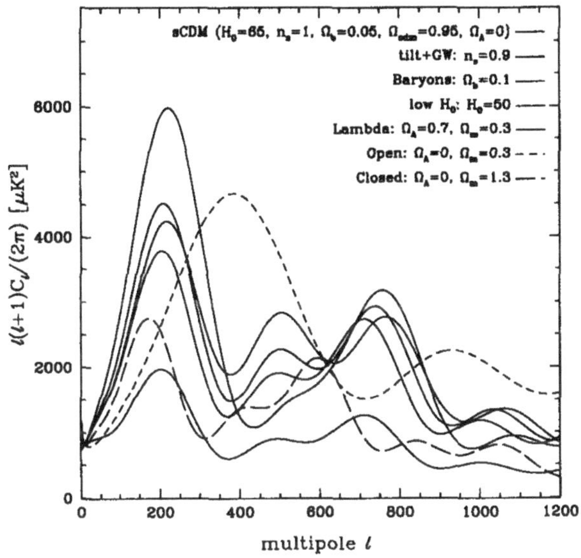

Figure 2. A sample of theoretical power spectra for various cosmological parameters, as marked.

3. CMB Experiments

The small-scale anisotropy of the CMB was first detected by the DMR instrument on the COBE satellite in the early 1990s.[20] DMR had an angular resolution of 7°, corresponding to a cutoff at $\ell \sim 20$. This scale is much larger than the horizon at last scattering and so the DMR signal is insensitive to the physical details of recombination; from Figure 2, we see that most structure in the spectrum is found at $\ell \gtrsim 100$.

The CMB community spent the 1990s refining the sensitivity and angular resolution of their detectors. By the end of the decade, there was a strong suggestion of a peak in C_ℓ at $\ell \sim 200$ corresponding to the predictions of a flat Universe.[21] These results were from a wide variety of experiments using a variety of experimental techniques: bolometers and radiometers, scanning and differencing, interferometers and telescopy.

In 2000, BOOMERANG[7] and MAXIMA[6] reported results with high signal to noise out to $\ell \sim 800$ showing an unambiguous peak at $\ell \sim 200$, interpreted to imply a flat Universe under the inflationary paradigm.[22,23,24] BOOMERANG and MAXIMA used similar experimental techniques: cooled bolometers flown from a balloon-borne telescope. BOOMERANG flew as a Long-Duration Balloon (LDB) over Antarctica in 1998–99; MAX-

IMA flew overnight from the US National Scientific Ballooning Facility in Palestine, TX. Both teams have subsequently reanalyzed their data at higher resolution and, in BOOMERANG's case, with considerably greater sky area.[1,2,25]

Since then, other groups using a significantly different technique have reported results over a similar range of angular scales. A crucial cross-check is provided by the results from the CBI, DASI and VSA experiments, interferometers working at lower frequencies than the bolometers, and with completely different technologies, and hence completely different (potential) systematic problems. All of these data are shown in Figure 1.

4. CMB Data Analysis

The simple physics underlying the observed CMB fluctuations allows us to characterize the statistical properties of both the signal and most sources of experimental noise. Although there are many competing (but to some extent consistent) approaches to the problem, here I outline a "Bayesian" solution. We start with the eponymous theorem of Bayes:

$$P(\theta|D) = \frac{P(\theta)P(D|\theta)}{\int d\theta\, P(\theta)P(D|\theta)} \tag{3}$$

where $P(x)$ is the probability density that some parameter's value lies in $(x, x+dx)$, $P(x|y)$ is the conditional probability of x given y, θ is the parameter we desire to measure, and D is the experimental data. Hence, $P(\theta)$ is the "prior" for θ, and $P(D|\theta)$ is the likelihood. The denominator just enforces the requirement that probabilities sum to one.

In our case, we start with the data in a raw form:

$$d_t = \sum_p A_{tp} T_p + n_t \tag{4}$$

where d_t represents the data taken at time t, T_p represents the underlying sky temperature at pixel p, already smoothed by the known experimental beam. A_{tp} represents the operation of observing; for a mapping experiment $A_{tp} = 1$ when observing pixel p at time t, and zero otherwise. (For other types of experiments, including interferometers, A_{tp} is more complicated, but the formal problem remains the same.) Finally, n_t gives the noise at time t. We assume that the noise has a known covariance structure $\langle n_t n_{t'} \rangle = N_{tt'}$ and that its properties are well represented by a zero-mean Gaussian with this covariance (with some conceptual and computational complications[26,27,28]).

We can then ask a series of questions of our data, embodied in different parameters θ. First, what is the pattern of CMB fluctuations on the sky (T_p)? The answer to this question is just a least-squares solution to Eq. 4. The posterior distribution for T_p is just another Normal (Gaussian) distribution with mean $\bar{T} = C_N A^T N^{-1} d$ and variance $C_N = (A^T N^{-1} A)^{-1}$ (using matrix notation).

We can then ask, what is the power spectrum underlying the map? The answer to this question depends only on the map, $\bar{T}$ and its noise covariance, rather than on the full timestream data—the map is a "sufficient statistic." The likelihood function is again given by a Gaussian,

$$P(\bar{T}|C_\ell) = \frac{1}{|2\pi C|^{1/2}} \exp\left[-\frac{1}{2}\bar{T}^T C^{-1} \bar{T}\right] \; ; \tag{5}$$

the map has zero mean and covariance given by $C = C_N + C_T$ with

$$C_{T,pp'} = \sum_\ell \frac{2\ell+1}{4\pi} C_\ell B_\ell^2 P_\ell(\cos\theta_{pp'}) \tag{6}$$

where B_ℓ is the Legendre transform of the experimental beam (assumed to be circularly symmetric[29]), the P_ℓ are the Legendre polynomials, and $\theta_{pp'}$ gives the angle between the pixels. Because our desired parameters appear in the covariance, C, there is no analytic expression for the likelihood maximum, $\bar{C}_\ell$. We instead use various iterative schemes.[30,31,32] For interferometric experiments, there are similar expressions; solutions are somewhat simpler since they effectively measure linear combinations of the $a_{\ell m}$ coefficients themselves in narrow bands of ℓ. There are also various 'frequentist' approaches which seem to find very similar values for the spectra.[33]

Finally, we wish to find the cosmological parameters. We need to calculate quantities like

$$\begin{aligned} &P(\Omega_m, \Omega_b, h, n_s, \ldots |d_t) \\ &\quad \propto P(\Omega_m, \Omega_b, h, n_s, \ldots) P(d_t | C_\ell[\Omega_m, \Omega_b, h, n_s, \ldots]) \,. \end{aligned} \tag{7}$$

As above, the first factor gives the prior probability for the cosmological parameters (codifying either some form of ignorance or, more usefully, other cosmological results), and the second gives the likelihood. As we saw above, we can replace the timestream, d_t, with the map, $\bar{T}_p$, with no information loss; the likelihood as a function of T is just a Gaussian. However, we cannot quite replace $\bar{T}_p$ with $\bar{C}_\ell$: because the latter appears in the covariance, C, the likelihood no longer takes such a simple form. Nonetheless, there are several approximations available which give an excellent fit to the full likelihood shape at minimal extra computational cost.[34] Thus, we can indeed calculate the likelihood $P(\bar{C}_\ell | C_\ell(\text{cosmology}))$; the computational costs are dominated by the calculation of the C_ℓ from the cosmological parameters.

Other complications arise from the need to characterize the posterior probability over a large-dimensional space. Various solutions to these problems have been outlined; the use of Markov Chain Monte Carlo (MCMC) along with various speedups to the Boltzmann solvers CMBFAST and CAMB seem the most promising direction.[35,36]

5. The CMB today

As we saw in Figure 1, in June, 2002, we now have high signal-to-noise measurements of the CMB power spectrum, with multiple experiments covering the range $50 \lesssim \ell \lesssim 1000$. Individual experiments also cover the low-ℓ ($\ell \lesssim 20$, COBE/DMR) and high-ℓ ($1000 \lesssim \ell \lesssim 3500$, CBI) ranges.

Several broad features are apparent in the data. The most significant is the presence of a narrow peak at $\ell \simeq 200$. This corresponds to the angular size of the sound horizon in a flat Universe. There is also evidence for further peaks near $\ell \sim 500$ and $\ell \sim 800$.

Without going further we can already make several crucial inferences. First among these is that the generic cosmogony we have described seems to obtain. In roughly increasing order of specificity, the hot big bang, an FRW universe, adiabatic initial conditions and a scale-invariant spectrum of perturbations are all consistent with the data. Of course, it is possible that other scenarios could generate similar data, but at present this is the simplest explanation (as it is for other cosmological data). In particular, a Universe with initially negligible fluctuations undergoing a phase transition generating cosmological defects such as Cosmic Strings seems to be strongly disfavored. Such a scenario generates perturbations incoherently, solely within the horizon, and cannot reproduce the sharp peak structure of the data.[37]

Thus, the presence of this peak indicates a flat Universe. More specific aspects of the data, in turn, allow us to measure the baryon density, $\Omega_b h^2 \simeq 0.020$, the value preferred by observations of light-element abundances. We can also measure the 'tilt' of the spectrum, encoded in the scalar spectral index, $n_s \simeq 1$. The general decline of power after the first peak also gives an indication that we understand the physics of recombination including the damping effects. In Figure 3, we show an example of a more detailed parameter estimation calculation.

CMB data do not exist in a vacuum; In Figure 3, we show an example of combining CMB data with other astrophysical information. The CMB spectrum is sensitive to Ω_k rather than Ω_m or Ω_Λ separately. Recent results using distant Supernovae as distance indicators give evidence for an accelerating expansion of the Universe and are themselves sensitive to the

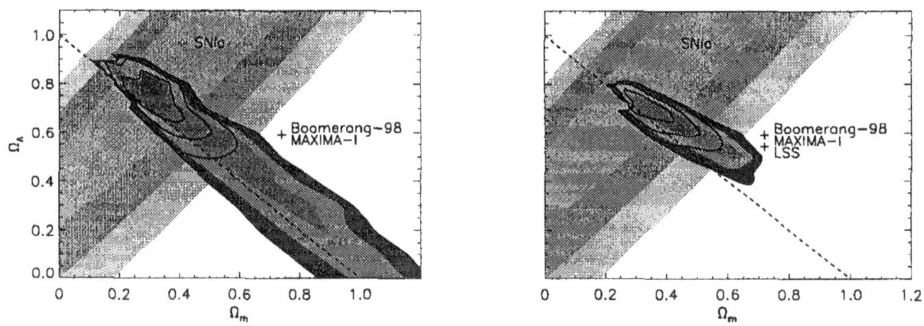

Figure 3. Probability distribution in the Ω_m, Ω_Λ plane. Left: 1-, 2- and 3-sigma contours for DMR, MAXIMA-I (Hanany et al 2000) and B98 (de Bernardis et al 2000) in Blue, the SNIa data in orange, and the combination (heavy lines). Right: CMB + Large-Scale structure (see text) in blue, CMB+Large-Scale Structure+SNIa (heavy lines). From Jaffe et al (2001).

approximate combination $\Omega_m - \Omega_\Lambda$. Thus in combination these data point to $\Omega_m \simeq 0.3$ and $\Omega_\Lambda \simeq 0.7$. At the same time, we can combine the CMB data with evidence from measurements of the distribution of matter in the Universe today. The amplitude of clustering is sensitive to the combination $\Omega_m^{0.6}\sigma_8$ and the distribution is sensitive to a combination of $\Omega_m h$ and n_s. In the figure, we see that these data combined with the CMB *also* point to a similar $\Omega_m \simeq 0.3$, $\Omega_\Lambda \simeq 0.7$ cosmology even without the SN data.

6. The future of the CMB: Polarization

In § 2 we considered polarization due to electrons moving with the fluid as it responds to the pressure and gravitational potential gradients. The polarization pattern that results is "curl-free"; it can be represented as the 2-d divergence (gradient) of some scalar on the sky. This is easy to understand if we heuristically identify the polarization pattern with the flow of the plasma. In linear perturbation theory this flow is a potential flow; any vorticity is damped by expansion and not driven to grow by gravity.

Gravitational waves at last–scattering can produce a "curl" pattern, since gravitational waves can produce non–potential flows in the plasma. For full-sky maps with high sensitivity and resolution, the "gradient" and "curl" components (also known as "Electric" and "Magnetic" or E and B from the obvious analogy to electromagnetic fields) can be separated completely; in more realistic situations statistical techniques are necessary.

The amplitude of such a curl pattern is proportional to the energy scale of inflation; indeed there appears to be no other way of determining this scale. Possibly this evidence would constitute a "smoking gun" for the occurrence of inflation, but there is no proof of inflation's uniqueness. One possible alternative for the generator of density perturbations, the 'cyclic Universe', produces a gravitational-wave background with a very low amplitude and bluer spectrum.[10] Other mechanisms for producing gravitational waves (such as first order phase transitions) result in redder power spectra.

Unfortunately, gravitational waves from inflation are not the only way to create polarization patterns with non–zero curl. Only if the inflationary scale is sufficiently high will we be able to disentangle the influence of the gravitational waves from astrophysical foregrounds such as lensing by local large-scale structure.

The next leap in knowledge of the CMB temperature power spectrum will likely come in 2002-03 with results from NASA's MAP satellite. MAP is the first all-sky CMB experiment since COBE/DMR, with a resolution comparable to current balloon and ground-based telescopes.

The Planck Surveyor, to be launched in 2007 and as discussed by Francois Bouchet at this meeting, will recover the *entire* primary temperature anisotropy. Neither Planck nor any other currently planned and funded mission will achieve the same for polarization. Only a small fraction of the E multipole moments will be significantly measured. For the B-mode, if it comes from inflationary gravitational waves, the situation is less certain, since the amplitude of the signal depends on the unknown energy scale of Inflation.

In this short review, we have not been able to examine the other directions CMB science will take over the next decade as microwaves become a waveband for extragalactic and galactic astrophysics, as we observe the Sunyaev-Zel'dovich effect, dust and synchrotron emission from our own and external galaxies. Satellites like MAP and Planck will be crucial to these observations, but so will further pioneering efforts from the ground and atmosphere.

Acknowledgments

The author thanks the MAXIMA, BOOMERANG and COMBAT collaborations with and by whom much of the work described here was done. He also wishes to thank Sarah Church and Lloyd Knox from which work with whom some of the text was adapted. This work was supported by the NSF and NASA in the USA and PPARC in the UK. Thanks also to the organizers for a wonderful meeting.

References

1. C. B. Netterfield *et al.*, *ApJ* **571**, 604 (2002), astro-ph/0104460.
2. A. T. Lee *et al.*, *ApJ* **561**, L1 (2001).
3. N. W. Halverson *et al.*, *ApJ* **568**, 38 (2002).
4. T. J. Pearson *et al.*, astro-ph/0205388.
5. P. F. Scott *et al.*, astro-ph/0205380.
6. S. Hanany *et al.*, *ApJ* **545**, L5 (2000).
7. P. de Bernardis *et al.*, *Nature* **404**, 955 (2000).
8. A. Linde, *Phys. Rep.* **333**, 575 (2000).
9. A. H. Guth, *Phys. Rep.* **333**, 555 (2000).
10. P. J. Steinhardt and N. Turok, *Science* **296**, 1436 (2002).
11. E. W. Kolb and M. S. Turner, *The Early Universe* (Frontiers in Physics, Addison-Wesley, Reading, MA, 1990).
12. P. J. E. Peebles, *Principles of Physical Cosmology* (Princeton Series in Physics, Princeton University Press, Princeton, NJ, 1993).
13. S. Church, A. Jaffe, and L. Knox, astro-ph/0111203.
14. W. Hu, D. Scott, and J. Silk, *Phys. Rev.* **D49**, 648 (1994).
15. U. Seljak and M. Zaldarriaga, *ApJ* **469**, 437 (1996).
16. http://physics.nyu.edu/matiasz/CMBFAST/cmbfast.html.
17. A. Lewis, A. Challinor, and A. Lasenby, *ApJ* **538**, 473 (2000).
18. http://camb.info/.
19. A. H. Jaffe, M. Kamionkowski, and L. Wang, *Phys. Rev.* **D61**, 083501 (2000).
20. G. F. Smoot *et al.*, *ApJ* **396**, L1 (1992).
21. L. Knox and L. Page, *Phys. Rev. Lett.* **85**, 1366 (2000).
22. A. E. Lange *et al.*, *Phys. Rev.* **D63**, 042001 (2001).
23. A. Balbi *et al.*, *ApJ* **545**, L1 (2000).
24. A. H. Jaffe *et al.*, *Phys. Rev. Lett.* **86**, 3475 (2001).
25. R. Stompor *et al.*, *ApJ* **561**, L7 (2001).
26. R. Stompor *et al.*, *Phys. Rev.* **D65**, 022003 (2002).
27. P. G. Ferreira and A. H. Jaffe, *MNRAS* **312**, 89 (2000).
28. O. Doré, R. Teyssier, F. R. Bouchet, D. Vibert, and S. Prunet, *A&A* **374**, 358 (2001).
29. J. H. P. Wu *et al.*, *ApJS* **132**, 1 (2001).
30. M. Tegmark, *Phys. Rev.* **D55**, 5895 (1997).
31. J. R. Bond, A. H. Jaffe, and L. Knox, *Phys. Rev.* **D57**, 2117 (1998).
32. M. P. Hobson and K. Maisinger, astro-ph/0201438.
33. E. Hivon *et al.*, *ApJ* **567**, 2 (2002).
34. J. R. Bond, A. H. Jaffe, and L. Knox, *ApJ* **533**, 19 (2000).
35. A. Lewis and S. Bridle, astro-ph/0205436.
36. M. Kaplinghat, L. Knox, and C. Skordis, astro-ph/0203413.
37. C. Contaldi, M. Hindmarsh, and J. Magueijo, *Phys. Rev. Lett.* **82**, 679 (1999).

The Sunyaev-Zel'dovich Effect: Surveys and Science

Mark Birkinshaw

THE SUNYAEV-ZEL'DOVICH EFFECT: SURVEYS AND SCIENCE

M. BIRKINSHAW
Department of Physics, University of Bristol
Tyndall Avenue,
Bristol BS8 1TL, UK
E-mail: Mark.Birkinshaw@Bristol.ac.uk

High-significance measurements of the Sunyaev-Zel'dovich effect in luminous X-ray clusters can now be made using interferometers, bolometers, and radiometers. Subject to some modelling uncertainties, these data can be used for a variety of cosmological purposes, including the study of the Hubble diagram and the measurement of the baryon fraction in clusters. Several groups are working on the next steps in the utilization of the Sunyaev-Zel'dovich effect, surveys for the effect in cluster samples and in blank fields, and improved measurements of the spectrum of the effect to extract its kinematic component. This article reviews the current state of Sunyaev-Zel'dovich effect research, and discusses its scientific aims, with particular attention to the AMiBA and OCRA projects.

1. The Sunyaev-Zel'dovich effects

Rich clusters of galaxies are the largest obvious concentrations of dark matter and baryons in the Universe. Their presence is most evident on X-ray images, where the high luminosities of their extensive hot atmospheres make them prominent sources. The well-known cluster CL 0016+16 (Figure 1) is a good example. Despite lying at redshift $z = 0.5455$ [1], the cluster is clearly seen as a bright, extended, X-ray source in a recent *XMM-Newton* observation [2], and the large number of X-ray counts from the cluster allows the properties of its gas to be well measured. For CL 0016+16, the intracluster gas is found to have a temperature of 9.1 ± 0.2 keV, a metal abundance of 0.22 ± 0.04 solar, and to be distributed roughly as an isothermal β model

$$n_{\rm e} = n_{\rm e0} \left(1 + \frac{r^2}{r_{\rm c}^2}\right)^{-\frac{3}{2}\beta} \tag{1}$$

with core radius $r_{\rm c} = 230 \pm 10$ kpc and $\beta = 0.70 \pm 0.01$, and central electron density $n_{\rm e0} = (8.8 \pm 0.5) \times 10^{-3}\ {\rm cm^{-3}}$ [2].

Figure 1. A vignetting-corrected $0.3 - 5$ keV image of CL 0016+16 from the combined MOS1, MOS2, and pn data obtained during a 36.8-ksec *XMM-Newton* observation of the cluster [2]. A logarithmic intensity scale is used, with the data binned into 3 arcsec pixels, and J2000 coordinates are shown. The boundaries of the MOS and pn chips are visible. The bright source 3 arcmin north of CL 0016+16 is quasar E 0015+162, which lies at a similar redshift [3]. The extended source 9 arcmin south-west of CL 0016+16 is RX J0018.3+1618, which is believed to be in the same supercluster [4].

Such a hot, extended, atmosphere provides a pool of electrons capable of scattering photons of the microwave background radiation. The scattering optical depth

$$\tau_{\rm e} \approx \overline{n_{\rm e}}\, \sigma_{\rm T}\, L \tag{2}$$

where $\overline{n_{\rm e}}$ is the average electron density on a path of length L though the cluster atmosphere and $\sigma_{\rm T}$ is the Thomson scattering cross-section, 6.65×10^{-25} cm^2. If the structure of a β-model atmosphere is taken into account, we find that the central optical depth through the CL 0016+16 is

$$\tau_{\rm e0} = \sigma_{\rm T} \int dz\, n_{\rm e}(r) = \sigma_{\rm T}\, n_{\rm e0}\, r_{\rm c}\, \sqrt{\pi}\, \frac{\Gamma\left(\frac{3}{2}\beta - \frac{1}{2}\right)}{\Gamma\left(\frac{3}{2}\beta\right)} \approx 1.2 \times 10^{-2} \tag{3}$$

so that about 1.2% of the photons passing through the cluster are scattered by electrons in the hot gas.

Since the electrons have higher average energy than the photons that they scatter, the photons will emerge from the cluster with higher energies than when they enter. The average fractional frequency change of a photon scattered by a hot electron in CL 0016+16 is

$$\frac{\Delta\nu}{\nu} = \frac{k_{\rm B}T_{\rm e}}{m_{\rm e}c^2} \approx 1.8 \times 10^{-2} \quad . \tag{4}$$

At frequencies in the Rayleigh-Jeans ($h\nu < k_{\rm B}T_{\rm rad}$) regime, the specific intensity of the microwave background radiation $I_\nu \propto \nu^2$, and so the net change in brightness of the microwave background radiation seen through the centre of CL 0016+16 will be

$$\left(\frac{\Delta I_\nu}{I_\nu}\right)_0 = -2\,\tau_{\rm e0}\frac{\Delta\nu}{\nu} \approx 4.3 \times 10^{-4} \quad . \tag{5}$$

This implies that the $T_{\rm rad} = 2.73$ K brightness temperature of the microwave background radiation will decrease by 1.2 mK towards the centre of the cluster. This decrease is the thermal Sunyaev-Zel'dovich effect (SZE) [5].

A second SZE arises if the cluster is in motion relative to the Hubble flow, since then the moving scattering medium induces a further frequency change on photons passing through it. This kinematic SZE is

$$\left(\frac{\Delta I_\nu}{I_\nu}\right)_0 = -\tau_{\rm e0}\,\frac{v_{\rm z}}{c} \approx -4 \times 10^{-5}\left(v_{\rm z}/1000\ {\rm km\,s^{-1}}\right) \tag{6}$$

for CL 0016+16, if its peculiar radial velocity is $v_{\rm z}$. This is small relative to the thermal SZE for plausible values of $v_{\rm z}$, but might be thought to be measurable from its different spectrum (Fig. 2), or from a single measurement at about 218 GHz where the thermal SZE is zero (the precise frequency of this zero depends on the gas temperature [6]). However, the spectrum of the kinematic SZE is indistinguishable from the spectrum of primordial fluctuations in the microwave background radiation, and so a strong confusing signal is present.

More detailed reviews of the physics of the scattering process are given by Rephaeli [6] and Birkinshaw [7], where details of the calculation of the spectrum of the thermal and kinematic SZEs are given.

2. Cluster gas distributions and recent X-ray studies

Eq. (1) describes the structures of clusters of galaxies as simple, isothermal, spherical, gas clouds. While studies of CL 0016+16 (Fig. 1) suggest that the gas is close to isothermal and distributed relatively smoothly, albeit with a non-circular shape, this is certainly not the case for all clusters of galaxies, as shown by recent studies with *Chandra* and *XMM-Newton*.

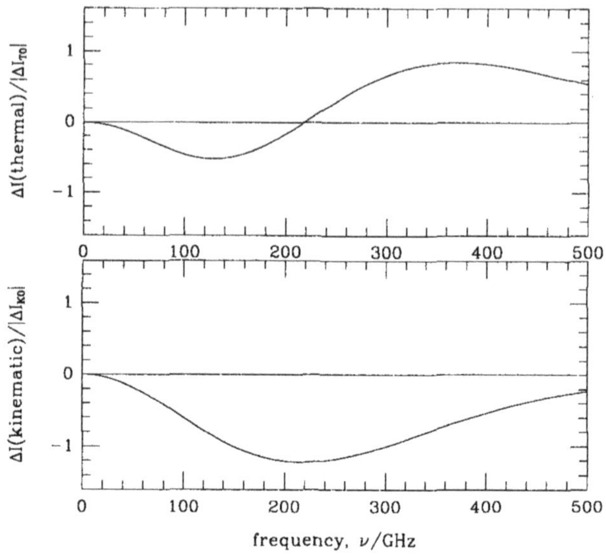

Figure 2. The spectra of the thermal (upper panel) and kinematic (lower panel) Sunyaev-Zel'dovich effects. The maximum kinematic effect occurs near the zero of the thermal effect.

Early cluster images with *Chandra* showed the presence of surprisingly sharp cold fronts within the intracluster medium. For Abell 2142, Markevitch et al. [8] found sharp temperature discontinuities in the central part of the cluster. A similar cold front has been described in Abell 3667 [9]. The distinct gas components, with their different temperatures, are presumably products of the continuing growth of the intracluster medium through accretion. The large temperature gradient across the interface implies that thermal conductivity in the intracluster medium is suppressed, probably by tangled magnetic fields.

Within the densest parts of clusters, where the most compact X-ray and SZE structures originate and central radio sources are common, considerable thermal and density complexity is seen. In the Perseus cluster [10] and near Hydra A [11] bubbles of plasma from the radio galaxy interact with the cluster gas. The radio galaxy may supply enough energy to disrupt the central part of a cooling flow, and will certainly make the central gas structure time-dependent.

Another type of structure is seen in Abell 1795 [12,13], where a trail of denser gas is created by the passage of a massive galaxy through the cluster. Though the X-ray enhancement is only a small fraction of the cluster's X-ray luminosity, it suggests that some level of clumping is maintained in the intracluster medium and should be included in models.

Some clusters show evidence of shocks where fast lumps of material are running quickly through the intracluster medium. A good example is 1E 0657-56, where *Chandra* clearly sees a gas "bullet" running through the cluster at about Mach 3, with the temperature and density jumps that it produces [14].

Another theme emerges from studies of the X-ray spectra of Abell 1795, Abell 1835, and other cooling flow clusters [15,16]. While some of the expected emission lines from the cooling gas are seen, the cooling appears to stop before lines characteristic of gas below 2 keV appear. The fate of cooling gas in clusters remains uncertain.

Finally, some clusters show a hard X-ray excess in their spectra. This can be attributed to a population of suprathermal electrons, which might also be responsible for the radio halos. Liang et al. [17] describe how this might work for the Coma cluster.

These recent results imply that the intracluster medium is more complicated than suggested by the simple isothermal β model of Eq. (1). Temperature, density, and velocity substructures are relatively common. Thus predictions of the brightness of the thermal and kinematic SZEs, especially predictions of the amplitudes of these effects on the smallest angular scales, are likely to be poor. The implication is that much can be learned about the physics of cluster atmospheres through detailed SZE mapping. However, we can expect difficulties in using cluster gas for cosmology (Sec. 4.1).

3. Sunyaev-Zel'dovich effect observations

Since the first reliable detection of the SZE in 1978 [18], there have been many campaigns to measure the SZEs in known X-ray clusters of galaxies. Single-dish radiometers and radiometer arrays, interferometers, and bolometers or bolometer arrays have all been used. In this article it is not possible to give more than a brief flavour of each method, with some representative references. A review of the observational state of affairs in 1999 [7] is now out of date because of the dramatic increase in the number and quality of observations over the past four or five years.

Perhaps the most effective method of observing cluster SZEs over the past few years has been interferometry. This has usually involved retrofitting existing interferometers to obtain an abundance of short antenna-antenna separations, either by installing lower-frequency receivers (as with the BIMA and OVRO arrays [19]) or by adding new close stations for the antennas (as with the Ryle telescope [20]). Cluster SZEs typically have angular sizes of an arcminute or more, so that much of the correlated power seen by an interferometer appears on antenna-antenna baselines less than about 2000λ. Such short spacings were not well populated in interferometers designed to make detailed maps of star-forming regions or radio galaxies, and hence have to be added by increasing the observing wavelength, λ, or altering the array geometry to convert the interferometer into an effective SZE instrument.

These retro-fitted interferometers have now detected a large number of clusters of galaxies at redshifts out to ~ 1 (recent examples include [21,22,23]). High signal/noise detections of cluster SZEs are possible because interferometers have an intrinsic ability to reject contaminating signals from their surroundings and the atmosphere. Extremely long interferometric integrations are therefore possible before residual systematic errors affect the quality of the data.

A limitation of these retro-fitted systems is that their range of antenna-antenna separations remains poor for mapping SZEs with high angular dynamic range. For this reason, a new generation of interferometers, better suited to this purpose, including AMI [24], SZA [25], and AMiBA [26], is under construction. These interferometers will have antenna-antenna spacings optimized for measuring SZEs in known X-ray clusters of galaxies (which has dominated the work to date), and will have sufficient sensitivity to perform deep blank-field searches for SZEs.

Single-dish radiometer systems provided the first detections of SZEs, and are still efficient for finding strong SZEs. Their large filled apertures can integrate the signal over much of the solid angle of a cluster. However, the systematic errors suffered by these systems, from spillover or residual atmospheric noise, has limited the length of useful integrations and hence their ultimate sensitivity (e.g., [27]). Since practical systems always involve a level of differencing between on-source and off-source radiometers, there is a maximum cluster angular size (a lower redshift limit) for which sensitive observations are possible, just as there is for an interferometer.

Radiometers also face the problem of contaminating radio sources. A large fraction of X-ray luminous clusters of galaxies contain strong radio sources whose flux densities fill in the negative hole of the SZE. While at higher frequencies these sources tend to be weaker, they are often variable, so that not only is the SZE reduced in amplitude, but historical knowledge of the brightness of a source (perhaps from an interferometric map) does not provide a reliable flux density. This problem is less important for interferometers, which can simultaneously measure the SZE on the short baselines and contaminating radio sources on long baselines.

Modern antennas provide new opportunities for radiometer systems since they have superior spillover characteristics and which can be equipped with radiometer arrays. The GBT is an obvious example, and the OCRA project (Sec. 5.2) is intending to exploit array technology to the full to make a fast SZE mapper. The Viper system at the South Pole, working at 40 GHz, is an example of a sensitive system able to detect the effects from large numbers of X-ray clusters and perform blank-field SZE surveys [28].

Bolometers typically operate at much shorter wavelengths than interferometers or radiometers, and are capable of measuring the spectrum of the SZE above $\sim$ 90 GHz. They therefore provide an opportunity to separate the thermal SZE from the kinematic SZE and primordial confusion. The intrinsic sensitivity of bolometers should allow fast measurements of the SZEs of targeted clusters, or high-speed surveys of large regions of sky. However, bolometers are exposed to a high level of atmospheric and other environmental signals, and so the differencing scheme used to extract sky signals from the noise must be of high quality.

Early bolometric attempts to measure the spectrum of the SZE have been reported using SuZIE [29,30]. Newer bolometer arrays, such as BOLOCAM [31] and ACBAR [32] should allow fast surveys for clusters and confusion-limited measurements of the cluster velocity if the flux scale near 1 mm is well calibrated.

There are about 50 published, reliable, detections of the SZE in known X-ray luminous clusters of galaxies, many with $> 10\sigma$ significance and confirmations with several telescopes (e.g., CL 0016+16). Interferometric maps, albeit with restricted angular dynamic range, exist for many clusters. But only rudimentary data on the spectra of cluster SZEs exist.

4. Science with the Sunyaev-Zel'dovich effects

4.1. *The distance scale*

The most popular use of the SZE has been in combination with X-ray data to measure the distance scale of the Universe [33]. The SZE, X-ray brightness, and X-ray measured electron temperature, T_e, from a cluster of galaxies can provide an absolute cluster distance (under assumptions about the shape and smoothness of the cluster atmosphere). This distance is combined with the cluster redshift to measure the Hubble constant. If this is done for many clusters over a range of redshifts, then several cosmological parameters can be measured [34].

A good understanding of the structure of the atmospheres of clusters of galaxies is essential if this method is to be used. For a simple isothermal β model the method is relatively straightforward. The X-ray datum might be the central surface brightness, the distance-weighted cluster emission measure

$$E = \frac{n_{e0}^2 \, r_c^3}{\eta \, d_L^2} \frac{\sqrt{\pi}}{4} \frac{\Gamma\left(3\beta - \frac{3}{2}\right)}{\Gamma\left(3\beta\right)} \tag{7}$$

from fitting the cluster X-ray spectrum (η is the ratio of the electron and proton number densities), or some other indication of the cluster's X-ray brightness. This is combined with the thermal Sunyaev-Zel'dovich flux density, the central SZE brightness temperature change at low frequency

$$\Delta T_0 = -2 \left(\frac{k_B \, T_e}{m_e \, c^2} \right) \tau_{e0} \quad , \tag{8}$$

or some other indication of the inverse-Compton optical depth through the cluster, to eliminate the unknown scale electron density, n_{e0}, and so to measure a scale size for the cluster

$$r_c \propto \frac{E \, T_e^2}{\Delta T_0^2} \quad . \tag{9}$$

This scale size is then compared with the angular scale of the cluster, θ_c, to obtain its angular diameter distance, $d_A = r_c/\theta_c$, and this, combined with the redshift, yields the Hubble constant. A recent application of this method for CL 0016+16 [2] found a Hubble constant $H_0 = 68 \pm 8 \text{ km s}^{-1}\,\text{Mpc}^{-1}$ for a cosmology with $\Omega_m = 0.3$ and $\Omega_\Lambda = 0.7$. Similar precision measurements for a set of about 70 clusters would allow good estimates of H_0 and the cosmic equation of state parameter, w [34]. At present we are far from that state: a Hubble diagram based on 14 clusters is shown in Fig. 3.

This approach requires excellent absolute calibration of the X-ray and SZE data, and precise temperatures for the cluster atmospheres. Any significant substructure (Sec 2) must be modelled, or systematic errors will appear in the distance scale. Since clusters are relatively young, and change rapidly with redshift, the amount of substructure, and so error in the distance scale, will vary with redshift unless excellent X-ray imaging and spectroscopy are available to map each cluster's internal structure.

The type of SZE data used also make a difference. It is possible to tune the technique so that uncertainties in the model of the gas structure have little effect on the distance estimate. This is an advantage of interferometric observations, which are most sensitive to the central regions, where the best structural information is available, and the distance scale derived is relatively insensitive to modelling uncertainties [35].

Important selection effects are associated with the set of clusters used as distance estimators. If clusters are chosen based on any surface brightness

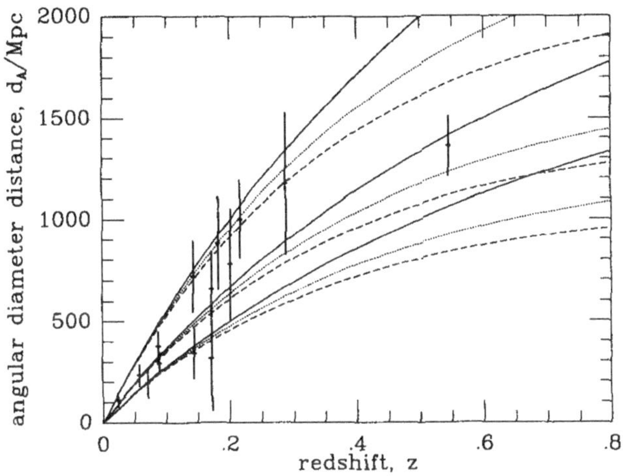

Figure 3. A Hubble diagram based on 14 clusters of galaxies with recent distance measurements. The curves show the distance relation for $H_0 = 50$, 75, and 100 $\mathrm{km\,s^{-1}}$ and a flat Universe with ($\Omega_m = 0.0, \Omega_\Lambda = 1.0$) (solid lines), ($\Omega_m = 0.3, \Omega_\Lambda = 0.7$) (dotted lines), and ($\Omega_m = 0.6, \Omega_\Lambda = 0.4$) (dashed lines). CL 0016+16 provides the point at $z = 0.5455$ [2].

criterion, then they will have preferred orientations: an ellipsoidal cluster has its highest surface brightness when its long axis is oriented towards the observer. This would cause a systematic bias in the length scale. Clusters should be selected on the basis of total X-ray luminosity or some other orientation-independent estimator [36].

4.2. *Surveys*

Blind surveys for SZEs provide another important cosmological test. Such surveys have the advantage over X-ray or optical surveys that the central SZE decrement is redshift-independent, and so readily visible to large redshift. This causes surveys with most instruments to be almost *mass-limited* and *orientation independent*, making them powerful indicators of the development of structure in the Universe. As shown by Fan and Chiueh [37], the redshift distribution and number counts of clusters can provide strong constraints on σ_8 and Ω_m.

Bolometer or radiometer arrays are likely to be efficient for surveys. Both techniques can survey the sky to high sensitivity relatively rapidly, though the radiometer technique is more subject to confusion from the non-thermal radio source population. Nearly-filled interferometric arrays, such as AMiBA (Sec. 5.1) can also be fast.

An interesting survey currently under way where this work will be possible is the *XMM-Newton* large-scale structure survey [38]. The 64 deg^2 survey should detect about 600 clusters to $z \sim 1$, and ~ 100 clusters at higher redshift, providing a measure of the evolution of large-scale structure as traced by the cluster population. A parallel SZE survey of this field would generate a well-populated Hubble diagram. The different cluster redshift distributions from the X-ray and SZE surveys will test structure formation theories, measuring σ_8 and $\Omega_{\rm m}$ and the changing properties of cluster atmospheres at $z > 1$.

4.3. *Baryon and thermal energy content*

The total SZE flux density of an object

$$S_{\rm RJ} \propto U_{\rm e} \propto N_{\rm e} T_{\rm e} \tag{10}$$

for an isothermal cluster, so that the total thermal energy contained in the electrons, $U_{\rm e}$, or the total electron count in the cluster, $N_{\rm e}$, can be deduced independent of the cluster's density structure. The gas mass of the cluster, which can be estimated if the temperature is known from an X-ray spectrum, can be compared with the total cluster mass deduced from the X-ray structure and temperature [39], or the weak gravitational lensing signal of the cluster, to measure the baryon fraction in intracluster gas, and to estimate the baryon fraction in the cluster as a whole.

The evolution, or otherwise, of the baryon fraction of clusters as a function of redshift would give much useful information on cluster formation if clusters are not fair samples of the total mass content of the Universe. Unfortunately, the baryon fractions measured are close to the value of $\Omega_{\rm b}/\Omega_{\rm m}$ for the Universe as a whole [40], and so we learn little about the development of clusters from the cluster baryon fraction.

4.4. *Velocities*

If cluster SZE spectra can be measured, then structure formation could be followed through the redshift evolution of cluster velocities. This measurement would have to be statistical, since we do not expect any one cluster to have a kinematic SZE above the level of confusion caused by primordial fluctuations in the microwave background radiation. I estimate that a set of 30 – 50 clusters in each redshift band, with confusion-limited spectra, would be needed to measure changes in cluster velocities at the level of about 200 km s^{-1} with cosmic epoch.

5. Dedicated Sunyaev-Zel'dovich effect instruments

Many dedicated microwave background telescopes are now operating (the CBI [41], VSA [42], MAP [43], MITO [44], Ryle [20], Viper + ACBAR [28], DASI [45] and others). Some of these are capable of serving as excellent SZE instruments. In the future, AMiBA [26], OCRA [46], AMI [24], Planck [47], BOLOCAM [31], and the SZA [25] will be operating and generating even better SZE detections, maps, and surveys. In this section I describe briefly some attributes of AMiBA and OCRA.

5.1. *AMiBA*

The properties of AMiBA, the Array for Microwave Background Anisotropy, are described in the article by Chiueh in this volume [48]. AMiBA is a platform-based system, operating at 95 GHz with a bandwidth of 20 GHz, with up to 19 1.2-m antennas for SZE work. It can reach a flux density limit of 1.3 mJy per beam in 1 hour, making it a highly effective interferometer for SZE surveys. AMiBA should be operational by 2004.

The high density of short antenna-antenna spacings and the 3-mm operating wavelength make AMiBA the most sensitive of the planned interferometers. However, AMI at 15 GHz [24], the SZA at 30 GHz [25], and AMiBA at 95 GHz have usefully complementary angular resolutions and operating frequencies. Although the spread of frequencies is inadequate for useful SZE spectral work, it is sufficient to provide a control against non-thermal radio sources, and an important cross-check on cluster counts.

The synthesized beam of AMiBA will be about 120 arcsec FWHM, so that high-redshift clusters appear almost unresolved to AMiBA. Each pointing will cover about a 10 arcmin diameter field, so that several thousand individual pointings would be needed for the planned shallow survey of the entire 64 deg^2 *XMM-Newton* survey field. While most of the $\sim 10^3$ SZE detections expected will correspond to X-ray clusters, a comparison of the *XMM-Newton* and AMiBA detection functions shows that AMiBA will be better at detecting clusters beyond $z = 0.7$, so that AMiBA detections with *XMM-Newton* non-detections are candidate $z > 0.7$ clusters with X-ray luminosities $> 4 \times 10^{43}$ $\text{erg}\,\text{s}^{-1}$ (in $0.5 - 10$ keV). Such clusters could be detected to $z > 2$, and detailed optical and infra-red follow-up work is planned.

A deeper survey of the central, higher-sensitivity, *XMM-Newton* survey field would push the mass limit for detectable clusters to lower levels, where confusion against the primordial microwave background structure will become important, although some of this confusing signal could be filtered out using a mosaic operating mode. Confusion from radio sources in the clusters is not expected to be a significant problem for either the deep or shallow surveys.

5.2. *OCRA*

While AMiBA is an interferometer and will have the advantages of such systems for rejecting contaminating environmental signals, radiometers provide another method of achieving high sensitivity. Accordingly, OCRA, the One Centimetre Radiometer Array [46], has been designed to make use of recent advances in radiometer technology.

The OCRA concept is that an array of $\sim$ 100 radiometers, housed in single cryostat at the secondary (or tertiary) focal plane of a large antenna, can achieve extremely high flux density sensitivity at 30 GHz. This allows a fast survey of a large area of sky for non-thermal radio sources (whose high-frequency population is little known) and SZEs.

The prototype, OCRA-p, is a two-beam system providing 1 arcmin FWHM beams separated by about 3 arcmin. It uses a receiver based on the Planck radiometers, and is nearing completion at Jodrell Bank. The system temperature should be less than 40 K, so that the receiver will achieve a flux density sensitivity of 5 mJy in 10 sec provided that the atmospheric noise is well controlled by the switching scheme. OCRA-p will be tested on the Torun 32-m radio telescope shortly, by making a restricted-area survey for radio sources and SZEs and pointed observations of known SZE clusters.

Following OCRA-p, the FARADAY project receiver with eight beams using MMICs on InP substrates, funded by a European Union grant, will be built and used at Torun. The collaboration then intends to proceed to the full 100-radiometer OCRA system.

For high-redshift objects, OCRA will be an extremely fast SZE instrument: a comparison of its mapping speed with some other systems is shown in Fig. 4. While OCRA is more susceptible to radio source confusion than AMiBA (indeed, the study of the radio source population at 30 GHz is one aim of OCRA), my recent work on the probability of detecting SZEs from clusters at 18.5 GHz suggests that the level of source confusion will not prevent OCRA from detecting most rich, distant, clusters.

6. Summary

The principal attribute of the SZEs that make them such powerful cosmological tools is their redshift-independence. The thermal SZE is, essentially,

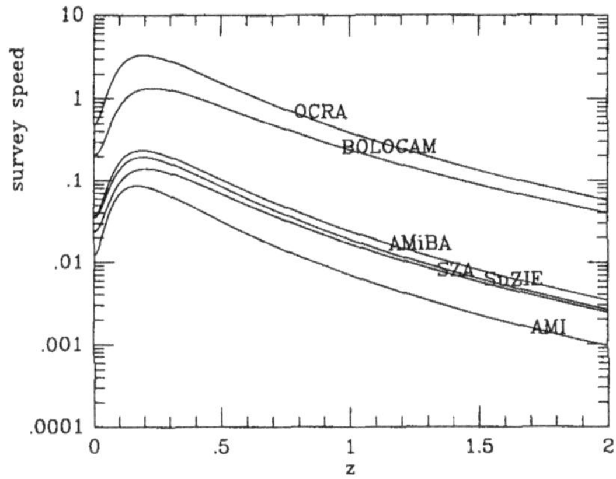

Figure 4. The relative survey speeds of several future SZE instruments, as a function of cluster redshift, for a model in which cluster gas evolves in density (as $(1+z)^{-0.5}$), in temperature (as $(1+z)^{-1}$) and in size (as $(1+z)^{-0.25}$). The decline in sensitivity at large redshift caused by this model is quite different from the predicted sensitivity increase in self-similar models: the shape of the number count as a function of SZE flux density will be a good indicator of cluster atmosphere evolution.

a *leptometer* — it counts the electron content of a cluster of galaxies, if the cluster temperature is known. The kinematic SZE is a *radial speedometer* — it can measure the changes in cluster peculiar velocities with cosmic epoch because of the changing (and linear) gravitational accelerations that the clusters encounter. The sum of the SZEs is an effective *hot mass finder* at all redshifts, not suffering from the $(1+z)^4$ dimming or K-correction of X-ray or optical studies of clusters. With new facilities for SZE work, such as AMiBA and OCRA, coming available, further exploitation of the SZEs will be a feature of microwave background work in the coming decade.

Acknowledgments

I thank to the Royal Society and the National Science Council for the award of a grant enabling collaborative work with Prof. Chiueh of the NTU.

References

1. A. Dressler and J.E. Gunn, *Astrophys. J. Suppl.* **78**, 1 (1992).
2. D.M. Worrall and M. Birkinshaw, *Mon. Not. R. astr. Soc.* submitted (2002).
3. B. Margon, R.A. Downes and H. Spinrad, *Nature* **301**, 221 (1983).
4. J.P. Hughes, M. Birkinshaw and J.P. Huchra, *Astrophys. J.* **448**, L93 (1995).
5. R.A. Sunyaev and Ya. B. Zel'dovich, *Comm. Astrophys. Sp. Phys.* **4**, 173 (1972).
6. Y. Rephaeli, *Ann. Rev. Astr. Astroph.* **33**, 541 (1995).

7. M. Birkinshaw, *Phys. Reports* **310**, 97 (1999).
8. M. Markevitch et al., *Astrophys. J.* **541**, 542 (2000).
9. A. Vikhlinin, M. Markevitch and S.S. Murray, *Astroph. J.* **551**, 160 (2001).
10. A.C. Fabian, A. Celotti, K.M. Blundell, N.E., Kassim and R.A. Perley, *Mon. Not. R. astr. Soc.* **331**, 369 (2002).
11. B. McNamara et al., *Astrophys. J.* **534**, L135 (2000).
12. S. Ettori, A.C. Fabian, S.W. Allen and R.M. Johnstone, *Mon. Not. R. astr. Soc.* **331**, 635 (2002).
13. A. Fabian et al., *Mon. Not. R. astr. Soc.* **321**, L33 (2001).
14. M. Markevitch et al., *Astrophys. J.* **567**, L27 (2002).
15. T. Tamura et al., *Astr. Astroph.* **365**, L87 (2001).
16. J.R. Peterson et al., `astro-ph/0202108` (2002).
17. H. Liang, V.A. Dogiel and M. Birkinshaw, *Mon. Not. R. astr. Soc.* in press (2002).
18. M. Birkinshaw, S.F. Gull and K.J.E. Northover, *Nature* **185**, 245 (1978).
19. J.E. Carlstrom, M. Joy and L. Grego, *Astrophys. J.* **456**, L75 and erratum **461**, L9 (1996).
20. M. Jones et al., *Nature* **365**, 320 (1993).
21. M. Joy et al., *Astrophys. J.* **551**, L1 (2001).
22. K. Grainge et al., *Mon. Not. R. astr. Soc.* **333**, 318 (2002).
23. G. Cotter et al., *Mon. Not. R. astr. Soc.* **334**, 323 (2002).
24. R. Kneissl et al. *Mon. Not. R. astr. Soc.* **328**, 783 (2001).
25. J.J. Mohr et al. *AMiBA 2001*, 43 (2002).
26. F.Y. Lo et al., *IAU Symp.* **201**, 51 (2001).
27. M. Birkinshaw and S.F. Gull, *Mon. Not. R. astr. Soc.*, **206**, 359 (1984).
28. G. Griffin et al., *Bull. A.A.S* **192**, 5803 (1998).
29. W. Holzapfel et al., *Astroph. J.* **481**, 35 (1997).
30. S. LaRoque et al., `astro-ph/0204134` (2002).
31. J. Glenn et al., *Proc. SPIE* **3357**, 326 (1998).
32. A.K. Romer et al., *Bull. A.A.S.* **199**, 1420 (2001).
33. M. Birkinshaw, *Mon. Not. R. astr. Soc.* **187**, 847 (1979).
34. S. Molnar, M. Birkinshaw and R. Mushotzky, *Astrophys. J.* **570**, 1 (2002).
35. E. Reese et al., *Astroph. J.* **533**, 38 (2002).
36. M. Birkinshaw, J.P. Hughes and K. Arnaud, *Astroph. J.* **379**, 466 (1991).
37. Z. Fan and T. Chiueh, *Astrophys. J.* **550**, 547 (2001).
38. M. Pierre et al., *ESO Messenger* **105**, 32 (2001).
39. D. Fabricant, M. Lecar and P. Gorenstein, *Astroph. J.* **241**, 552 (1980).
40. L. Grego et al., *Astroph. J.* **552**, 2 (2001).
41. T.J. Pearson et al., *IAU Symp.* **201**, 2 (2001).
42. R.A. Watson et al., `astro-ph/0205378` (2002)
43. E.L. Wright, *New Astr. Rev.* **43**, 257 (1999).
44. L. D'Alba et al., *New Astr. Rev.* **43**, 297 (1999).
45. N.W. Halverson, J.E. Carlstrom, M. Dragovan, W.L. Holzapfel and J. Kovac, *Proc. SPIE* **3357**, 416 (1998).
46. I.W.A. Browne et al., *Proc. SPIE* **4015**, 299 (2000).
47. J. Trauber, *IAU Symp.* **204**, 40 (2000).
48. T. Chiueh, this volume (2002).

AMiBA and Galaxy Cluster Survey via Thermal Sunyaev-Zel'dovich Effects

Tzihong Chiueh

AMIBA AND GALAXY CLUSTER SURVEY VIA THERMAL SUNYAEV-ZEL'DOVICH EFFECTS

TZIHONG CHIUEH

Department of Physics, National Taiwan University,
& Institute of Astronomy and Astrophysics, Academia Sinica,
Taipei, Taiwan
E-mail: chiuehth@phys.ntu.edu.tw

Array for Microwave Background Anisotropy (AMiBA) is a 19-element, platform-mounted interferometry telescope operating at 90 GHz. With a 10' field of view and 2' angular resolution, the designed sensitivity of AMiBA for real-space imaging of galaxy clusters via the Sunyaev-Zel'dovich effect can achieve 150 $\mu K/\sqrt{sec}$. I present our preliminary tests for the quality of AMiBA cluster images via mosaic mock observations on the simulated sky produced by cosmological N-body/SPH simulations. I also discuss the issue of confusion caused by the synchrotron and dust emission from galaxies residing in clusters.

1. AMiBA

The full AMiBA telescope is a 19-element interferometry array[1] sited at Mauna Loa, Hawaii, and to be completed in late 2004. In the interim period before late 2004, AMiBA will begin the observation with 7 dish elements in 2003 and with 13 dish elements in 2004.

1.1. *Science Targets:*

The AMiBA project aims at measuring (1) the polarization of the cosmic microwave background (CMB) anisotropy at the angular scale from $l =$ 1000 to $l = 5000$ in the spherical harmonics space, and (2) the high-redshift galaxy clusters at the angular scale from $l = 3000$ to $l = 16000$ via the thermal Sunyaev-Zel'dovich (SZ) effect[2,3]. To achieve the two science goals, AMiBA is equipped with two separate sets of dish antenna of diameter 60cm and 120cm respectively to probe physics at different angular scales. Figure (1) shows the AMiBA angular scale overlaid upon relevant scales of CMB physics.

The AMiBA telescope is designed to achieve a sensitivity $260\mu K/\sqrt{t_{sec}}$ for the polarization measurement at the most sensitive scale $l = 1200$.

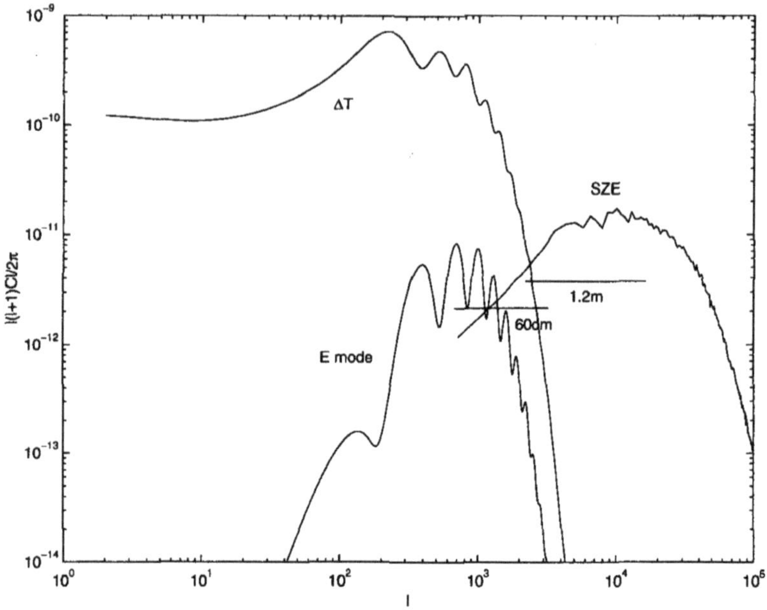

Figure 1. Relevant CMB scales in comparison with the scales sensitive to AMiBA

The experiment, with 60cm dishes, is expected to detect the 1σ E-mode polarization within few-hour integration. AMiBA may even detect the polarization features of scales down to $l = 600$ by adopting the drift-scan observing strategy.[4] Moreover, AMiBA, being an interferometry array, can also observe the CMB polarization in the single-dish mode. With 19 single dishes that scan a sky annulus, AMiBA can detect the polarization E-mode and B-mode separately with relatively high efficiencies.[5] Some details of the AMiBA CMB polarization measurements can be found in the presentation given by K.W. Ng in this proceeding.

The AMiBA telescope with 120cm dishes also has superb imaging capability. The field of view of AMiBA is about 10' and the synthesized beam, or angular resolution, about 2'. The 2' resolution is chosen to match the typical size of galaxy clusters located at $z = 0.5 - 1$. Imaging with an interferometry array requires the full use of all baselines. With 19 dish elements, AMiBA has 171 baselines and is expected to have a sensitivity $150\mu K/\sqrt{t_{sec}}$.

1.2. *Specifications*

AMiBA is equipped with *dual*-polarization wide-band receivers and designed to have a system temperature 70 K that includes the atmospheric temperature. The receiver operates at 90 GHz frequency, with a bandwidth $\Delta\nu = 20$ GHz. Within the 20 GHz band, the signal is further divided into *two* channels of 10 GHz wide, thus defining the frequency resolution. The interference signals of a given baseline are obtained for each channel as well as for each polarization.

All AMiBA dish elements are mounted on a flat platform, permitting all dishes to stare at the same patch of sky without mutual shadowing and without changes of baseline lengths as the sky rotates. The platform is in turn supported by a hexapod mount, allowing for platform maneuvers with 3 translational and 3 rotational degrees of freedom. The platform diameter is 6 meters; this aperture defines the angular resolution of AMiBA, which amounts to 2 arcminutes.

Two separate sets of dishes, of 60cm and 120cm diameters, are to be in use for different science purposes. The interferometry array has the highest sensitivity in the shortest baselines, where the dish diameter equals the dish separation and the closest dish pair nearly touch each other. For the CMB polarization measurement, the closest pairs of 60cm dishes are optimized to measure angular patterns around $l = 1200$. With 13 dish elements, AMiBA has 24 such 60cm baselines. For the interferometric imaging of SZ clusters, the 120cm dishes are chosen so as to fill out the 6 meter platform by 19 dish elements in a closely packed hexagonal configuration and also to avoid picking up the CMB primary anisotropy. A 70% antenna efficiency is expected from this dish-receiver system in the AMiBA design.

2. AMiBA Sunyaev-Zel'dovich Cluster Survey

The Sunyaev-Zel'dovich effect primarily involves the physics of inverse Compton scattering of the cold CMB photons by the optically thin, intra-cluster hot gases (Sunyaev & Zel'dovich 1972; also Birkinshaw 1999 for a review). Unlike thermalized photons, upon Compton scattering the cold photons gain only their energies but not their number density. Therefore the CMB spectrum deviates from black-body. The amount by which the spectrum deviates from the thermal one depends on a dimensionless parameter, the Compton y parameter. For typical parameters in galaxy clusters, the Compton y is much less than unity and the deviation from thermal is proportionally small. In the Rayleigh-Jeans limit, the change of surface brightness ΔT is related to the Compton y parameter as $\Delta T = 2yT_{cmb}$, where T_{cmb} is the present CMB temperature 2.73K. For AMiBA which operates at *90* GHz, the correction to the Rayleigh-Jeans limit amounts to replacing the factor 2 by 1.6.

The SZ effect is redshift-independent, in the sense that the total SZ flux of a given cluster is independent of the redshift of this cluster. This nice feature arises from the fact that the observed light is the bright CMB, and the SZ signal is nothing but the shadow of any object that blocks the CMB light on the light path. Moreover, the SZ flux is approximately proportional to $M^{5/3}$ where M is the mass of the light-blocking baryonic object, and hence the SZ measurements directly probe the clumpy baryons throughout the universe. If baryons trace the dark matter, as are commonly believed, the SZ flux can further serve to probe the clumpy dark matter distribution of the universe.

Of course, in reality the early light-blocking objects tend to be of smaller size than the late objects, according to the prevailing hierarchical formation theory, where small objects formed earlier than big ones. Hence the SZ flux does depend on the redshift. But the dependence enters in a clean way, in that it depends only on the dynamics of how the structures grew during the past history of the universe and not *directly* on the kinematics of how the universe expands, though the latter also controls the structure growth and enters the problem in an indirect way. Together with the advantage mentioned in the last paragraph, the SZ measurements serve to provide unique information about the evolution of matter clumpiness.

2.1. *Determination of* σ_8

Among the various avenues in assessing the evolution of matter clumpiness, the most useful one in the present context is the Press-Schechter theory for the formation of nonlinear, virialized bound objects. This is because the dense objects are most efficient in producing local changes of CMB surface brightness. As a result, the SZ measurement can serve to test the hierarchical structure formation theory in a direct way. On the other hand, if the structure formation theory is indeed a valid one, one may further combine the Press-Schechter theory of bound-object formation with the SZ measurements to investigate various factors that govern the rate at which the bound objects formed as a function of their mass. Assuming that the universe is flat, the obvious governing factors are Ω_0, which anti-correlates with the background accelerating force, and σ_8, the (linearly extrapolated) amplitude of present density fluctuations averaged over a sphere of 8Mpc h^{-1}. For example, with a fixed σ_8, Fig.(2) shows the predictions of the Press-Schechter theory for the distributions of galaxy clusters over the redshift space for different flat CDM cosmological models in a flux-limited sample. (Fig.(2) also shows the distribution for the open CDM cosmology as a comparison.)

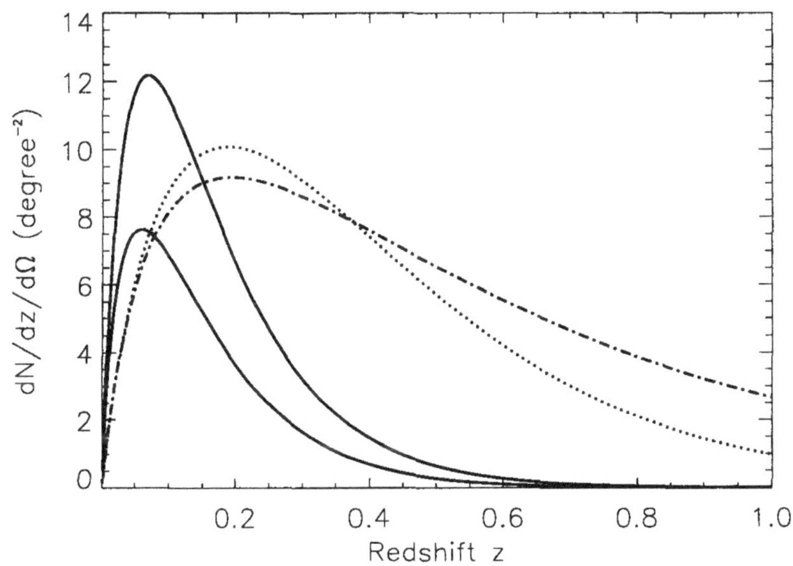

Figure 2. The flux-limited distributions of cluster number over the redshift space for the cosmological models of τCDM (solid), ΛCDM (dotted), SCDM (triple dot-dashed) and open CDM (dot-dahsed).

Of these two relevant factors it turns out that the result of SZ measurements is much more sensitive to σ_8 than to Ω_0. Plotted in Fig.(3) are the constraints to be derived from the cluster number counts measured by the SZ effect for two different flux-limited samples at 1 mJy and 6 mJy. These constraints lie almost horizontally in the $\Omega_0 - \sigma_8$ diagram,[6] indicative of that the SZ cluster count is almost independent of Ω_0, as along as the universe is flat with $\Omega_0 + \Omega_\Lambda = 1$. A similar conclusion has also been predicted for the measurements of the SZ surface-brightness power spectrum.[7,8] By contrast, we also plot in Fig.(3) the constraint given by the cluster number count measured by x-ray[9]. The x-ray constraint is sensitive to both σ_8 and Ω_0. To see the difference between the SZ and x-ray detected clusters, we note that when the SZ flux limit is raised to a higher value, the SZ constraint can approach the x-ray constraint. This tendency illustrates that the x-ray measurements tend to pick up nearby or massive clusters, whereas the low-flux-limit SZ measurements tend to pick up high-redshift ($z > 0.5$), low-to-medium mass clusters.

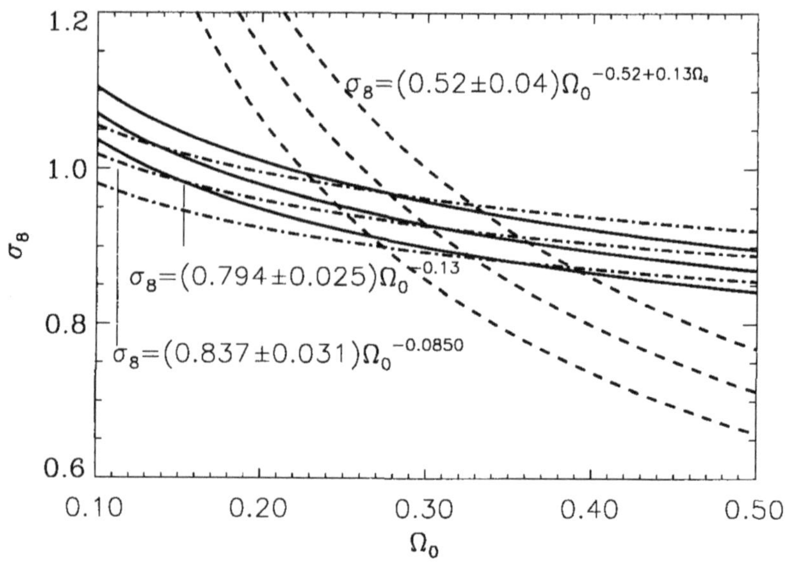

Figure 3. Bands of constraints set by the 1mJy (dot-dashed lines) and 6mJy (solid lines) flux-limited SZ surveys, as well as by the x-ray survey (dashed lines). The band widths indicate the 3-σ confidence level set in a survey area of 50 deg^2 for the 6-mJy SZ and x-ray surveys, whereas the 3-σ band width of the 1-mJy SZ survey is for a survey area of 1.5 deg^2.

2.2. *AMiBA Images of SZ Clusters*

We test the AMiBA imaging capability by performing mock observations on a simulated SZ sky. The parallel CDM/SPH code, GADGET, was run on our AMD/dual-cpu PC cluster of 32 nodes to conduct the cosmological N-body/hydrodynamics simulation. It yields a dozen of 1-square-degree SZ maps, which represent independent patches of a simulated SZ sky. The simulation was conducted with 256^3 dark matter particles and 256^3 gas particles in a 100 Mpc h^{-3} periodic box. The cosmological parameters chosen for the simulation were $\Omega_0 = 0.3$, $\Omega_\Lambda = 0.7$, $\Omega_b = 0.05$, $h = 0.65$ and $\sigma_8 = 0.75$. (This simulation does not incorporate preheating.) Shown on the left of Fig.(4) is one patch of the simulated one-deg SZ sky. (The CMB primary anisotropy has not been included in the simulated sky.)

MIRIAD, an analysis package for radio astronomy, was then used to do mock observations on the simulated sky, taking into account the specifications of AMiBA. Shown on the right of Fig.(4) is a mosaic image of the simulated sky for 500-hour AMiBA observation. Faint circles in the AMiBA map indicate 4$-\sigma$ detections of 6 clusters at redshifts, $z = 0.37, 0.4, 0.46, 0.56, 0.7$ and 1.1, respectively. The relatively low cluster number count in this mock

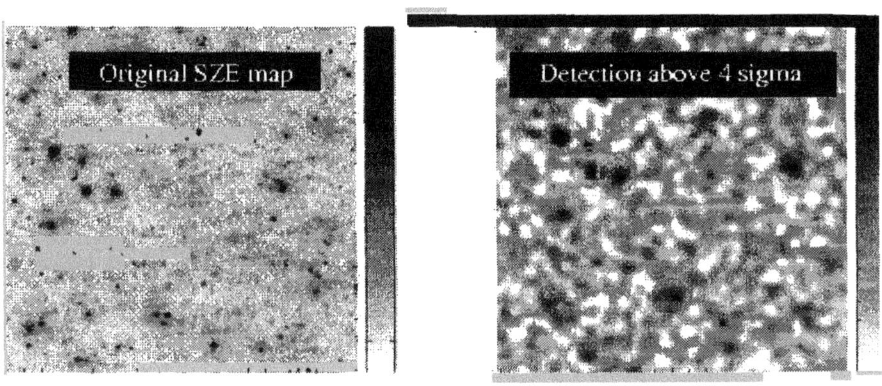

Figure 4. The simulated 1-square-deg sky on the left and the mosaic map produced by AMiBA on the right. Faint circles in the map indicate 4-σ detection.

observation is due to the relatively low σ_8 adopted in our simulation. A slight increase of σ_8 by 15%, or including preheating, can enhance the detections by at least a factor of 2.

These mock observations have been quite useful for testing the AMiBA hardware designs. For example, the 10 GHz frequency channel of AMiBA had once been thought to be too wide that it creates severe band smearing in the $u - v$ space and deteriorates the image quality. We find from the mock observations that when the mosaic strategy is adopted, the $u - v$ smearing caused by the poor frequency resolution creates little problems for the final map, thus verifying the AMiBA specifications.

3. Synchrotron/Dust Confusion Sources

It is known that most radio-emission sources in the universe reside in galaxies. On the other hand, galaxy clusters contain several tens to few hundreds of galaxies. As the SZ survey intends to measure galaxy clusters, the radio emission from cluster galaxies can therefore potentially contaminate the inverse-Compton scattered CMB photons severely. The dominant radio emission arises both from relativistic electrons that emit via synchrotron mechanisms and from dust particles that process the ultra-violet photons received from young stars and re-emit at the far-infrared. The synchrotron emission dominates the low-frequency flux and the dust emission dominates the high-frequency flux. Fortunately, the frequency at which the two emission fluxes become comparable and the total emission flux reaches the minimum turns out to be rather universal, around 100GHz, a frequency near which AMiBA operates.

However, the 100 GHz window refers to the rest-frame frequency, and the choice of 90 GHz observing frequency is good only for local galaxies. As the SZ measurements tend to pick up clusters of $z \gtrsim 0.5$, this window moves to lower observing frequencies. We are so motivated to examine how serious the confusion caused by galaxy emission affects the SZ cluster measurement.

Our strategy for investigating this problem takes advantage of the available 1.4 GHz cluster radio luminosity function[10] (CRLF) for local sources in clusters of $z < 0.1$, completed up to power 10^{22} W/Hz, as well as the power index distribution of synchrotron sources[11,12] and the spectral index distribution[13] in between 1.4 GHz and 350 GHz. The latter distribution is obtained by comparing the synchrotron flux at 1.4 GHz and the dust flux at 350 GHz; the flux ratio follows a narrow distribution as a result of the well-known strong radio/far-infrared correlation. The former distribution is obtained from two existing samples, one for sources in the average local galaxies and the other for sources in medium-redshift and very massive clusters. The higher-frequency CRLF can then be computed by convolving the 1.4 GHz CRLF with these power/spectral index distributions. Shown in Fig.(5) are the CRLFs at various higher frequencies.

The 1.4 GHz CRLF adopted here refers to the fraction of cluster galaxies with the 1.4 GHz emission power $P_{1.4}$ in between $\log(P_{1.4}) - 0.5$ and $\log(P_{1.4}) + 0.5$ among all cluster galaxies brighter than the optical r-band absolute magnitude $M^r = -20.5$. In order to relate the cluster mass to the number of 1.4 GHz radio sources, one needs the cluster mass-to-light ratio for cluster galaxies of $M^r \leq -20.5$. Such information is also available from data of the CNOC survey[14]. It was found that the mass per galaxy of $M^r \leq -20.5$ amounts to $1.5 \times 10^{13} M_\odot$. Given a cluster mass, we can thus obtain the number of such galaxies N_{gal} in the cluster.

Note that the CRLF so far mentioned is for a single cluster galaxy. The CRLF for a cluster of N_{gal} galaxies differs from that for a single galaxy; in fact the distribution of N-galaxy CRLF should approach a Gaussian as N_{gal} becomes larger as a result of the central limit theorem. Shown in Fig.(6) is an example of the 15 GHz N-galaxy CRLFs for various N_{gal}. Similar curves were also found for the N-galaxy CRLFs of other frequencies, except for the horizontal shifts in the $\log P_\nu$ axis.

After the above issues being properly taken care, the confusion flux at various frequencies and redshifts can be calculated by assuming the null-evolution hypothesis, a hypothesis that is likely a good one for sources below $z = 1$ after quasars were turned off. The N-galaxy CRLF allows

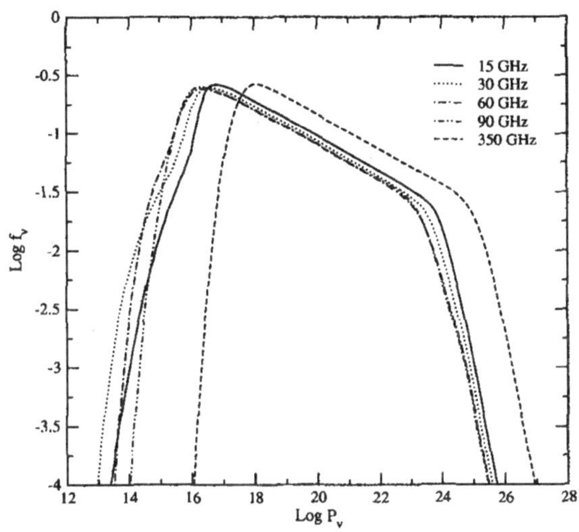

Figure 5. The cluster radio luminosity function at various frequencies.

us to compute not only the mean flux but also the variance. Plotted in Fig.(7) are the ratios of the confusion flux to the SZ flux for 30 GHz and 90 GHz as functions of cluster mass at various redshifts. In each panel, there are two curves, and they correspond to the confusion flux derived from the power-index distribution given by Slee et al.[11] for sources in average local galaxies (dark circular dots) and that given by Cooray et al.[12] for sources in galaxies (light square dots) of very massive clusters ($> 10^{15} M_\odot$). The predicted CRLFs derived from the two differ by almost a constant factor of 2 in the synchrotron-dominant regime, i.e., rest-frame frequencies < 60

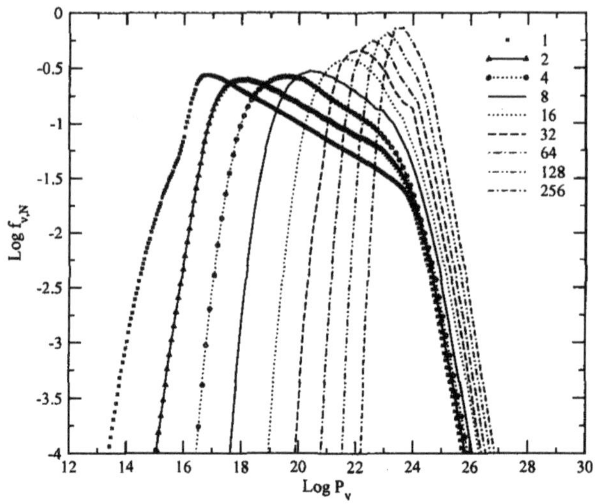

Figure 6. The N-galaxy cluster radio luminosity function at 15 GHz for various galaxy numbers N_{gal}.

GHz. But the difference begins to diminish as the rest-frame frequency exceeds 100 GHz in the dust-dominant regime.

From Fig.(7), it is clear that the 90 GHz SZ measurements need not concern about the radio confusion, except for those measurements on nearby low-mass clusters. As a result, subtraction of the confusion point sources with the aid of a larger telescope is generally not necessary for the AMiBA cluster survey. On the other hand, the lower-frequency, e.g., 30 GHz, SZ measurements are likely to be plagued by radio confusion, except for those measurements of high-redshift massive clusters.

In this presentation, I summarize the key results relevant to AMiBA from our recent work[15]. The kinematic Sunyaev-Zel'dovich effect,[16] though un-related directly to AMiBA, has been a subject of current theoretical interests, and the radio confusion to this effect can be the deciding factor that tells whether the effect is realistically detectable. Readers are referred to the original work for the details.

4. Conclusion

The AMiBA project is an international collaborative project among the Academia Sinica of Taiwan, National Taiwan University, Australia Telescope National Facilities, Carnegie-Mellon University and Canadian Institute for Theoretical Astrophysics. The prototype AMiBA, with two dish elements, is currently under testing as of mid 2002; the CMB polarization

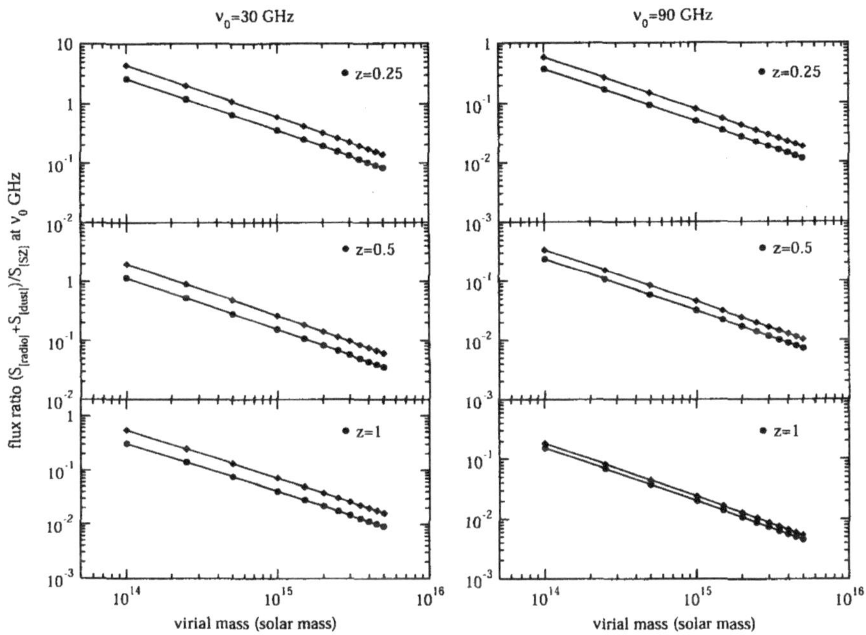

Figure 7. A comparison of the confusion flux ratios for clusters of various virial masses and located at different redshifts when observed at 30 GHz and at 90 GHz.

measurement will begin in 2003 with 7, and subsequently 13, dish elements; the SZ project will begin in 2004 with a complete 19-element array.

The AMiBA SZ project aims primarily at the galaxy cluster survey. Due to the nature of SZ effect, the SZ cluster survey has the advantage of detecting clusters of redshift $z \geq 0.5$ up to $z \sim 1.5$, characteristic of the SZ observations unparalleled by optical or x-ray cluster surveys. Hence the AMiBA cluster survey will focus on deep-field observations, with an average of 15-hour exposure per 10' field. Within a realistic operation time span of a couple of years, AMiBA will survey about 10 square degrees for this deep-survey mode. The field of this deep survey has not yet been selected, but it will likely coincide with the common fields of x-ray and optical surveys.

Acknowledgments

I would like to thank Z.Fan, H.Lin, K.Y.Lin, C.J.Ma, K.Umetsu, T.B.Woo and the AMiBA team for productive collaboration that yields the results presented here.

References

1. K.Y. Lo et al. astro-ph/0012282 (2000).
2. R.A. Sunyaev and Ya.B. Zel'dovich, *Comm. Astrophys. Space Phys.* **4**, 173 (1972).
3. M. Birkinshaw, *Phys. Rep.*, **310**, 97 (1999).
4. U.L. Pen et al., www.cita.utoronto.ca/~pen/download/AMiBA, "drift.ps" (2001).
5. T. Chiueh and C.J. Ma, *ApJ*, Oct.20 issue (2002); astro-ph/0101205.
6. Z. Fan and T. Chiueh, *ApJ* **550**, 547 (2001).
7. E. Komatsu and U. Seljak, astro-ph/0205468 (2002).
8. P. Zhang, U.L. Pen and B. Wang, astro-ph/0201375 (2002).
9. V.R. Eke, S. Cole and C.S. Frenk, *MNRAS* **282**, 263 (1996).
10. M.J. Ledow and F.N. Owen, *AJ* **112**, 9 (1996).
11. O.B. Slee, A.L. Roy and H. Andernach, *Australian J. Phys.* **49**, 977 (1996).
12. R.A. Cooray et al., *AJ* **115**, 1388 (1998).
13. C.L. Carilli and M.S. Yun, *ApJ* **530**, 618 (2000).
14. R.G. Calberg et al., *ApJ* **462**, 32 (1996).
15. H. Lin, T. Chiueh, and X.P. Wu, astro-ph/0202174 (2002).
16. R.A. Sunyaev and Ya.B. Zel'dovich, *Ann. Rev. Astron.& Astrophys.* **18**, 537 (1980).

AMiBA Observation of CMB Anisotropies

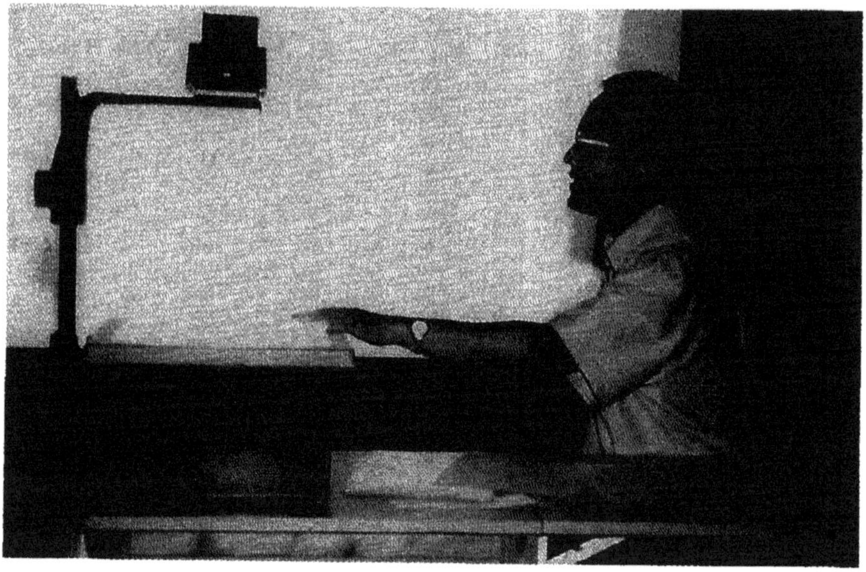

Kin-Wang Ng

AMIBA OBSERVATION OF CMB ANISOTROPIES

KIN-WANG NG*
Institute of Physics, Academia Sinica, Taipei, Taiwan 11529
E-mail: nkw@phys.sinica.edu.tw

The Array for Microwave Background Anisotropies (AMiBA), a 13-element dual-channel 85-105 GHz interferometer array with full polarization capabilities, is being built to search for high redshift clusters of galaxies via the Sunyaev-Zel'dovich effect as well as to probe the polarization properties of the cosmic microwave background (CMB). We discuss several important issues in the observation of the CMB anisotropies such as observing strategy, l space resolution and mosaicing, optimal estimation of the power spectra, and ground pickup removal.

1. Introduction

CMB experiments are commonly single-dish chopping instruments, whose scanning strategy and data analysis procedure are well developed. In the last decade, interferometers were introduced to study the microwave sky. Recent advancement in low-noise, broadband, GHz amplifiers, in addition to mature synthesis imaging techniques, has made interferometry a particularly attractive technique for detecting CMB anisotropies. An interferometric array is intrinsically a high-resolution instrument well suited for observing small-scale intensity fluctuations, while being flexible in coverage of a wide range of angular scales with the resolution and sensitivity determined by the aperture of each element of the array and the baselines formed by the array elements. A desirable feature of the interferometer for CMB observation is that it directly measures the power spectrum of the sky, and that the polarimetry is a routine. In addition, many systematic problems that are inherent in single-dish experiments, such as ground and near field atmospheric pickup, and spurious polarization signal, can be reduced or avoided in interferometry. A brief account of the interferometric CMB observations can be found in White et al.[1] Recently, the Cosmic Background Imager (CBI)[2], the Degree Angular Scale Interferometer (DASI)[3], and the Very Small Array (VSA)[4] have reported results on the detections

*in collaboration with the amiba science team

of the CMB anisotropy. The AMiBA,[5] a 13-element dual-channel 85-105 GHz interferometer array, to be sited on Mauna Loa in Hawaii, will reach a sensitivity of 10 μK in 1 hour. The construction of AMiBA is scheduled to starting operating in early 2004. As for the polarization capability, the AMiBA will have full polarizations, whereas the DASI and the CBI will not be polarization sensitive initially. The AMiBA, when used with the 0.6 m apertures, will be sensitive to CMB polarization over the range $1000 < l < 2750$. A 4-σ detection of the CMB polarization at $l \sim 1250$ needs 51 days with a single baseline.

The output of an interferometer is the visibility that is the Fourier transform of the intensity fluctuations on the sky, thus offering a sampling of the power spectrum and allowing for direct analysis of the power spectrum on the visibility plane.[6] Both CBI and DASI experiments consisting of closely packed dishes have used the visibility correlation function to extract the CMB power spectrum from the visibility data. The results have shown that in the estimation of the low-l power spectrum the uncertainty mainly attributes to sample variance. This is because in the close-packed configuration the short baselines can only have limited independent samplings on the visibility plane (see below). Nevertheless, in order to maximize the size of the dish to obtain optimal sensitivity for signal detection, the AMiBA will adopt a close-packed hexagonal configuration with the dishes almost in touching. We will discuss the observing strategy and data analysis for such a close-packed configuration.

2. Complex Visibility

The output of the interferometer is the time-averaged correlation of the electric field measured by two antennae pointing in the same direction to the sky but at two different locations:

$$V(\vec{u}, \hat{e}) = \int d\hat{e}' A(\hat{e}', \hat{e}) T(\hat{e}') e^{2\pi i \vec{u}\cdot\hat{e}'}, \tag{1}$$

where $\vec{u}$ is the separation vector (baseline) of the two antennae measured in units of the observation wavelength λ, A denotes the primary beam with the phase tracking center pointing along the unit vector $\hat{e}$, and T is the intensity fluctuations.

In typical interferometric measurements, we have $\lambda \ll D$, where D is the dish size, dictating a small field of view. Thus, for a single pointing, it is very good to make the flat-sky approximation by decomposing

$$\hat{e}' = \hat{e} + \mathbf{x}, \quad \text{with} \quad \mathbf{x} \cdot \hat{e} = 0, \quad \text{and} \quad |\mathbf{x}| \ll 1, \tag{2}$$

meaning that $\mathbf{x}$ is a two-dimensional vector lying in the plane of the sky. Hence, the complex visibility is reduced to the two-dimensional Fourier transform of the sky intensity multiplied by the primary beam:

$$V(\mathbf{u}) \equiv V(\vec{u}, \hat{e})e^{-2\pi i \vec{u}\cdot\hat{e}} \simeq \int d\mathbf{x} A(\mathbf{x})T(\mathbf{x})e^{2\pi i \mathbf{u}\cdot\mathbf{x}}, \tag{3}$$

where $\mathbf{u}$ is the two-dimensional projection vector of $\vec{u}$ in the $\mathbf{x}$ plane. In case of CMB anisotropy, we can expand

$$T(\mathbf{x}) = \int d\mathbf{u} a_T(\mathbf{u})e^{-2\pi i \mathbf{u}\cdot\mathbf{x}}, \quad \langle a_T^*(\mathbf{u})a_T(\mathbf{w})\rangle = C_T(l)\delta(\mathbf{u}-\mathbf{w}), \tag{4}$$

where $C_T(l)$ is the anisotropy power spectrum with $l = 2\pi|\mathbf{u}|$. In case of CMB polarization, T is replaced by the Q and U fields:

$$Q(\mathbf{x}) = \int d\mathbf{u}\,(a_E(\mathbf{u})\cos 2\theta_u - a_B(\mathbf{u})\sin 2\theta_u)\, e^{-2\pi i \mathbf{u}\cdot\mathbf{x}},$$
$$U(\mathbf{x}) = \int d\mathbf{u}\,(a_E(\mathbf{u})\sin 2\theta_u + a_B(\mathbf{u})\cos 2\theta_u)\, e^{-2\pi i \mathbf{u}\cdot\mathbf{x}}, \tag{5}$$

where the E-polarization, B-polarization, and ET correlation power spectra are

$$\langle a_{E,B}^*(\mathbf{u})a_{E,B}(\mathbf{w})\rangle = C_{E,B}(l)\delta(\mathbf{u}-\mathbf{w}), \tag{6}$$
$$\langle a_T^*(\mathbf{u})a_E(\mathbf{w})\rangle = C_{TE}(l)\delta(\mathbf{u}-\mathbf{w}).$$

3. Visibility Correlation Matrix

The visibility correlation matrix is then given by [6]

$$C_{ij}^T \equiv \langle V^*(\mathbf{u}_i)V(\mathbf{u}_j)\rangle = \int d\mathbf{w}\tilde{A}^*(\mathbf{u}_i-\mathbf{w})\tilde{A}(\mathbf{u}_j-\mathbf{w})C_T(2\pi|\mathbf{w}|), \tag{7}$$

where

$$A(\mathbf{x}) = \int d\mathbf{u}\tilde{A}(\mathbf{u})e^{-2\pi i \mathbf{u}\cdot\mathbf{x}}. \tag{8}$$

Different from single-dish experiments that usually scan a significant fraction of the sky and extract the CMB power spectrum from the sky map, the interferometer traces one point of the sky for a sufficiently long integration time. For a given set of measured visibilities per single pointing one can estimate the band powers of the power spectrum $C_T(l)$ by using the quadratic estimator method[7] for the correlation matrix of total visibilities $C_{ij} = C_{ij}^T + C_{ij}^N$, where C_{ij}^N is the noise correlation matrix. To reduce the sample variance, one can repoint the telescope to uncorrelated patches of the sky to measure more independent sets of visibilities. This method has been used by the CBI and the DASI. However, several problems have to be solved. The uncertainty in low-l power spectrum is dominated by sample variance. This implies that a large sky coverage is needed. The resolution which we have in l space for a single pointing of the close-packed interferometer is equal to the size of the primary beam. That is not adequate to resolve the feature of the power spectrum. The dominant foreground contamination is due to ground spillover. Although it can be eliminated by marginalizing over a common component across different fields in a given observation for each visibility, we have not taken the advantage of the drift-scan mode for a ground-based telescope that can effectively remove the ground contamination.

It has been pointed out that a close-packed interferometer is similar to a single-beam antenna undergoing chopping and wobbling.[8] So when we attempt to construct a two-point correlation function (7) with short spacings $u_i \sim u_j \sim D/\lambda$, we obtain a pure phase function which does not contain any useful information. The reason is simply that the correlation over the domain in the $\mathbf{u}$ space spanned by $\mathbf{u}_i$ and $\mathbf{u}_j$, whose size is still comparable to the size of the dish, is almost a constant. Therefore, we should sample the visibility at different parts of the sky by repointing the entire telescope, and analyze the data in the same way as in the single-dish observation. In addition, we can increase the l resolution by combining several contiguous pointings of the telescope.[1] This is known as mosaicking in interferometry. However, for long baselines, more independent samplings in $\mathbf{u}$ plane can be made, thus allowing us to use the visibility correlation method to estimate the band powers in the beam.[8]

4. Mosaicking

For a close-packed interferometer such as AMiBA, mosaicking is needed to reduce the sample variance as well as to increase the l resolution. If the sky coverage is small, one can still use the flat-sky approach. Otherwise, we would have to take the curved space into account. Furthermore, if we adopt the drift-scan mode, the convolution of the instrument beam with the sky signal in interferometry and an appropriate observing strategy will be needed.

4.1. *Flat-sky Limit*

As long as the mosaicked map is small, the flat-sky approximation is still applicable and the visibility is measured as a function of position $\mathbf{y}$ from the map,

$$V(\mathbf{u}, \mathbf{y}) = \int d\mathbf{x} A(\mathbf{x}) T(\mathbf{x} + \mathbf{y}) e^{2\pi i \mathbf{u}\cdot\mathbf{x}}. \tag{9}$$

Hence, the correlation matrix evaluated over the $\mathbf{u}$ space as well as the flat sky is given by

$$\begin{aligned} C_{ij}^T &\equiv \langle V^*(\mathbf{u}_i, \mathbf{y}_i) V(\mathbf{u}_j, \mathbf{y}_j) \rangle \\ &= \int d\mathbf{w} \tilde{A}^*(\mathbf{u}_i - \mathbf{w}) \tilde{A}(\mathbf{u}_j - \mathbf{w}) e^{2\pi i \mathbf{w}\cdot(\mathbf{y}_i - \mathbf{y}_j)} C(2\pi|\mathbf{w}|). \end{aligned} \tag{10}$$

The mosaicked data that contain information in the correlations between visibilities can be analysed using the same correlation matrix method as mentioned above but allowing different pointing centers for each visibility.[1] Recently, the CBI has used this method to extract the anisotropy power spectrum from mosaicked data (see Pearson *et al.*[2]). Polarization visibility correlation matrices can be similarly constructed and used to estimate the E and B power spectra.

4.2. *All-sky Convolution*

We discuss all-sky convolution of the instrument beam with the CMB sky signal in interferometry experiments with dishes mounted on a single tracking platform, such as the CBI, the DASI, and the AMiBA. This is necessary in the case that the curvature of the sky is non-negligible due to a large sky coverage, and the formalism can be applicable to dealing with the time-stream data obtained in the drift-scan mode of an interferometer.

Let us begin with equation (1). In Figure 1, we set up a spherical coordinate system to define the various angles. The baseline vector $\mathbf{u}$ lying on the platform is perpendicular to the pointing direction $\hat{e}(\theta, \phi)$ and is making an angle ψ with the longitude. We rewrite $\vec{u} = \mathbf{u}$, and for a small field of view we have $\mathbf{u} \cdot \hat{e}' = \mathbf{u} \cdot \mathbf{x}$ with $|\mathbf{x}| \simeq \beta$. Then, expanding the anisotropy field

$$T(\hat{e}') = \sum_{lm} a_{lm} Y_{lm}(\hat{e}'), \quad \text{where} \quad \langle a_{l'm'}^* a_{lm} \rangle = C_l \delta_{l'l} \delta_{m'm}, \tag{11}$$

and using the decomposition formula [8]

$$Y_{lm}(\hat{e}') = \sqrt{\frac{4\pi}{2l+1}} \sum_{m'} Y_{lm'}(\beta, \alpha) \, {}_{-m'}Y_{lm}(\hat{e}), \tag{12}$$

the visibility $V(\mathbf{u}, \hat{e})$ can be rewritten as

$$V(\theta, \phi, \psi) = \sum_{lmm'} a_{lm} b_{lm'}(u) \, {}_{-m'}Y_{lm}(\theta, \phi) e^{im'\psi},$$

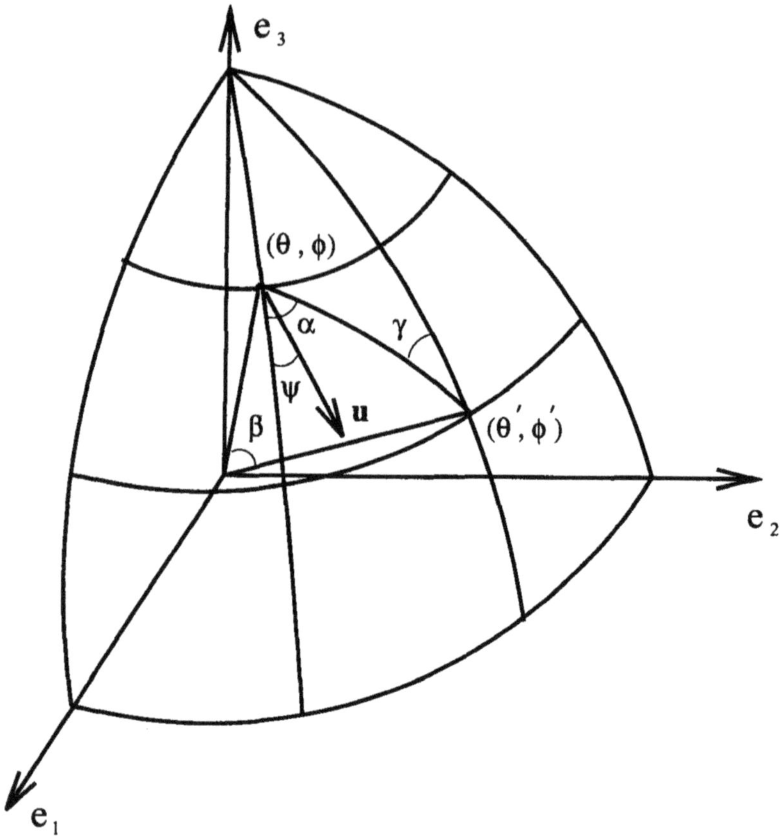

Figure 1. Spherical coordinates showing two unit vectors $\hat{e}'(\theta', \phi')$ and $\hat{e}(\theta, \phi)$ with separation angle β. The angles between the great arc connecting the two points and the longitudes are γ and α. $\mathbf{u}$ is the baseline vector tangential to the sphere with the orientation angle ψ.

$$b_{lm'}(u) = \sqrt{\frac{(l-m')!}{(l+m')!}} 2\pi i^{m'} \int_0^{\pi} d\beta\, \beta A(\beta) P_l^{m'}(\cos\beta) J_{m'}(2\pi u \beta), \quad (13)$$

where we have assumed an axisymmetric beam and integrated out α. The window function $b_{lm'}$ is a function of the length of the baseline. This result is actually a special case of the all-sky convolution in single-dish CMB experiments with asymmetric beams.[9,10] The function $V(\theta, \phi, \psi)$ can be rapidly evaluated by the method used in Challinor et al.[10] Therefore, we can use the common pipeline in single-dish experiments to analyse the mosaicked data. In constructing the two-point correlation function, we can either consider a fixed orientation angle, or sum up the orientation angles corresponding to all different baselines of the same length.

The all-sky convolution can be easily extended to polarimetry experiments by replacing the anisotropy field $T(\hat{e})$ by the Q and U Stokes parameters

$$(Q - iU)(\hat{e}) = \sum_{lm} a_{2,lm}\ {}_2Y_{lm}(\hat{e}),$$
$$(Q + iU)(\hat{e}) = \sum_{lm} a_{-2,lm}\ {}_{-2}Y_{lm}(\hat{e}), \tag{14}$$

where

$$\langle a^*_{2,l'm'} a_{2,lm} \rangle = (C_{El} + C_{Bl})\delta_{l'l}\delta_{m'm},$$
$$\langle a^*_{2,l'm'} a_{-2,lm} \rangle = (C_{El} - C_{Bl})\delta_{l'l}\delta_{m'm}, \tag{15}$$

where C_{El} and C_{Bl} are respectively E-polarization and B-polarization angular power spectra. Then we can derive the Q and U visibility functions similar to equation (13) but replaced with the spin-2 window function ${}_2b_{lm'}$.[9,10,8]

4.3. *Drift Scanning*

Drift scanning is the simplest way to remove the ground contamination because an interferometer is insensitive to the ground emission that is a dc signal. When the interferometer is in the drift-scan mode, the visibility is a function of time through a scan path described by an ordered set of tuples $(\theta(t), \phi(t), \psi(t))$. For a given scan path, we can apply the algorithms proposed by Wandelt & Górski[11] for convolving the full sky with an asymmetric beam pattern (also see Challinor et al.[10]) to analyse the visibility data. If the drift-scan map is small, we can pixelize the map and use the analysis method in the flat-sky limit in Section 4.1.

5. Mock Observations and Estimation of Power Spectra

We use the CMBFAST package[12] to obtain the power spectra $C_T(l)$, $C_E(l)$, and $C_{TE}(l)$ in a flat ΛCDM model with $\Omega_\Lambda = 0.6$, $\Omega_M = 0.4$, $h = 0.6$, and $\Omega_b h^2 = 0.0125$ in which the B-polarization vanishes. To generate small patches of T, Q, and U fields, each $10^o \times 10^o$, we use equations (4,5,7) by adding a Gaussian random number to each power spectrum with the variance predicted in the theory. According to equation (3), these fields are then multiplied by the primary beam and Fourier transformed to give a regular array of visibilities.

The AMiBA instrumental noise on the data is simulated by adding a random complex number to each visibility whose real and imaginary parts are drawn from a Gaussian distribution with the variance of the noise predicted in a real observation. Here we use the sensitivity per baseline per polarization defined as [8]

$$s_b = \frac{1}{\eta_s \eta_a} \frac{2k_B T_{sys}}{A_{phy}} \frac{1}{\sqrt{2\Delta_\nu}}, \tag{16}$$

where k_B is the Boltzmann constant, T_{sys} is the system temperature, η_a and η_s are respectively the antenna and system efficiencies, Δ_ν is the bandwidth, t_{int} is the integration time, and A_{phy} is the physical area of the dish. Here, we use $\eta_a = \eta_s = 0.8$, $T_{sys} = 70K$, and the total bandwidth $\Delta_\nu = 16$ GHz.

We have measured the CMB power spectra using the quadratic estimator method[7] from the simulated visibilty data. We wrote a FORTRAN code based on BLAS, LAPACK, and LINPACK routines, which are very efficient for matrix operations and also appropriate for parallel computing, to compute the Fisher information matrix,

$$F_{bb'} = \frac{1}{2} \mathrm{Tr} \left(C^{-1} \frac{\partial C_T}{\partial C_b} C^{-1} \frac{\partial C_T}{\partial C_{b'}} \right), \tag{17}$$

where C and C_T are respectively the total and the CMB visibility correlation matrices, and C_b is the band power in each flat band. For a simulated set of visibilities Δ and an initial guess of the band powers C_b, we compute the correction

$$\delta C_b = \frac{1}{2} F_{bb'}^{-1} \mathrm{Tr} \left[(\Delta\Delta^T - C) \left(C^{-1} \frac{\partial C_T}{\partial C_{b'}} C^{-1} \right) \right], \tag{18}$$

and then repeat for the new values of C_b. Usually, we obtain converged C_b's after 5 iterations. The error bars in the estimated band powers are simply given by the square root of the diagonal components of the inverse Fisher matrix, $(F_{bb}^{-1})^{1/2}$.

Figure 2 shows the measured anisotropy power spectrum from simulated visibility data on 10 independent fields each of which is a 19 hexagonal mosaic with overlapping of 10 arcminutes. In each pointing, we have used 7 dishes with 0.6 m apertures, 3 bands with $\Delta_\nu = 5.33$ GHz, and the integration time is 24 hours. So, the total integration time is 190 days. The horizontal bars represent the widths of the flat bands, while the vertical bars are the uncertainties due to detector noise and sample variances. Figure 3 shows the measured E-polarization and TE correlation power spectra.

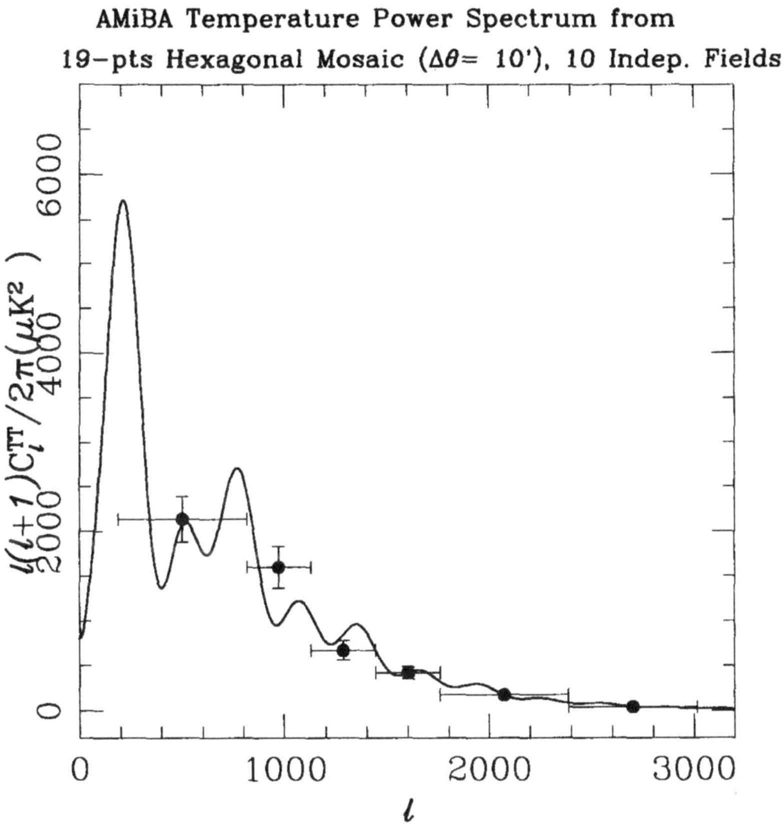

Figure 2. Estimated anisotropy power spectrum in an observation for 190 days with 7 dishes.

6. Conclusions

In practical situation, the parallactic angle changes with time as the telescope is staring at a point on the sky. In the above simulations of the mosaicking of the 0.6 m 7-element AMiBA, we have assumed an active rotation of the platform so as to keep the baselines fixed. We are developing a routine without parallactification, mainly based on the pixelization of the visibility plane. Drift scanning is the simplest way to remove the ground contamination. When the interferometer is in the drift-scan mode, the visibility is a function of time described by an time-ordered set of data through a scanning path. We have given a formalism on how to deal with the data and a pipeline to analyse this time-ordered data set is under progress.

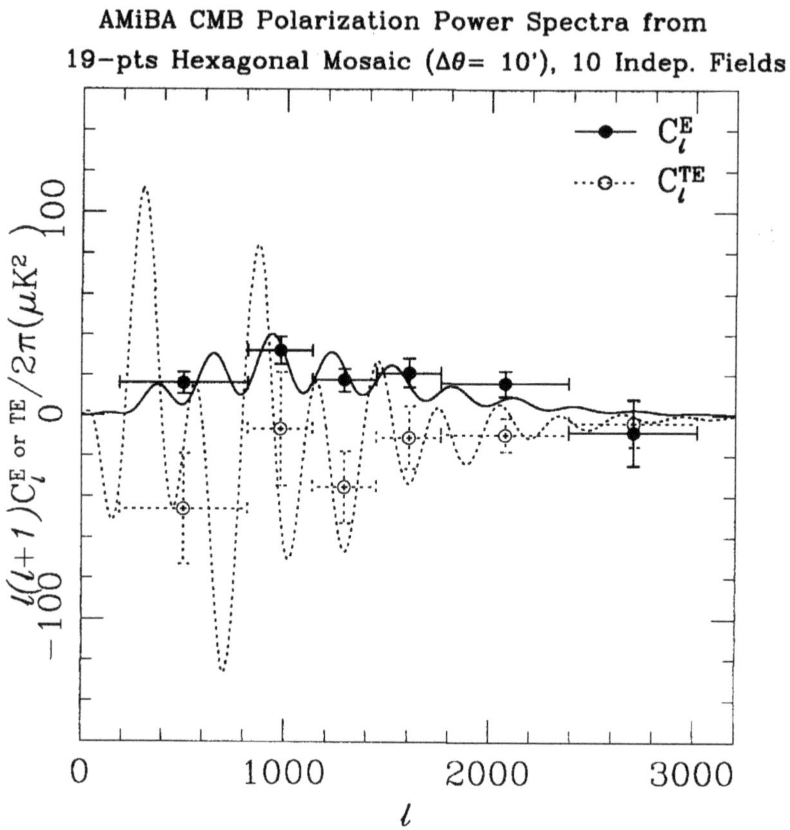

Figure 3. Estimated anisotropy power spectrum in an observation for 190 days with 7 dishes.

References

1. M. White *et al.*, *Astrophys. J.* **514**, 12 (1999).
2. S. Padin *et al.*, *Astrophys. J.* **549**, L1 (2001); B. S. Mason *et al.*, astro-ph/0205384; T. J. Pearson *et al.*, astro-ph/0205388.
3. E. M. Leitch *et al.*, *Astrophys. J.* **568**, 28 (2002); N. W. Halverson *et al.*, *Astrophys. J.* **568**, 38 (2002).
4. P. F. Scott *et al.*, astro-ph/0205380; A. C. Taylor *et al.*, astro-ph/0205381.
5. K. Y. Lo *et al.*, in *New Cosmological Data and the Values of the Fundamental Parameters*, IAU Symp. 201, ed. A. Lasenby and A. Wilkinson (Astronomical Society of the Pacific, San Francisco, CA, 2001).
6. M. P. Hobson, A. N. Lasenby, and M. Jones, *Mon. Not. R. Astron. Soc.* **275**, 863 (1995).
7. J. R. Bond, A. H. Jaffe, and L. Knox, *Phys. Rev.* **D57**, 2117 (1998).
8. K.-W. Ng, *Phys. Rev.* **D63**, 123001 (2001).
9. K.-W. Ng and G.-C. Liu, *Int. J. Mod. Phys.* **D8**, 61 (1999).

10. A. Challinor *et al.*, *Phys. Rev.* **D62**, 123002 (2000).
11. B. D. Wandelt and K. M. Górski, *Phys. Rev.* **D63**, 123002 (2001).
12. U. Seljak and M. Zaldarriaga, *Astrophys. J.* **469**, 437 (1996).

If the Universe is Finite

J. H. Proty Wu

IF THE UNIVERSE IS FINITE

JIUN-HUEI PROTY WU*

Department of Physics, National Taiwan University,
No.1 Sec.4 Roosevelt Road, Taipei, Taiwan
E-mail: jhpw@phys.ntu.edu.tw

If our universe is finite, there will be several distinct intrinsic properties in the Cosmic Microwave Background that we can detect. Based on the observations to date, previous studies conclude that the size of our universe is at least of the order of the size of the last-scattering surface. However, by including more physics in the analysis, such as a cosmological constant or gravitational waves, our results overturn this conclusion. We show that models with a small universe are still alive as well as those with a large universe. Better observational data and analysis methods are needed for us to learn more about this subject.

1. Introduction

With the development of modern technology, cosmology has shifted from a philosophical or a metaphysical subject in the old days to an experimental science today. So far we have been able to estimate with unprecedented precision several intrinsic properties of our universe, such as the age, the total energy density, the spatial geometry, etc.[1,2,3,4,5,6] Among these, one question that has not been answered, but soon will be, is 'How big is our universe?'

Similar to the question in the old ages that 'Is there an edge with the sea?' or 'How big is the earth?', we can now ask 'Is our universe finite?' or 'How big is our universe?' Just like the surface of the earth being two-dimensional but curved in a three-dimensional space and thus closed and finite, it is possible that our three-dimensional universe is curved in a four-dimensional space and thus closed and finite. This means that in such a universe, when we travel along one direction we are able to come back to the origin due to the periodic boundaries. Depending on how the periodic boundaries are identified, there are different possible topologies for a finite

*homepage: http://jhpw.phys.ntu.edu.tw

universe. Therefore, in the future we hope to know not only the size of the universe, if finite, but also its topology.

There are uncountable number of possible topologies for a finite universe. Fortunately, the observational evidence that the spatial geometry of our universe is very close to flat, if not exactly flat,[1,2,3,4,5,6] has allowed us to confine ourselves to the cases where the universe is flat. There are only seventeen possible topologies under this condition. For cases where the universe is close to but not exactly flat, the size of the universe that makes the geometry so close to flat will be too large for us to detect using observations, and therefore out of our interest here. The simplest topology in a flat finite universe is based on a cube with pairs of opposite faces identified as periodic boundaries. This is the so-called hypertorus topology. In the literature, six topologies based on either a parallelepiped or a hexagonal prism in a flat space have been studied[7,8,9,10,11,12,13] (see Figure 1).

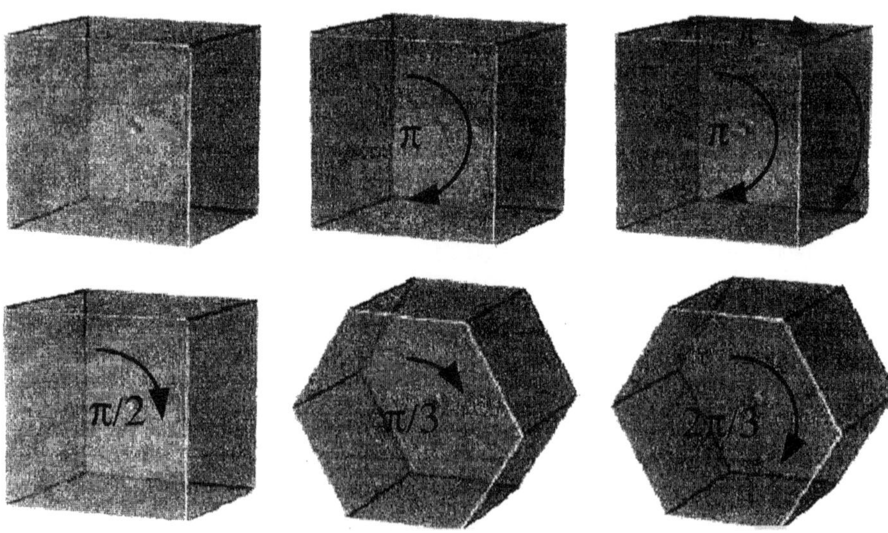

Figure 1. Six topologies based on either a parallelepiped or a hexagonal prism. Indicated are the directions and angles of rotation of one surface before being identified with its counter part on the opposite side. For a geometrically flat universe, these are the only six topologies that have been studied in the literature.

The Cosmic Microwave Background (CMB) has been used in these studies to constrain the size and topology of the universe. The CMB is chosen because it is arguably the cleanest cosmic signal on the largest scales that we can reach today. If the universe is finite, then there are several interesting properties that we can use to probe the size and topology of our universe. First, if the size of the universe L is smaller than twice of the radius y of the last-scattering surface, then the last-scattering surface will intersect with the periodic boundary of the universe, resulting in pairs of circles of matching patterns[12,13] (see Figure 2 and 3). By determining the

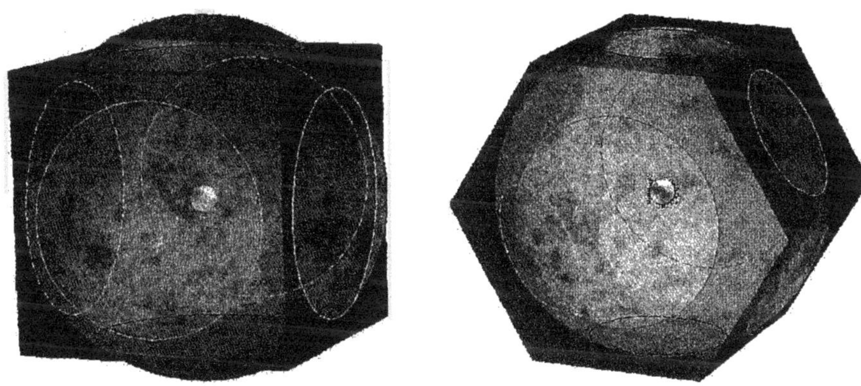

Figure 2. The last-scattering surface (the sphere) intersects with the periodic boundary of the universe, forming pairs of matching circles. The cube (left) and the hexagonal prism (right) indicate the finite volume of the universe, whose boundaries are periodic.

size of these matching circles and how they are matched, we shall be able to find the size and topology of the universe. Second, if L is not much larger than y, then the two-point correlation function of the CMB anisotropy will be anisotropic due to the anisotropic topology of the universe. This can be illustrated by the pair of small squares and the pair of small circles in Figure 3. Both pairs correspond to the same angular scale. In the figure, the large circle indicates the last-scattering sphere while the large square represents the periodic boundary of the universe. We can see that the pair of circles are essentially at the same location in the universe, while the pair of squares are not. This apparently causes some anisotropy in the two-point correlation function. Third, the CMB perturbation on one scale in the three-dimensional space may correspond to different angular scales on the two-dimensional CMB sphere (the last-scattering surface). This will result in scale-scale correlation in the CMB anisotropy. As shown in Figure 3, the physical length d actually corresponds to at least two different angular scales on the last-scattering surface, one being π and the other much smaller than π. Finally, the Fourier transform of the distribution of the contents in the universe, radiation or matter, will have discrete Fourier modes and lack

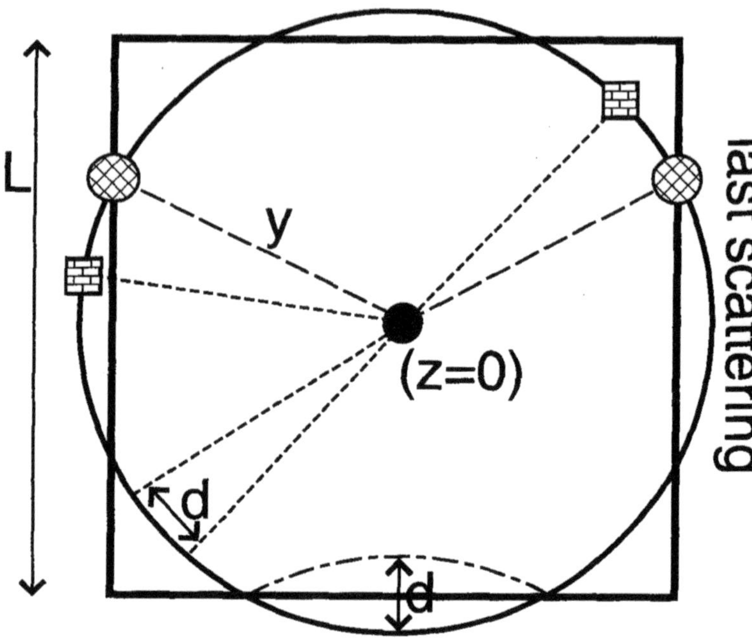

Figure 3. A finite universe with the hypertorus topology and $L < 2y$. In this case the last-scattering surface (the large circle) intersects with the periodic boundary of the universe (the large square).

of low-frequency (large-scale) power. For CMB photons, converting this consequence in the three-dimensional space to the two-dimensional CMB sphere will result in a relatively low power on large angular scales in the power spectrum of CMB anisotropy. The smaller the universe, the smaller the large-angular-scale power. Thus a comparison in the CMB power spectrum between the theoretical prediction and the observation will give a constraint on the size of the universe.

In the literature, no matching circles have been found in the current observational data,[12,13] neither the anisotropy in the two-point correlation function nor the scale-scale correlations.[10,11] Nevertheless, by investigating the power spectrum of CMB anisotropy on large angular scales[7,8,9] and by searching for symmetric patterns in the CMB,[10,11] it has been concluded that the size of the universe L is at least of the order of y. This means that the current observations require the size of the universe to be at least of the order of the size of the last-scattering surface. However, by including more physics that has been overlooked in all previous studies, in this article we shall show some preliminary results that are strongly against the previous conclusion. That is, we shall show that the current CMB observations are actually consistent with a small universe. A more detailed work will be presented elsewhere.[14]

2. Loopholes of the problems in the literature

The observed CMB today is contributed from several different physical processes at different epochs. Thus we can decompose it linearly as

$$\left.\frac{dT}{T}\right|_{\rm obs} = \left.\frac{dT}{T}\right|_{\rm ls} + \left.\frac{dT}{T}\right|_{\rm ISW} + \left.\frac{dT}{T}\right|_{\rm SZ} + \left.\frac{dT}{T}\right|_{\rm len} + \ldots, \tag{1}$$

where the subscripts 'ls', 'ISW', 'SZ', and 'len' denote respectively the contribution from the last-scattering epoch, the integrated Sachs-Wolfe effects, the Sunyaev-Zeldovich effects, and the gravitational lensing effects. On large angular scales, the first two terms dominate. So far all previous work in the literature has ignored the second term, as well as the scenarios where the gravitational waves exist.

The ignorance of these two factors may have significantly biased the current constraint that the universe should be at least about the size of the last-scattering surface. First, the second term is effectively zero if the universe is flat without a cosmological constant Λ. However, since recent observations have favored a non-zero Λ,[1,2,3,4,5,6] the second term may have contributed significantly to the observed total. If it has, it is arguable that it mainly contributes to the large angular scales. Second, gravitational waves do exist in some inflationary models. This tensor mode can result in extra CMB anisotropy on large angular scales. Even in a finite universe, because both factors exist between the last-scattering surface and the observer, us, they may induce significant CMB anisotropy on angular scales larger than the angular size of the universe on the last-scattering surface. Therefore, the inclusion of these two factors may compensate for the lack of large-angular-scale power that has been seen in models with a small universe. This may make a small universe consistent with the current observations. In addition, these two factors may have also obscured the intrinsic property expected in models with a finite universe, resulting in no detection of a finite universe. In this article, we shall take the hypertorus model as an example, to demonstrate how the inclusion of the gravitational waves can alter the previous conclusion that the universe must be large.

3. The effect of gravitational waves

We consider the CMB as contributed from both the scalar and tensor modes. The scalar mode is the usual component that has been intensively studied in the literature.[8,9,10] It results in the density contrasts in the contents of the universe that we see. The tensor mode manifests itself as the gravitational waves, and is allowed to exist in many inflationary models. Thus the theoretical prediction of the CMB power spectrum can be decomposed as

$$C_{\ell(\text{thy})} = (1-R)C_\ell^{\text{S}} + RC_\ell^{\text{T}}, \tag{2}$$

where the superscripts 'S' and 'T' denote the scalar and tensor modes respectively, $0 \leq R \leq 1$, and $C_{25}^{\text{S}} = C_{25}^{\text{T}}$. The C_ℓ^{S} and C_ℓ^{T} are simulated using the cosmological parameters of the current best-fit model. This simulation takes into account that the size of the universe is finite. For C_ℓ^{T}, we have taken the spectral index $n_{\text{T}} = -0.5$. This value will result in rising power towards low ℓ, instead of a constant power for $n_{\text{T}} = 0$. We note that it is commonly assumed that $n_{\text{T}} = n_{\text{S}} - 1$, where n_{S} is the spectral index of the scalar mode. This implies that if $n_{\text{S}} \approx 1$, as required by the observation,[1,2,3,4,5,6] then $n_{\text{T}} \approx 0$. However, this n_{S}-n_{T} relation is only a common assumption but not a requirement. In many inflationary models, the n_{S} and n_{T} actually deviate dramatically from the above relation, depending on the form of the inflaton potential. As an example for demonstrating how the C_ℓ^{T}, which has been completely ignored in the literature, can alter the previous conclusion, we shall fix R to 0.2 for simplicity. A more detailed work will be presented elsewhere.[14]

Figure 4 shows the C_ℓ^{S} (the usual component studied in the literature) and the $C_{\ell(\text{thy})}$ (including a tensor mode with $R = 20\%$ and $n_{\text{T}} = -0.5$), compared with the observed $C_{\ell(\text{obs})}$.[15] We note that all curves are normalized at $\ell = 24$, so when compared with the observation their amplitudes are not the best fit in the figure. It is clear that the inclusion of C_ℓ^{T} boosts the power at low ℓ. When compared with the observation, this compensates for the lack of the low-ℓ power in C_ℓ^{S} if the universe is small. Thus we see that a small universe, such as the model with $L/y = 0.1$, is no longer inconsistent with the observation. This means that the earlier constraint $L/y >\approx 1$ is no longer valid.

To further quantify how well the inclusion of gravitational waves can resolve the earlier problem, we compute the likelihood for C_ℓ^{S} and $C_{\ell(\text{thy})}$, given the CMB observations. Figure 5 shows the results. The result for C_ℓ^{S} (dashed line) is consistent with previous studies,[8,9,10,11] suggesting that the universe is inconsistent with $L/y <\approx 1$. On the other hand, with the inclusion of the tensor mode, the solid line indicates that the universe is consistent with almost the gamut of L/y, with a maximum likelihood at $L/y \approx 0.2$. This confirms again, in a more quantitative way, the result

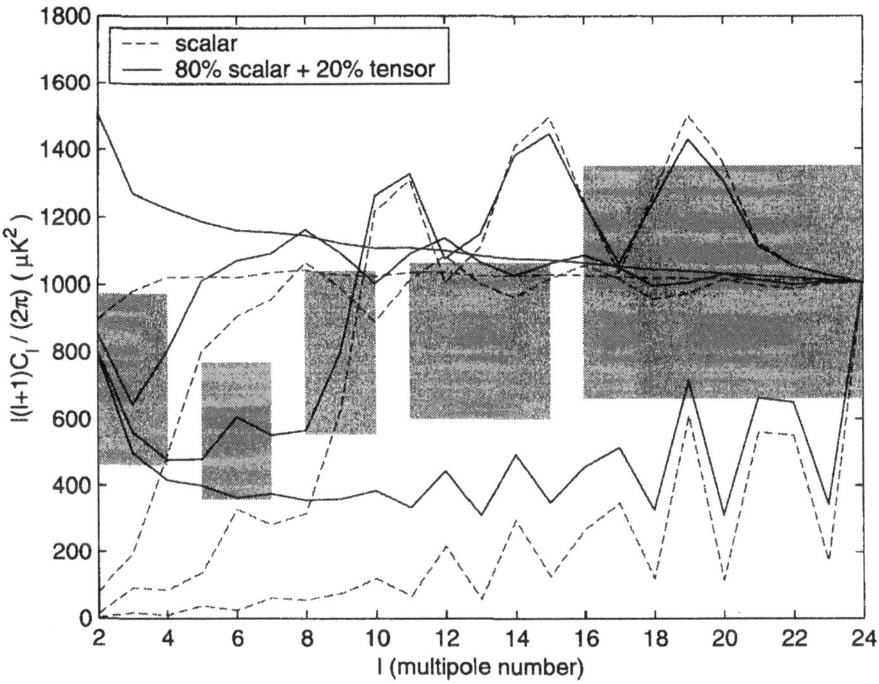

Figure 4. The theoretical predictions of $C_\ell^{\rm S}$ (dashed) and $C_{\ell({\rm thy})}$ (solid; including a tensor mode with $R = 20\%$ and $n_{\rm T} = -0.5$), compared with the observed $C_{\ell({\rm obs})}$ (shaded squares).[15] Here we have used $L/y = 0.1, 0.5, 1, 3$ (from bottom up at the left ends of the curves), so there are four pairs of $C_\ell^{\rm S}$ and $C_{\ell({\rm thy})}$. All curves are normalized at $\ell = 24$.

observed in Figure 4. Thus we see that the inclusion of gravitational waves in the theoretical modeling may allow a small universe when confronting the current observational data.

4. Future and conclusion

We consider the ISW effects and the gravitational waves as the loophole of the problem that was earlier seen in models with a small universe. Taking the gravitational waves as an example, the preliminary result presented here has overturned the earlier conclusion about the observational constraint on the size and topology of the universe. We show that a small universe is still consistent with the current observation, provided that the gravitational waves may exist. The ignorance of the ISW effects and the gravitational waves in the past are understandable, because these two factors do not play an important role in the cosmological models when the topology problem

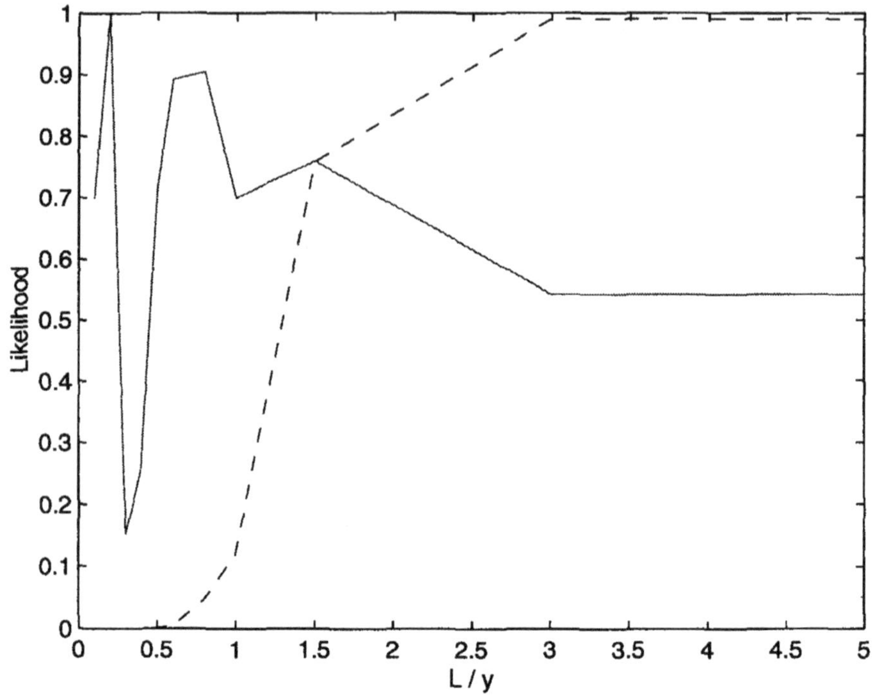

Figure 5. The likelihood of the $C_\ell^{\rm S}$ alone (dashed) and and that of the $C_{\ell(\rm thy)}$ (solid; $R = 0.2$ and $n_{\rm T} = -0.5$).

was raised about ten years ago. They became important only after the arrival of the new high-quality CMB data and the supernova data a couple of years ago, and after people had paid more attention to the search for gravitational waves.

The low resolution and the low signal to noise ratio in the current observational data have limited what we can say about the question whether or not the universe is finite. Nevertheless, the flood of high-quality full-sky data in the near future should teach us much more. We need the full-sky data because the size of the universe may not be small enough for us to detect using the current small-sky high-quality data. In the face of these future data, we need to improve our capability in several aspects. In future, the resolution of the full-sky data will be boost from the currently thousands of pixels from the COBE DMR to several millions from the MAP or Planck, both with unprecedented precision (see Figure 6 for comparison). Our current capability is apparently insufficient for dealing with such data. Thus we are devoted to much more efficient simulation and analysis for

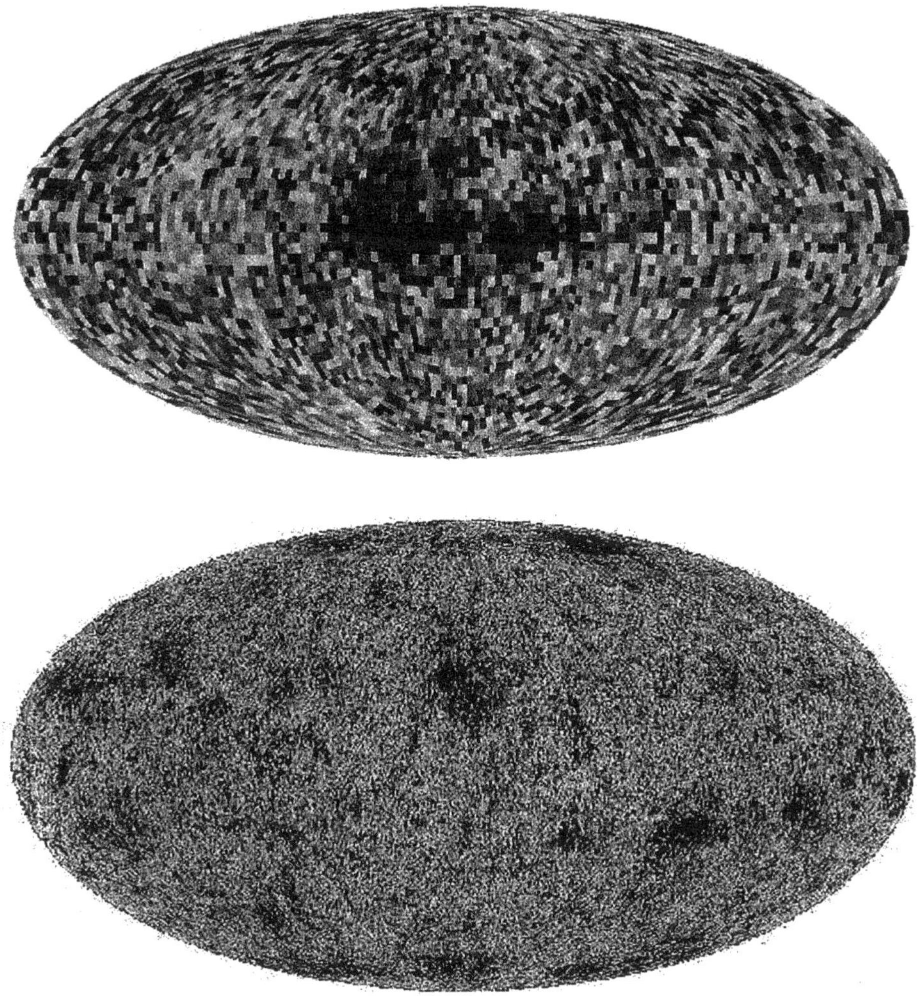

Figure 6. The COBE DMR observation (top) and the simulation for MAP and Planck (bottom). The former is currently the best observational result, which has several thousand pixels, while the latter will be available in the near future, which will have millions of pixels.

high-quality and high-resolution data, with more physics included. On the other hand, we also need to improve our methods of detecting the size and topology of the universe. If our analysis method is powerful enough, it is not impossible that we can detect the size and topology of the universe even when its size is larger than the size of the last-scattering surface. In summary, the models with a small universe are still alive, and so are those with a large universe. We need both better data and better analysis method to investigate the size and topology of our universe.

References

1. S. Hanany et al., *Astrophys. J.* **545**, L5 (2000).
2. P. de Bernardis et al., *Nature* **404**, 955 (2000).
3. N. W. Halverson et al., *Astrophys. J.* **568**, 38 (2002).
4. A. H. Jaffe et al., *Phys. Rev. Lett.* **86**, 3475 (2001).
5. T. J. Pearson et al., astro-ph/0205388 (2002).
6. R. A. Watson et al., astro-ph/0205378 (2002).
7. D. Stevens, D. Scott & J. Silk, *Phys. Rev. Lett.* **71**, 20 (1993).
8. J. Levin, E. Scannapieco & J. Silk, *Phys. Rev.* **D58**, 103516 (1998).
9. E. Scannapieco, J. Levin & J. Silk, *Mon. Not. Roy. Ast. Soc.* astro-ph/9811226 (1999).
10. A. de Oliveira-Costa & G. Smoot, *Astrophy. J.* **448**, 477 (1995).
11. A. de Oliveira-Costa, G. Smoot & A. Starobinsky, *Astrophy. J.* **468**, 457 (1996).
12. N. J. Cornish, D. N. Spergel & G. Starkman, *Phys. Rev.* **D57**, 5982 (1998).
13. B. F. Roukema, *Mon. Not. Roy. Ast. Soc.* **312**, 712 (2000).
14. J. H. P. Wu, in preparation (2002).
15. G. Smoot et al., *Astrophys. J.* **396**, L1 (1992).

Trans-Planckian Physics and Inflationary Cosmology

Robert Brandenberger

TRANS-PLANCKIAN PHYSICS AND INFLATIONARY COSMOLOGY

ROBERT H. BRANDENBERGER
Physics Dept., Brown University,
Providence, R.I. 02912, USA
E-mail: rhb@het.brown.edu

Due to the quasi-exponential redshifting which occurs during an inflationary period in the very early Universe, wavelengths which at the present time correspond to cosmological lengths are in general sub-Planckian during the early stages of inflation. This talk discusses two approaches to addressing this issue which both indicate that the standard predictions of inflationary cosmology - made using classical general relativity and weakly coupled scalar matter field theory - are not robust against changes in the physics on trans-Planckian scales. One approach makes use of modified dispersion relations for a usual free field scalar matter theory, the other uses some properties of space-time noncommutativity - a feature expected in string theory. Thus, it is possible that cosmological observations may be used as a window to explore trans-Planckian physics. We also speculate on a novel way of obtaining inflation based on modified dispersion relations for ordinary radiation.

1. Introduction

Inflationary cosmology[1] is an elegant paradigm of the very early Universe which solves several conceptual problems of standard cosmology and leads to a predictive theory of the origin of cosmological fluctuations. The basic idea of inflation is to replace the time line of Standard Big Bang (SBB) cosmology (a late time phase of matter domination preceded by a period in which radiation is dominant and which begins with a cosmological singularity) by a modified time line for which during some time interval $I = [t_i, t_R]$ - long before the time of nucleosynthesis - the Universe is accelerating, often involving quasi-exponential expansion of the scale factor $a(t)$ (to set our notation, we will be writing the space-time metric for our homogeneous and isotropic background cosmology in the form

$$ds^2 = dt^2 - a(t)^2[dx^2 + dy^2 + dz^2], \tag{1}$$

where t denotes physical time and x, y, z are the three Euclidean comoving spatial coordinates - we are neglecting the spatial curvature). The time t_i

stands for the beginning of the inflationary phase, the time t_R stands for the end, the time of "reheating".

Inflationary cosmology solves several of the conceptual problems of SBB cosmology. In particular, it resolves the homogeneity problem, the inability of SBB cosmology to address the reason for the near isotropy of the cosmic microwave background (CMB), it explains the spatial flatness of the Universe, and it provides the first ever mechanism of cosmological structure formation based on causal physics.

Let us briefly recall how inflationary cosmology leads to the existence of a mechanism for using causal microphysics for producing fluctuations on cosmological scales (when measured today), which at the time of equal matter and radiation had a physical wavelength larger than the Hubble radius $l_H(t) = H^{-1}(t)$, where $H = \dot{a}/a$ is the expansion rate (for references to the original literature see textbook treatments[2,3]). The relevant space-time sketch is shown in Fig. 1. If we trace a fixed comoving length scale (corresponding e.g. to CMB anisotropies on a fixed large angular scale) back into the past, then in SBB cosmology this scale is larger than the horizon (which in standard cosmology is H^{-1} - up to a numerical coefficient of order one) at early times. The time when such scales "cross" the horizon is in fact later than t_{rec}. Thus, it appears impossible to explain the origin of the primordial fluctuations measured in the CMB without violating causality.[a]

In an inflationary Universe (taking exponential inflation to be specific), during the period of exponential expansion the physical wavelength corresponding to a fixed comoving scale decreases exponentially as we go back in time, whereas the Hubble radius $l_H(t)$ is constant.[b] Thus, as long as the period of inflation is sufficiently long, all scales of cosmological interest today originate inside the Hubble radius in the early stages of inflation. As will be shown in the next section, the squeezing of quantum vacuum fluctuations when they exit and propagate outside the Hubble radius is the *mechanism for the origin of fluctuations in inflationary cosmology.*

The same background dynamics which yields the causal generation mechanism for cosmological fluctuations, the most spectacular success of inflationary cosmology, bears in it the nucleus of the trans-Planckian problem. This can also be seen from Fig. 1. If inflation lasts only slightly longer than the minimal time it needs to last in order to solve the horizon problem and to provide a causal generation mechanism for CMB fluctuations, then

---

[a]Topological defect models are a counterexample to this "standard dogma"

[b]The Hubble radius measures the distance over which microphysics can act coherently at a fixed time t, and in inflationary cosmology differs in a crucial way from the causal horizon (the forward light cone, which becomes exponentially larger).

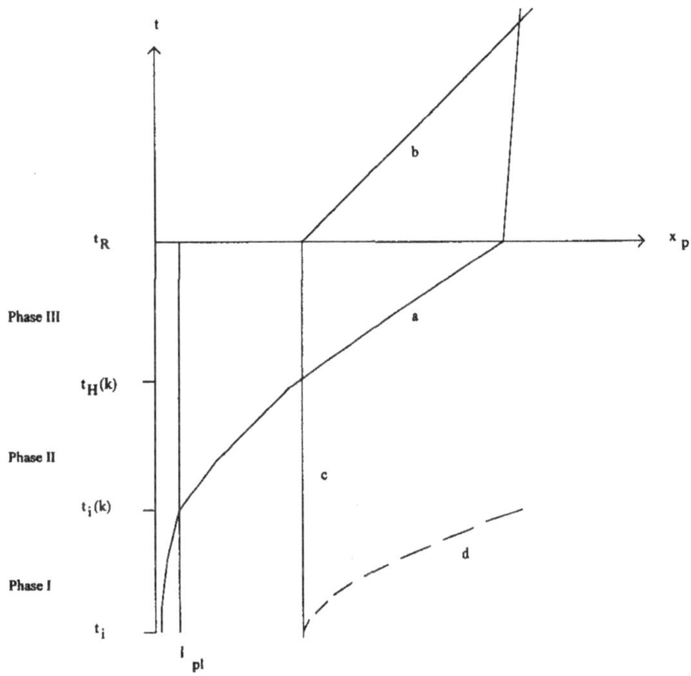

Figure 1. Space-time diagram (physical distance vs. time) showing how inflationary cosmology provides a causal mechanism for producing cosmological fluctuations. The line labeled a) is the physical wavelength associated with a fixed comoving scale k. The line b) is the Hubble radius or horizon in SBB cosmology. Note that at t_{rec} the fluctuation is outside the Hubble radius. Curve c) shows the Hubble radius during inflation. As depicted, at early times during inflation the curve a) is inside the Hubble radius, thus allowing for a causal generation mechanism for fluctuations on the corresponding scale. The horizon in inflationary cosmology is shown in curve d). The graph also demonstrates the trans-Planckian problem of inflationary cosmology: at very early times, the wavelength is smaller than the Planck scale ℓ_{Pl} (Phase I), at intermediate times it is larger than ℓ_{Pl} but smaller than the Hubble radius H^{-1} (Phase II), and at late times during inflation it is larger than the Hubble radius (Phase III).

the corresponding physical wavelength of these fluctuations is smaller than the Planck length at the beginning of the period of inflation. The theory of cosmological perturbations is based on classical general relativity coupled to a weakly coupled scalar field description of matter. Both the theory of gravity and of matter will break down on trans-Planckian scales, and this immediately leads to the trans-Planckian problem: are the predictions of standard inflationary cosmology robust against effects of trans-Planckian physics?[4] To answer this question, we need to give the reader a short review of the standard theory of cosmological fluctuations.

2. Fluctuations in Inflationary Cosmology

In the following we give an overview of the quantum theory of cosmological perturbations (see review articles[5] for details and references to the original literature). Since gravity is a purely attractive force, and since the fluctuations on scales of the CMB anisotropies were small in amplitude when the anisotropies were generated, the fluctuations had to have been very small in the early Universe. Thus, a linearized analysis of the fluctuations is justified. In this case, the Fourier modes of the cosmological fluctuations evolve independently.

The basic idea of the theory of cosmological perturbations is to quantize the linear fluctuations about a classical background cosmology. The starting point is the full action of gravity plus matter

$$S = \int \mathrm{d}^4x\sqrt{-g}R + S_{\mathrm{m}}, \tag{2}$$

where the first term is the usual Einstein-Hilbert action for gravity, R being the Ricci scalar and g the determinant of the metric, and S_{m} is the matter action. For the sake of simplicity, and since it is the usual assumption in simple inflationary Universe models, we take matter to be described by a single minimally coupled scalar field φ. Then, we separate the metric and matter into classical background variables $g_{\mu\nu}^{(0)}, \varphi^{(0)}$ which depend only on time, and fluctuating fields $\delta g_{\mu\nu}, \delta\varphi$ which depend on space and time and have vanishing spatial average:

$$g_{\mu\nu} = g_{\mu\nu}^{(0)}(\eta) + \delta g_{\mu\nu}(\eta, \mathbf{x}), \quad \varphi = \varphi^{(0)}(\eta) + \delta\varphi(\eta, \mathbf{x})\,. \tag{3}$$

We will focus on *scalar metric fluctuations*, the fluctuating degrees of freedom which couple to matter perturbations.[c]

The description of scalar metric perturbations is complicated by the fact that some perturbation modes correspond to space-time reparametrizations of a homogeneous and isotropic cosmology. A simple way to address this issue of gauge is to work in a system of coordinates which completely fixes the gauge. A simple choice is the *longitudinal* gauge, in which the metric takes the form[5]

$$\mathrm{d}s^2 = a^2(\eta)\left[(1+2\Phi)d\eta^2 - (1-2\Psi)\delta_{ij}\mathrm{d}x^i\mathrm{d}x^j\right], \tag{4}$$

where the space- and time-dependent functions Φ and Ψ are the two physical metric degrees of freedom which describe scalar metric fluctuations, and η is conformal time (related to physical time t via $dt = a(\eta)d\eta$). The

[c]Greek variables run over space-time indices, Latin variables only over spatial indices.

fluctuations of matter fields give additional degrees of freedom for scalar metric fluctuations. In the simple case of a single scalar matter field, the matter field fluctuation is $\delta\varphi$.

In the absence of anisotropic stress, it follows from the $i \neq j$ Einstein equations that the two metric fluctuation variables Φ and Ψ coincide. Due to the Einstein constraint equation, the remaining metric fluctuation Ψ is determined by the matter fluctuation $\delta\varphi$. It is clear from this analysis of the physical degrees of freedom that the action for scalar metric fluctuations must be expressible in terms of the action of a single free scalar field v with a time dependent mass (determined by the background cosmology). As shown in Ref. 6 (see also Ref. 7), this field is

$$v = a\left(\delta\varphi + \frac{\varphi^{(0)\prime}}{\mathcal{H}}\Psi\right) = z\mathcal{R}\,, \tag{5}$$

where $\mathcal{H} = a'/a$ (a prime denoting the derivative with respect to η),

$$z \equiv a\frac{\varphi^{(0)\prime}}{\mathcal{H}}\,, \tag{6}$$

and $\mathcal{R}$ denotes the curvature perturbation in comoving gauge.[8] The action for scalar metric fluctuations is [9]

$$\delta S_{\mathcal{R}} = \frac{1}{2}\int \mathrm{d}^4\mathbf{x}\left[(v_{\mathbf{k}}')^2 - \delta^{ij}v_{\mathbf{k},i}v_{\mathbf{k},j} + \frac{z''}{z}v_{\mathbf{k}}^2\right], \tag{7}$$

which leads to the equation of motion

$$v_{\mathbf{k}}'' + \left(k^2 - \frac{z''}{z}\right)v_{\mathbf{k}} = 0\,, \tag{8}$$

the equation of motion of a harmonic oscillator with time-dependent mass, the time dependence being given by $z(\eta)$ which is a function of the background cosmology. Note that if $a(\eta)$ is a power of η, then φ_0' and $\mathcal{H}$ scale with the same power of η, and the variable z is then proportional to a.

At this point we can summarize the mechanism by which cosmological fluctuations are generated in inflationary cosmology: we canonically quantize the linearized metric/matter fluctuations about a classical background cosmology, solve the resulting dynamical problem as a standard initial value problem, setting off all Fourier modes of the scalar field v in their vacuum state at the initial time (e.g. the beginning of the period of inflation).

The equation (8) is a harmonic oscillator equation with a time-dependent mass given by z''/z. On scales smaller than the Hubble radius ($t < t_{\mathrm{H}}(k)$), the mass term is negligible, and the mode functions oscillate with constant amplitude. On scales larger than the Hubble radius, however, the mass term dominates and the k^2 term can be neglected. The mode functions no longer oscillate. In an expanding background, the dominant mode of $v_{\mathbf{k}}(\eta)$ scales as $z(\eta)$. Note that it is incorrect to assume that fluctuations are created at the time of Hubble radius crossing (an impression one could get from reading Ref. 10). Rather, the role of the time $t_{\rm H}(k)$ of Hubble radius crossing is to set the time when the classical mode functions cease to oscillate and begin to increase in amplitude (squeezing).

The quantum mechanical interpretation of the two phases $t < t_{\rm H}(k)$ and $t > t_{\rm H}(k)$ is the following: on sub-Hubble scales we have oscillating quantum vacuum fluctuations and there is no particle production. Once the scales cross the Hubble radius, the mode functions begin to grow and the fluctuations get frozen. The initial vacuum state then becomes highly squeezed for $t \gg t_{\rm H}(k)$.[d] The squeezing leads to the generation of effectively classical cosmological perturbations.

For cosmological applications, it is particularly interesting to calculate the power spectrum of the curvature perturbation $\mathcal{R}$, defined as

$$\mathcal{P}_{\mathcal{R}}(k) = \frac{k^3}{2\pi^2 z^2}|v_k|^2 . \tag{9}$$

This last quantity can be estimated very easily. From the fact that on scales larger than the Hubble radius the mode functions are proportional to $a(\eta)$, we find

$$\mathcal{P}_{\mathcal{R}}(k) \sim \frac{k^3}{2\pi^2}\frac{1}{2k}\frac{1}{a^2[\eta_{\rm H}(k)]}\,, \tag{10}$$

where $\eta_{\rm H}(k)$ is the conformal time of Hubble radius crossing for the mode with comoving wavenumber k. Note that the second factor on the r.h.s. of (10) represents the vacuum normalization of the wave function.

As is evident from Figure 1, the standard theory of cosmological fluctuations summarized in this section relies on extrapolating classical general relativity and weakly coupled scalar matter field theory to length scales smaller than the Planck length. Thus, it is legitimate to ask whether the predictions resulting from this theory are sensitive to modifications of physics on physical length scales smaller than the Planck length. There are various ways in which such physics could lead to deviations from the standard predictions. First, new physics could lead to a non-standard evolution of the initial vacuum state of fluctuations in Period I, such that at Hubble radius crossing the state is different from the vacuum state. A

[d] The equation of motion for gravitational waves in an expanding background cosmology is identical to (8) with $z(\eta)$ replaced by $a(\eta)$, and thus the physics is identical. The case of gravitational waves was first discussed in Ref. 11.

model of realizing this scenario is summarized in the following section. Secondly, trans-Planckian physics may lead to different boundary conditions, thus resulting in a different final state. A string-motivated example for this scenario is given in Section 4.

3. Trans-Planckian Analysis I: Modified Dispersion Relations

The simplest way of modeling the possible effects of trans-Planckian physics, while keeping the mathematical analysis simple, is to replace the linear dispersion relation $\omega_{\rm phys} = k_{\rm phys}$ of the usual equation for cosmological perturbations by a non standard dispersion relation $\omega_{\rm phys} = \omega_{\rm phys}(k)$ which differs from the standard one only for physical wavenumbers larger than the Planck scale. This method was introduced[12,13] in the context of studying the dependence of thermal spectrum of black hole radiation on trans-Planckian physics. In the context of cosmology, it has been shown[14,15,16] that this amounts to replacing k^2 appearing in (8) with $k^2_{\rm eff}(n,\eta)$ defined by

$$k^2 \rightarrow k^2_{\rm eff}(k,\eta) \equiv a^2(\eta)\omega^2_{\rm phys}\left[\frac{k}{a(\eta)}\right]. \qquad (11)$$

For a fixed comoving mode, this implies that the dispersion relation becomes time-dependent. Therefore, the equation of motion of the quantity $v_k(\eta)$ takes the form

$$v_k'' + \left[k^2_{\rm eff}(k,\eta) - \frac{a''}{a}\right]v_k = 0\,. \qquad (12)$$

A more rigorous derivation of this equation, based on a variational principle, has been provided[17] (see also Ref. 18).

The evolution of modes thus must be considered separately in three phases, see Fig. 1. In Phase I the wavelength is smaller than the Planck scale, and trans-Planckian physics can play an important role. In Phase II, the wavelength is larger than the Planck scale but smaller than the Hubble radius. In this phase, trans-Planckian physics will have a negligible effect (this statement can be quantified[10]). Hence, by the analysis of the previous section, the wave function of fluctuations is oscillating in this phase,

$$v_k = B_1 \exp(-ik\eta) + B_2 \exp(ik\eta) \qquad (13)$$

with constant coefficients B_1 and B_2. In the standard approach, the initial conditions are fixed in this region and the usual choice of the vacuum state leads to $B_1 = 1/\sqrt{2k}$, $B_2 = 0$. Phase III starts at the time $t_{\rm H}(k)$ when the mode crosses the Hubble radius. During this phase, the wave function is squeezed.

One source of trans-Planckian effects[14,15] on observations is the possible non-adiabatic evolution of the wave function during Phase I. If this occurs, then it is possible that the wave function of the fluctuation mode is not in its vacuum state when it enters Phase II and, as a consequence, the coefficients B_1 and B_2 are no longer given by the standard expressions above. In this case, the wave function will not be in its vacuum state when it crosses the Hubble radius, and the final spectrum will be different. In general, B_1 and B_2 are determined by the matching conditions between Phase I and II. By focusing only [10] on trans-Planckian effects on the local vacuum wave function at the time $t_{\rm H}(k)$, one misses this important potential source of trans-Planckian signals in the CMB. If the dynamics is adiabatic throughout (in particular if the a''/a term is negligible), the WKB approximation holds and the solution is always given by

$$v_k(\eta) \simeq \frac{1}{\sqrt{2k_{\rm eff}(k,\eta)}} \exp\left[-i\int_{\eta_{\rm i}}^{\eta} k_{\rm eff}\,{\rm d}\tau\right], \tag{14}$$

where η_i is some initial time. Therefore, if we start with a positive frequency solution only and use this solution, we find that no negative frequency solution appears. Deep in Region II where $k_{\rm eff} \simeq k$ the solution becomes

$$v_k(\eta) \simeq \frac{1}{\sqrt{2k}} \exp(-i\phi - ik\eta), \tag{15}$$

i.e. the standard vacuum solution times a phase which will disappear when we calculate the modulus. To obtain a modification of the inflationary spectrum, it is sufficient to find a dispersion relation such that the WKB approximation breaks down in Phase I.

A concrete class of dispersion relations for which the WKB approximation breaks down is

$$k_{\rm eff}^2(k,\eta) = k^2 - k^2|b_m|\left[\frac{\ell_{\rm Pl}}{\lambda(\eta)}\right]^{2m}, \tag{16}$$

where $\lambda(\eta) = 2\pi a(\eta)/k$ is the wavelength of a mode. If we follow the evolution of the modes in Phases I, II and III, matching the mode functions and their derivatives at the junction times, the calculation[14,15,19] demonstrates that the final spectral index is modified and that superimposed oscillations appear.

However, the above example suffers from several problems. First, in inflationary models with a long period of inflationary expansion, the dispersion relation (16) leads to complex frequencies at the beginning of inflation for scales which are of current interest in cosmology. Furthermore, the initial conditions for the Fourier modes of the fluctuation field have to be set in a region where the evolution is non-adiabatic and the use of the usual vacuum prescription can be questioned. These problems can be avoided in a toy model in which we follow the evolution of fluctuations in a bouncing cosmological background which is asymptotically flat in the past and in the future. The analysis[20] shows that even in this case the final spectrum of fluctuations depends on the specific dispersion relation used.

4. Trans-Planckian Analysis II: Space-Time Uncertainty Relation

A justified criticism against the method summarized in the previous section is that the non-standard dispersion relations used are completely ad hoc, without a clear basis in trans-Planckian physics. There has been a lot of recent work[21,22,23,24] on the implication of space-space uncertainty relations[25,26] on the evolution of fluctuations. The application of the uncertainty relations on the fluctuations lead to two effects.[27] Firstly, the equation of motion of the fluctuations in modified. Secondly, for fixed comoving length scale k, the uncertainty relation is saturated before a critical time $t_i(k)$. Thus, in addition to a modification of the evolution, trans-Planckian physics leads to a modification of the boundary condition for the fluctuation modes. The upshot of this work is that the spectrum of fluctuations is modified.

We have recently studied[28] the implications of the stringy space-time uncertainty relation[29,30]

$$\Delta x_{\rm phys} \Delta t \, \geq \, l_s^2 \tag{17}$$

on the spectrum of cosmological fluctuations. Again, application of this uncertainty relation to the fluctuations leads to two effects. Firstly, the coupling between the background and the fluctuations is nonlocal in time, thus leading to a modified dynamical equation of motion. Secondly, the uncertainty relation is saturated at the time $t_i(k)$ when the physical wavelength equals the string scale l_s. Before that time it does not make sense to talk about fluctuations on that scale. By continuity, it makes sense to assume that fluctuations on scale k are created at time $t_i(k)$ in the local vacuum state (the instantaneous WKB vacuum state).

Let us for the moment neglect the nonlocal coupling between background and fluctuation, and thus consider the usual equation of motion for fluctuations in an accelerating background cosmology. We distinguish two ranges of scales. Ultraviolet modes are generated at late times when the

Hubble radius is larger than l_s. On these scales, the spectrum of fluctuations does not differ from what is predicted by the standard theory, since at the time of Hubble radius crossing the fluctuation mode will be in its vacuum state. However, the evolution of infrared modes which are created when the Hubble radius is smaller than l_s is different. The fluctuations undergo *less* squeezing than they do in the absence of the uncertainty relation, and hence the final amplitude of fluctuations is lower. From the equation (10) for the power spectrum of fluctuations, and making use of the condition

$$a(t_i(k)) = kl_s \tag{18}$$

for the time $t_i(k)$ when the mode is generated, it follows immediately that the power spectrum is scale-invariant

$$\mathcal{P}_{\mathcal{R}}(k) \sim k^0 . \tag{19}$$

In the standard scenario of power-law inflation the spectrum is red ($\mathcal{P}_{\mathcal{R}}(k) \sim k^{n-1}$ with $n < 1$). Taking into account the effects of the nonlocal coupling between background and fluctuation mode leads [28] to a modification of this result: the spectrum of fluctuations in a power-law inflationary background is in fact blue ($n > 1$).

Note that, if we neglect the nonlocal coupling between background and fluctuation mode, the result of (19) also holds in a cosmological background which is NOT accelerating. Thus, we have a method of obtaining a scale-invariant spectrum of fluctuations without inflation. This result has also been obtained in Ref. 31, however without a microphysical basis for the prescription for the initial conditions.

5. Non-Commutative Inflation

Another important problem with the method of modified dispersion relations is the issue of back-reaction.[32,33] If the mode occupation numbers of the fluctuations at Hubble radius crossing are significant, the danger arises that the back-reaction of the fluctuations will in fact prevent inflation. This issue is currently under investigation.

Surprisingly, it has been realized [34] that back-reaction effects due to modified dispersion relations (which in turn are motivated by string theory) might it fact yield a method of obtaining inflation from pure radiation. In this section, p and ω will denote physical wavenumber and frequency, respectively.

One of the key features expected from string theory is the existence of a minimal length, or equivalently a maximal wavenumber $p_{\max}$. Thus, if we consider the dispersion relation for pure radiation in string theory, the dispersion relation must saturate or turn over at p_{max}. Some cosmological consequences of a dispersion relation which saturates at p_{max}, i.e. for which the frequency ω diverges as $k \to p_{max}$ were explored recently.[36] It was found that a realization of the varying speed of light scenario can be obtained.

Recent work[34] has focused on the case when the dispersion relation turns over at p_{max}, i.e. if the frequency is increased, then before $p(\omega)$ reaches p_{max}, the wavenumber begins to decrease again. This implies that for each value of p there are two states corresponding to two different frequencies, i.e. that there are two branches of the dispersion relation. A class of such dispersion relations is given by

$$\omega^2 - p^2 f^2 = 0 \ , \ f(\omega) = 1 + (\lambda E)^\alpha \, , \tag{20}$$

where α is a free parameter.

Let us now consider an expanding Universe with such a dispersion relation, and assume that in the initial state both branches are populated up to a frequency much larger than the frequency at which the dispersion relation turns over. As the Universe expands, the physical wavenumber of all modes decreases. However, this implies - in contrast to the usual situation - that the energy of the upper branch states *increases*. This is what one wants to be able to achieve an inflationary cosmology. Eventually, the high energy states will decay into the lower branch states (which have the usual equation of state of radiation), thus leading to a graceful exit from inflation.

To check whether the above heuristic scenario is indeed realized, one must compute the equation of state corresponding to the dispersion relation (20). The spectrum is deformed, and a thermodynamic calculation[34] yields an equation of state which in the high density limit tends to $P = 1/[3(1 - \alpha)]\rho$ where P stands for the pressure. Thus, we see that there is a narrow range of values of α which indeed give the correct equation of state for power-law inflation.

In such an inflationary scenario, the fluctuations are of thermal origin. Taking the initial r.m.s. amplitude of the mass fluctuations on thermal length scale T^{-1} to be order unity, assuming random superposition of such fluctuations on larger scales to compute the amplitude of fluctuations when a particular comoving length scale exits the Hubble radius, and using the usual theory of cosmological fluctuations to track the fluctuations to the time when the scale enters the Hubble radius, one finds a spectrum of fluctuations with the same slope as in regular power-law inflation, and with an amplitude which agrees with the value required from observations if the fundamental length scale l_s at which the dispersion relation turns over is about an order of magnitude larger than the Planck length.

6. Conclusions

Due to the exponential redshifting of wavelengths, present cosmological scales originate at wavelengths smaller than the Planck length early on during the period of inflation. Thus, Planck physics may well encode information in these modes which can now be observed in the spectrum of microwave anisotropies. Two examples have been shown to demonstrate the existence of this "window of opportunity" to probe trans-Planckian physics in cosmological observations. The first method makes use of modified dispersion relations to probe the robustness of the predictions of inflationary cosmology, the second applies the stringy space-time uncertainty relation on the fluctuation modes. Both methods yield the result that trans-Planckian physics may lead to measurable effects in cosmological observables. An important issue which must be studied more carefully is the back-reaction of the cosmological fluctuations (see e.g. Ref. 35 for a possible formalism). As demonstrated in the final section, it is possible that trans-Planckian physics can in fact lead to dramatic changes even in the background cosmology.

Acknowledgments

I am grateful to the organizers of CosPA2002 for their wonderful hospitality, and to S. Alexander, P.-M. Ho, S. Joras, J. Magueijo and J. Martin for collaboration. The research was supported in part by the U.S. Department of Energy under Contract DE-FG02-91ER40688, TASK A.

References

1. A. H. Guth, *Phys. Rev.* **D23**, 347 (1981).
2. A. Linde, "*Particle Physics and Inflationary Cosmology*", Harwood Academic, Chur, (1990).
3. A. Liddle and D. Lyth, "*Cosmological inflation and large-scale structure*, Cambridge Univ. Press, Cambridge, (2000).
4. R. H. Brandenberger, arXiv:hep-ph/9910410.
5. V. F. Mukhanov, H. A. Feldman and R. H. Brandenberger, *Phys. Rept.* **215**, 203 (1992).
6. V. F. Mukhanov, *JETP Lett.* **41**, 493 (1985) [*Pisma Zh. Eksp. Teor. Fiz.* **41**, 402 (1985)].
7. V. Lukash, *Sov. Phys. JETP* **52**, 807 (1980).
8. D. H. Lyth, *Phys. Rev.* **D31**, 1792 (1985).
9. V. F. Mukhanov, *Sov. Phys. JETP* **67** (1988) 1297 [*Zh. Eksp. Teor. Fiz.* **94N7** (1988 ZETFA,94,1-11.1988) 1].

10. N. Kaloper, M. Kleban, A. E. Lawrence and S. Shenker, arXiv:hep-th/0201158.
11. L. P. Grishchuk, *Sov. Phys. JETP* **40**, 409 (1975) [*Zh. Eksp. Teor. Fiz.* **67**, 825 (1974)].
12. W. G. Unruh, *Phys. Rev.* **D51**, 2827 (1995).
13. S. Corley and T. Jacobson, *Phys. Rev.* **D54**, 1568 (1996) [arXiv:hep-th/9601073].
14. J. Martin and R. H. Brandenberger, *Phys. Rev.* **D63**, 123501 (2001) [arXiv:hep-th/0005209].
15. R. H. Brandenberger and J. Martin, *Mod. Phys. Lett.* **A16**, 999 (2001) [arXiv:astro-ph/0005432].
16. J. C. Niemeyer, *Phys. Rev.* **D63**, 123502 (2001) [arXiv:astro-ph/0005533].
17. M. Lemoine, M. Lubo, J. Martin and J. P. Uzan, *Phys. Rev.* **D65**, 023510 (2002) [arXiv:hep-th/0109128].
18. T. Jacobson and D. Mattingly, *Phys. Rev.* **D63**, 041502 (2001) [arXiv:hep-th/0009052].
19. J. Martin and R. H. Brandenberger, *Phys. Rev.* **D65**, 103514 (2002) [arXiv:hep-th/0201189].
20. R. H. Brandenberger, S. E. Joras and J. Martin, arXiv:hep-th/0112122.
21. R. Easther, B. R. Greene, W. H. Kinney and G. Shiu, *Phys. Rev.* **D64**, 103502 (2001) [arXiv:hep-th/0104102].
22. A. Kempf and J. C. Niemeyer, *Phys. Rev.* **D64**, 103501 (2001) [arXiv:astro-ph/0103225].
23. R. Easther, B. R. Greene, W. H. Kinney and G. Shiu, arXiv:hep-th/0110226.
24. F. Lizzi, G. Mangano, G. Miele and M. Peloso, *JHEP* **0206**, 049 (2002) [arXiv:hep-th/0203099].
25. D. Amati, M. Ciafaloni and G. Veneziano, *Phys. Lett.* **B197**, 81 (1987).
26. D. J. Gross and P. F. Mende, *Nucl. Phys.* **B303**, 407 (1988).
27. A. Kempf, *Phys. Rev.* **D63**, 083514 (2001) [arXiv:astro-ph/0009209].
28. R. Brandenberger and P. M. Ho, *Phys. Rev.* **D66**, 023517 (2002) [arXiv:hep-th/0203119].
29. T. Yoneya, *Mod. Phys. Lett.* **A4**, 1587 (1989).
30. M. Li and T. Yoneya, arXiv:hep-th/9806240.
31. S. Hollands and R. M. Wald, arXiv:gr-qc/0205058.
32. T. Tanaka, arXiv:astro-ph/0012431.
33. A. A. Starobinsky, *Pisma Zh. Eksp. Teor. Fiz.* **73**, 415 (2001) [*JETP Lett.* **73**, 371 (2001)] [arXiv:astro-ph/0104043].
34. S. Alexander, R. Brandenberger and J. Magueijo, arXiv:hep-th/0108190.
35. L. R. Abramo, R. H. Brandenberger and V. F. Mukhanov, *Phys. Rev.* **D56**, 3248 (1997) [arXiv:gr-qc/9704037].
36. S. Alexander and J. Magueijo, arXiv:hep-th/0104093.

CMB Constraints on Cosmic Quintessence and its Implication

Wolung Lee

CMB CONSTRAINTS ON COSMIC QUINTESSENCE AND ITS IMPLICATION

WO-LUNG LEE

Institute of Physics, Academia Sinica,
Taipei, Taiwan 11529, ROC
E-mail: leewl@phys.sinica.edu.tw

The equation of state of the hypothetical dark energy component, which constitutes about two thirds of the critical density of the universe, may be very different from that of a cosmological constant. Employing a phenomenological model, we investigate the constraints imposed on the scalar quintessence by supernovae observations, and by the acoustic scale extracted from recent CMB data. We show that a universe with a quintessence-dominated phase in the dark age is consistent with the current observational constraints. This may have effects on the evolution of density perturbations and the subsequent structure formation. Furthermore, we explore the possibility of coupling the quintessence to electromagnetism, and discuss its implication to the generation of primordial magnetic fields.

1. Introduction

Recent astrophysical and cosmological observations such as dynamical mass, Type Ia supernovae (SNe), gravitational lensing, and cosmic microwave background (CMB) anisotropies, concordantly prevail a spatially flat universe containing a mixture of matter and a dominant smooth component, which provides a repulsive force to accelerate the cosmic expansion.[1] The simplest candidate for this invisible component carrying a sufficiently large negative pressure is a true cosmological constant. The current data, however, are consistent with a somewhat broader diversity of such a repulsive "dark energy" as long as its equation of state w_X approaches that of the cosmological constant, $w_\Lambda = -1$, at recent epoch. A dynamically evolving scalar field called "quintessence" (Q) is probably the most popular scenario so far to accommodate the dark energy component.

Principally there are two different approaches to explore the nature of the quintessential dark energy. Many efforts have been put forth to reconstruct the potential of the dynamical scalar field ϕ based on various reasonable physical motivations. They include pseudo Nambu-Goldstone boson (PNGB), inverse power law, exponential, tracking characteristics, oscillating feature, and others.[2] Unfortunately, all quintessence models up to the minute fail to overcome two well-known theoretical difficulties, namely, "the cosmological constant problem" in which some unknown deep symmetry is required to evanesce the vacuum energy,[3] and "the coincidence problem", where an explanation to the timely concurrence of dark matter and dark energy is called for.[4] Accordingly, it seems premature at this stage to perform a detailed data fitting to a particular quintessence model.

An alternative model-independent approach, which is adopted in this study, is to constrain the character of dark energy by means of experimental data with minimal underlying theoretical assumptions. Since the scalar potential $V(\phi)$ of the Q-field is scarcely known, it is convenient to discuss the evolution of ϕ through its equation of state, $p_\phi = w_\phi \varrho_\phi$. Physically, $-1 \leq w_\phi \leq 1$, where the former equality holds for a pure vacuum state. Here, we assume a phenomenological form for the time dependent w_ϕ to accommodate all known observational results and then use it to unfold the evolution of dark energy in a standard big bang universe up to the epoch of the last scattering surface. In particular, we focus on the CMB constraints that apply to such a generic quintessence (GQ) scenario in the hope of distinguishing the Q field from the true cosmological constant case.

It has been pointed out[5] that the time variations of the equation of state with redshift z are essentially undetectable. Furthermore, the study on the complementary observables connecting to the expansion history provides little help in practice to determine $w_\phi(z)$ since the Q field becomes dynamically important only at low redshift in any observationally viable model. This conclusion is drawn on the assumption that the scalar quintessence evolves slowly from time to time. The situation can be significantly altered if one take into account the possibility of fast motion of Q along the evolution. By considering the rapid transition in the equation of motion, the GQ scenario thus offers more information than other conventional quintessence models. We will not address the detectability of time-varying w_ϕ here but provide an implication to illuminate the ample application the scenario may achieve.

2. Current Observational Findings

Lately some progress has been made in constraining the behavior of quintessential fields from observational data. A combined large scale structure (LSS), SNe, and CMB analysis has set an upper limit on Q-models with a constant $w_\phi < -0.7$,[6] and a more recent analysis of CMB observations gives $w_\phi = -0.82^{+0.14}_{-0.11}$.[7] Furthermore, the SNe data and measurements of the position of the acoustic peaks in the CMB anisotropy spectrum have been used to put a constraint on the present $w_\phi^0 \leq -0.96$.[8] The apparent brightness of the farthest SN observed to date, SN1997ff at redshift $z \sim 1.7$, is consistent with that expected in the decelerating phase of the flat ΛCDM model with $\Omega_\Lambda \simeq 0.7$,[9] inferring $w_\phi = -1$ for $z < 1.7$. In addition, several attempts have been made to test different Q-models.[10] As mentioned previously, it is nevertheless primitive to differentiate between the variations, and the reconstruction of $V(\phi)$ would require next-generation observations.

2.1. *Constraints Imposed by the CMB Acoustic Peaks*

The theory of CMB anisotropies is well developed by the end of last century.[11] The tightly coupled baryon-photon cosmic soup experienced a serial acoustic oscillation just before the recombination epoch. The acoustic scale (the angular momentum scale of the acoustic oscillation) sets the locations of the peaks in the power spectrum of the CMB anisotropies,[12] and is characterized by

$$l_A = \pi \frac{d_*}{h_s} = \pi \frac{\eta_0 - \eta_{\rm dec}}{\int_0^{\eta_{\rm dec}} c_s d\eta}, \tag{1}$$

where η_0 and $\eta_{\rm dec}$ are respectively the conformal time today and at the last scattering, defined by $\eta = H_0 \int dt a^{-1}(t)$ with the scale factor a and the Hubble constant H_0. The quantity d_* represents the comoving distance to the last scattering surface, and h_s denotes the sound horizon at the decoupling epoch with the sound speed c_s. Both d_* and h_s are affected in the presence of the quintessential component.

On the other hand, the locations of the m−th peak can be parametrized in practice by the empirical fitting formula, $l_m = l_A(m - \varphi_m)$, where the phase shifts φ_m caused by the plasma driving effect are solely determined by pre-recombination physics.[13] It was shown [14] that the shift of the third peak is relatively insensitive to cosmological parameters. Consequently, by assuming $\varphi_3 = 0.341$, the value of the acoustic scale derived from the analysis of BOOMERANG data[15] lies in the range [16]

$$l_A = 316 \pm 8, \tag{2}$$

which is estimated to within one percent if the location l_3 is measured.

3. The Generic Quintessence Scenario

Consider a flat universe in which the total density parameter of the universe today is represented by $\Omega_0 = \Omega_m^0 + \Omega_r^0 + \Omega_\phi^0 = 1$ with a negligible Ω_r^0 and

$\Omega_m^0/\Omega_\phi^0 \sim 1/2$. We define an Ω_ϕ-weighted average [17]

$$\langle w_\phi \rangle = \int_{\eta_{\rm dec}}^{\eta_0} \Omega_\phi(\eta) w_\phi(\eta) d\eta \times \left(\int_{\eta_{\rm dec}}^{\eta_0} \Omega_\phi(\eta) d\eta \right)^{-1} . \tag{3}$$

Assuming a spatially homogeneous ϕ field, the evolution of the cosmic background is governed by

$$\frac{d\rho_\phi}{d\eta} = -3a\mathcal{H}\,(1+w_\phi)\,\rho_\phi, \tag{4}$$

$$\frac{d\mathcal{H}}{d\eta} = -\frac{3}{2}a\mathcal{H}^2 - \frac{a}{2}(w_r\rho_r + w_\phi\rho_\phi), \tag{5}$$

where we have used $\dot{\phi}^2 = (1+w_\phi)\varrho_\phi$ and $V(\phi) = (1-w_\phi)\varrho_\phi/2$, and rescaled the energy density of the i-th component by the reduced Planck mass $M_p \equiv (8\pi G)^{-1/2}$ and H_0, i.e. $\rho_i \equiv \varrho_i/(M_p H_0)^2$. Accordingly, the dimensionless Hubble parameter is given by

$$\mathcal{H}^2 \equiv \left(\frac{H}{H_0}\right)^2 = \Omega_m^0\, a^{-3} + \Omega_r^0\, a^{-4} + \Omega_\phi^0\, \mathcal{H}^2. \tag{6}$$

The generic quintessence scenario assumes a phenomenological form for the equation of state w_ϕ to accommodate as many observational outcomes as possible. For example, taking $\Omega_m^0 = 0.3$, $\Omega_r^0 = 9.89 \times 10^{-5}$, with $w_r = 1/3$, a simple square-wave form of w_ϕ gives rise to the background evolution as shown in Figure 1. In order to satisfy the above-mentioned observational constraints on w_ϕ, we have chosen $w_\phi = -1$ for $z < 2$, and a width of the square-wave such that $\langle w_\phi \rangle \simeq -0.7$. Note that at the present time $\Omega_\phi = \rho_\phi/(3\mathcal{H}^2) \approx 0.7$ and $H_0 t = 0.95$. We have also plotted $d\theta/d\eta = a\sqrt{(1+w_\phi)\rho_\phi}$, where $\theta \equiv \phi/M_p$.

To impose the constraint from the peak separation of CMB, one needs to calculate the acoustic scale l_A [eq.(1)] for different phenomenological GQ models. Since the sound speed in the pre-recombination plasma is characterized by [12]

$$c_s = \frac{1}{\sqrt{3(1+R)}} \quad \text{with} \quad R \equiv \frac{3\rho_b}{4\rho_\gamma} \approx 30233 \left(\frac{T_0}{2.728}\right)^{-4} \frac{\Omega_b^0 h^2}{1+z}, \tag{7}$$

where the baryon-photon momentum density ratio R sets the baryon loading to the acoustic oscillation of CMB, the sound horizon h_s at the decoupling epoch can be determined by the differential equation

$$\frac{dh_s}{d\eta} = c_s \tag{8}$$

along the background evolution. Using $\Omega_b^0 = 0.05$, $h = 0.65$, and fixing the last scattering surface at $z_{\rm dec} = 1100$, we have calculated the acoustic scales

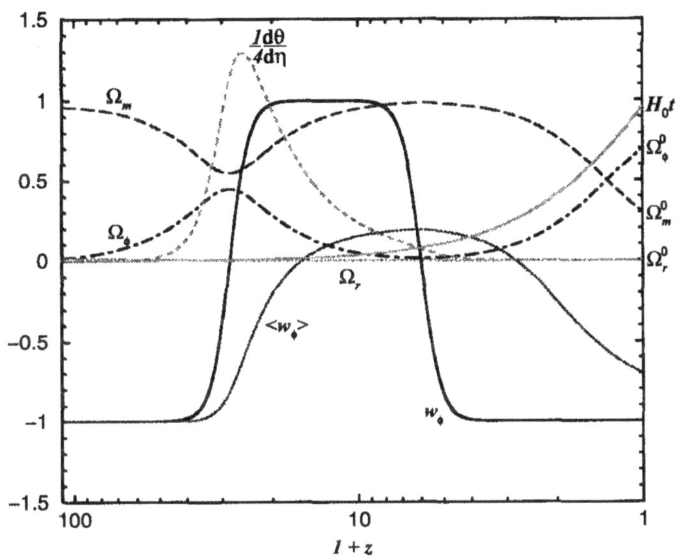

Figure 1. An example of GQ scenario

of different GQ-models associated with $\langle w_\phi \rangle = -0.6, \; -0.7, \; -0.8, \; -0.9$, and the case equivalent to the cosmological constant model $\langle w_\phi \rangle = -1$. Figure 2 plots the results against the quintessence density at decoupling $\Omega_\phi(z_{\rm dec})$. In general, when $\langle w_\phi \rangle$ is fixed, the acoustic scale is proportional to the quintessence density at the last scattering surface. On the other hand, as Ω_ϕ at photon decoupling is fixed, the acoustic scale will decrease when the averaged $\langle w_\phi \rangle$ increases.

The criterion for the acoustic scale derived from the BOOMERANG data [eq.(2)] has also been imposed where two horizontal lines are drawn as the upper- and lower-bounds of l_A. Apparently, all models encircled within the triangle-like area formed by the curve of $\langle w_\phi \rangle = -0.7$, the upper-bound, and the vertical axis of the graph are consistent with the current experimental data. As an example, let us consider the model located at the upper right corner of the allowed region, i.e. $(\langle w_\phi \rangle, \; \Omega_\phi^{\rm dec}) = (-0.7, \; 0.34)$. Figure 3 shows the detailed background evolution of the model, which bears a salient feature that the Q-field overwhelms the matter component during the dark age. Consequently, the coincidence problem is relaxed in such a GQ scenario. Meanwhile, a strong integrated Sachs-Wolfe effect (ISW),

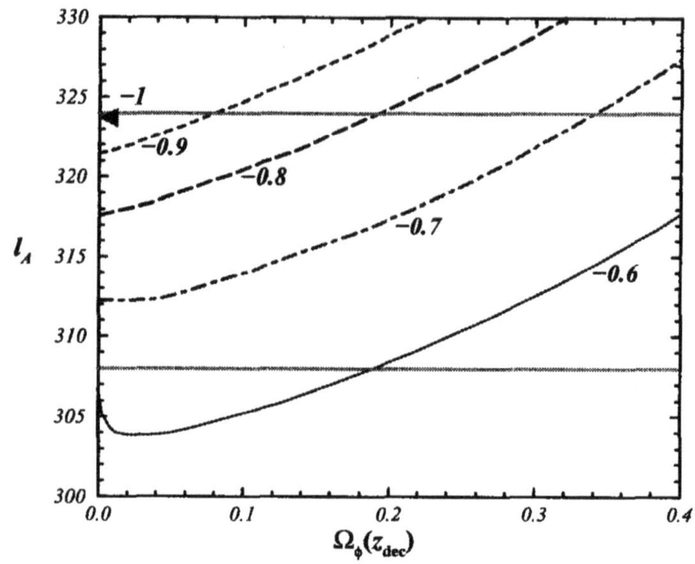

Figure 2. The acoustic scales of GQ-models are plotted as a function of the quintessence density at decoupling while keeping the Ω_ϕ-weighted averaged equation of state fixed. The model equivalent to the cosmological constant case has $l_A \approx 323.71$ and is indicated by a triangle. The criterion for l_A [eq.(2)] has also plotted as the two horizontal lines signifying the upper- and the lower-bounds permitted by the BOOMERANG data.

the most notable secondary anisotropy of CMB, is anticipated to arise since a significant portion of quintessence at decoupling would inevitably crush the Newtonian potential after the last scattering surface. Thus, a complete numerical study of the power spectrum is called for. Moreover, the matter-radiation equality time has shifted much ahead comparing to the prediction in the standard big bang model. This may affect the growing process of the large scale structure.

4. The Origin of the Primordial Magnetic Fields

As an implication of the GQ scenario, let us consider the primordial magnetic fields of the universe. As we know, the issue of the origin of the observed cluster and galactic magnetic fields of about a few μG [18] remains a puzzle.[19] These magnetic fields could have been resulted from the amplification of a seed field of $B_{seed} \sim 10^{-23}$G, or correspondingly $\rho_B \simeq 10^{-34}\rho_\gamma$, on a comoving scale larger than Mpc via the so-called galactic dynamo

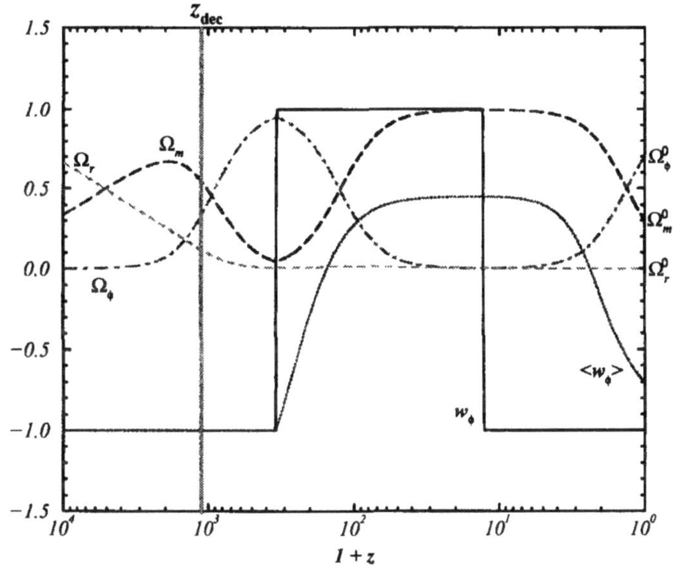

Figure 3. The background evolution of an extreme GQ-model allowed by the current observations. The last scattering surface is marked as the vertical line at $z_{\rm dec} = 1100$.

effect.

It has been pointed out by Carroll that the existence of an approximate global symmetry would allow a coupling of the Q field, ϕ, to the pseudoscalar $F_{\mu\nu}\tilde{F}^{\mu\nu}$ of electromagnetism.[20] As long as an ultralight ϕ field couples to photon and the mass m_ϕ is comparable to H_0, it is conceivable to have very long-wavelength electromagnetic fields generated via spinodal instabilities[21] from the dynamics of ϕ as a possible source of seed magnetic fields for the galactic dynamo.

Considering the ϕ-photon coupling in a flat universe,

$$L_{\phi\gamma} = \frac{c}{M_p}\phi\, \epsilon^{\alpha\beta\mu\nu} F_{\alpha\beta} F_{\mu\nu}, \tag{9}$$

where $F_{\mu\nu} = \partial_\mu A_\nu - \partial_\nu A_\mu$, and c is a coupling constant which we treat as a free parameter, the mode equation of the comoving magnetic field can be written as $(q = k/H_0)$ [22]

$$\frac{d^2}{d\eta^2} V_{\pm q} + \left(q^2 \mp 4cq\frac{d\theta}{d\eta}\right) V_{\pm q} = 0, \tag{10}$$

and the comoving energy density of the magnetic field is given by $\rho_B =$

$\langle B^2 \rangle / 8\pi = \int dq (d\rho_B/dq)$ with

$$\frac{d\rho_B}{dq} = \frac{H_0^4}{32\pi^3} q^3 \coth\left[\frac{qH_0}{2T_0}\right] \sum_{\lambda=\pm} |V_{\lambda q}|^2, \tag{11}$$

where the coth term is the Bose-Einstein enhancement factor due to the presence of the CMB with current temperature T_0 and energy density $\rho_\gamma = \pi^2 T_0^4/15$. With proper initial conditions, we have numerically solved the mode equations (10) using $c = 130$ and the background solution as shown in Fig. 1, and plotted the ratio $(d\rho_B/dq)/\rho_\gamma$ in Fig. 4. Although photons are being produced as the scalar field starts rolling at $z \sim 60$, we have counted the photons produced only after $z = 10$ when the universe has presumably entered the non-linear regime. The result shows that a sufficiently large seed magnetic field of 10 Mpc scale has been produced before $z \sim 4$. Evidently, through the spinodal instability, the coupling of quintessential dark energy and electromagnetism can effectively produce the large scale cosmic magnetic fields.

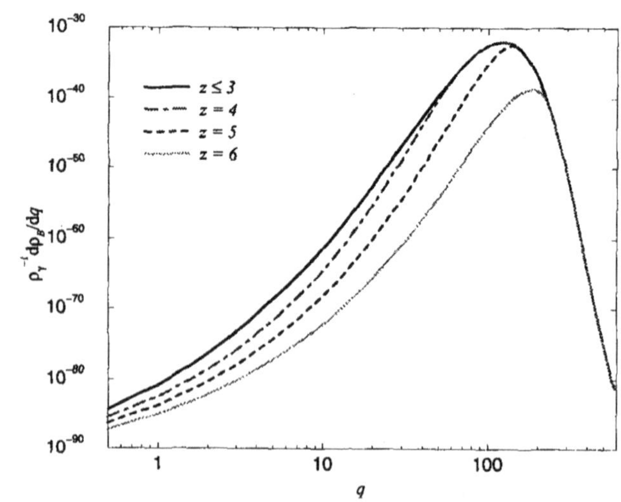

Figure 4. Ratios of the spectral magnetic energy density to the present CMB energy density at various redshifts. The present wavelength of the magnetic field is given by $2\pi/(qH_0)$.

Acknowledgments

The author thanks Kin-Wang Ng and Da-Shin Lee for the cooperation. This work was supported in part by the National Science Council under Contract NSC90-2112-M-001-028.

References

1. See, e.g., L. Wang, R. R. Caldwell, J. P. Ostriker, and P. J. Steinhardt, *Astrophys. J.* **530**, 17 (2000).
2. I. Waga and J. A. Frieman, *Phys. Rev.* **D62**, 043521 (2000); S. Dodelson, M. Kaplinghat, and E. Stewart, *Phys. Rev. Lett.* **85**, 5276 (2000); L. A. Boyle, R. R. Caldwell, M. Kamionkowski, astro-ph/0105318; T. Chiba, *Phys. Rev.* **D64**, 103503 (2001), and references therein.
3. S. M. Carroll, *Liv. Rev. Rel.* **4**, 1 (2001).
4. M. S. Turner, astro-ph/0202007.
5. J. Kujat, A. M. Linn, R. J. Scherrer, and D. H. Weinberg, *Astrophys. J.* **572**, 1 (2001); I. Maor, R. Brustein, J. McMahon, and P. J. Steinhardt, *Phys. Rev.* **D65**, 123003 (2002).
6. J. R. Bond *et al.*, astro-ph/0011379.
7. C. Baccigalupi *et al.*, *Phys. Rev.* **D65**, 063520 (2002).
8. P. S. Corasaniti and E. J. Copeland, *Phys. Rev.* **D65**, 043004 (2002).
9. A. G. Riess *et al.*, *Astrophys. J.* **560**, 49 (2001); M. S. Turner and A. G. Riess, astro-ph/0106051.
10. P. Brax, J. Martin, and A. Riazuelo, *Phys. Rev.* **D62**, 103505 (2000); C. Baccigalupi, S. Matarrese, and F. Perrotta, *Phys. Rev.* **D62**, 123510 (2000); A. Balbi *el al.*, *Astrophys. J.* **547**, L89 (2001); L. Amendola, *Phys. Rev. Lett.* **86**, 196 (2001); M. Pavlov, C. Rubano, M. Sazhin, and P. Scudellaro, *Astrophys. J.* **566**, 619 (2002); B. F. Roukema, G. A. Mamon, and S. Bajtlik, *Astron. Astrophys.* **382**, 397 (2002); M. Yahiro *et al.*, *Phys. Rev.* **D65**, 063502 (2002).
11. W. Hu, N. Sugiyama and J. Silk, *Nature* **386**, 37 (1997).
12. W. Hu and N. Sugiyama, *Astrophys. J.* **444**, 489 (1995).
13. W. Hu, M. Fukugita, M. Zaldarriaga and M. Tegmark, *Astrophys. J.* **549**, 669 (2001).
14. M. Doran and M. Lilley, *Mon. Not. Roy. Ast. Soc.* **330**, 965 (2001).
15. P. de Bernardis et al., *Astrophys. J.* **564**, 559 (2002).
16. M. Doran, M. Lilley, and C. Wetterich, *Phys. Lett.* **B528**, 175 (2002).
17. G. Huey *et al.*, *Phys. Rev.* **D59**, 063005 (1999).
18. P. P. Kronberg, *Rep. Prog. Phys.* **57**, 325 (1994).
19. For reviews, see A. V. Olinto, in Proceedings of the 3rd RESCEU Symposium, Tokyo, Japan, 1997, edited by K. Sato, T. Yanagida, and T. Shiromizu; D. Grasso and H. R. Rubinstein, astro-ph/0009061.
20. S. M. Carroll, *Phys. Rev. Lett.* **81**, 3067 (1998).
21. D. Boyanovsky, D.-S. Lee, and A. Singh, *Phys. Rev.* **D48**, 800 (1993); D.-S. Lee and K.-W. Ng, *Phys. Rev.* **D61**, 085003 (2000).
22. D.-S. Lee, W.-L. Lee, and K.-W. Ng, astro-ph/0109184, *Phys. Lett.* **B** in press.

A Way to the Dark Side of the Universe through Extra Dimensions

Je-An Gu

A WAY TO THE DARK SIDE OF THE UNIVERSE THROUGH EXTRA DIMENSIONS

JE-AN GU

Department of Physics, National Taiwan University,
Taipei 106, Taiwan, R.O.C.
E-mail: jagu@phys.ntu.edu.tw

As indicated by Einstein's general relativity, matter and geometry are two faces of a single nature. In our point of view, extra dimensions, as a member of the *geometry face*, will be treated as a part of the *matter face* when they are beyond our poor vision, thereby providing dark energy sources effectively. The geometrical structure and the evolution pattern of extra dimensions therefore may play an important role in cosmology. Various possible impacts of extra dimensions on cosmology are investigated. In one way, the evolution of homogeneous extra dimensions may contribute to dark energy, driving the accelerating expansion of the universe. In the other way, both the energy perturbations in the ordinary three-space, combined with homogeneous extra dimensions, and the inhomogeneities in the extra space may contribute to dark matter. In this paper we wish to sketch the basic idea and show how extra dimensions lead to the dark side of the universe.

1. Introduction

It is strongly suggested by observational data that our universe has the critical energy density and consists of 1/3 of dark matter and 2/3 of dark energy (see e.g., Ref. 1 and references therein), where "dark" indicates the invisibility. Even though it is generally not an elegant way to explain data via something we cannot see, the avalanche of data, including those from type Ia supernova measurements,[2,3] cosmic microwave anisotropies,[4] galactic rotation curves, and surveys of galaxies and clusters (providing the power spectrum of energy density fluctuations), make it more and more convincing. Nevertheless, we accordingly need to ask a question: *Why are dark matter and dark energy so dark?*

This question reminds us another "dark" stuff, extra dimensions. The existence of extra dimensions is required in various theories beyond the standard model of particle physics, especially in the theories for unifying gravity and other forces, such as superstring theory. Extra dimensions should be "hidden" (or "dark") for consistency with observations. This common feature, "invisible existence", of dark energy, dark matter, and extra dimensions provides us a hint that there may be some deep relationship among them.

In this paper we show how extra dimensions may manifest themselves as a source of energy in the ordinary three-space and lead to the dark side of the universe. Basically homogeneous extra dimensions will contribute to dark energy and may also provide some sort of dark matter effectively if combined with the effects of inhomogeneities in the ordinary three-space, and inhomogeneities in the extra space will contribute to dark matter effectively. The basic idea is sketched in the next section, and then we discuss in Sec. 3 how homogeneous extra dimensions provide "effective" dark energy and influence the evolution of the ordinary three-space, especially, producing the accelerating expansion of the universe. The extra dimensions employed throughout this paper are small and compact, as introduced in the Kaluza-Klein theories.[a]

2. A Sketch of the Idea

We consider a $(3+n+1)$-dimensional space-time where n is the number of extra spatial dimensions. The unperturbed metric tensor $g_{\alpha\beta}$ $(\alpha, \beta = 0, 1, \ldots, 3+n)$, which describes a universe with homogeneous, isotropic ordinary three-space and extra space, is defined by

$$ds^2 = dt^2 - a^2(t)\left(\frac{dr_a^2}{1-k_a r_a^2} + r_a^2 d\Omega_a^2\right) - b^2(t)\left(\frac{dr_b^2}{1-k_b r_b^2} + r_b^2 d\Omega_b^2\right), \quad (1)$$

where $a(t)$ and $b(t)$ are scale factors, and k_a and k_b relate to curvatures of the ordinary 3-space and the extra space, respectively. The value of r_b is set to be within the interval $[0, 1)$ for the compactness of extra dimensions. The perturbed metric describing a lumpy universe is defined by

$$ds^2 = [g_{\mu\nu} + \delta g_{\mu\nu}(x)]\, dx^\mu dx^\nu + [1 + \delta_b(x^\mu)]^2\, g_{pq} dx^p dx^q, \quad (2)$$

where $g_{\mu\nu}$ and g_{pq} are unperturbed metric tensors, while $\delta g_{\mu\nu}(x)$ and $\delta_b(x^\mu)$ corresponding to perturbations, of the ordinary $(3+1)$-dimensional space-time and the extra space, respectively. As a convention, $x^{\mu,\nu}$ and $x^{p,q}$ denote the coordinates of the ordinary space-time and the extra space, respectively, while x denotes all of the coordinates. For the sake of simplicity

[a]Various scenarios for hidden extra dimensions have been proposed, for example, a brane world with large compact extra dimensions in factorizable geometry proposed by Arkani-Hamed *et al.*,[5,6] a brane world with extra dimensions in warped nonfactorizable geometry proposed by Randall and Sundrum,[7] and small compact extra dimensions in factorizable geometry as introduced in the Kaluza-Klein theories.[8]

the cross terms $dx^\mu dx^p$ are abandoned by requiring the symmetry with respect to extra space inversion, i.e., $x^p \to -x^p$. We note that the extra space is kept to be homogeneous and isotropic after introducing perturbations, so that we only need $\delta_b(x^\mu)$, the perturbation of the scale factor $b(t)$ as a function of the coordinates of the ordinary space-time, to represent perturbations of the extra space. On the contrary we have no symmetry requirement for the perturbed ordinary space-time, and hence the metric perturbations $\delta g_{\mu\nu}$ in general is a function of all the coordinates $\{x^\gamma, \gamma = 0, 1, \ldots, 3+n\}$.

Assuming that both the metric perturbations $\delta g_{\mu\nu}$ and δ_b are small, such that the Einstein equations from the perturbed metric can be expanded with respect to these perturbations, we obtain

$$G_{\alpha\beta} = 8\pi\bar{G}T_{\alpha\beta} = 8\pi\bar{G}\left[T^{(0)}_{\alpha\beta}(t) + \delta T_{\alpha\beta}(x)\right] \tag{3}$$

$$= G^{(0)}_{\alpha\beta}\left[g_{\mu\nu}(t)\right] + G^{(1)}_{\alpha\beta}\left[g_{\mu\nu}(t), \delta g_{\mu\nu}(x)\right] + G^{(2)}_{\alpha\beta}\left[g_{\mu\nu}(t), b(t)\right] \tag{4}$$

$$+G^{(3)}_{\alpha\beta}\left[g_{\mu\nu}(t), \delta g_{\mu\nu}(x), b(t)\right] + G^{(4)}_{\alpha\beta}\left[g_{\mu\nu}(t), \delta g_{\mu\nu}(x), b(t), \delta_b(x^\mu)\right],$$

where $\bar{G}$ is the gravitational constant in the higher-dimensional space-time, and $T_{\alpha\beta}$ denotes the energy-momentum tensor, $T^{(0)}_{\alpha\beta}(t)$ the unperturbed, and $\delta T_{\alpha\beta}(x)$ the perturbed one. The first two terms in the above expansion of the Einstein tensor, $G^{(0)}_{\alpha\beta}$ and $G^{(1)}_{\alpha\beta}$, are exactly the unperturbed and the perturbed Einstein tensor, respectively, of the ordinary $(3+1)$-dimensional space-time. In contrast, $G^{(2)}_{\alpha\beta}$, $G^{(3)}_{\alpha\beta}$, and $G^{(4)}_{\alpha\beta}$ are additional terms coming from extra dimensions. In our point of view, if observers are too blind to see extra dimensions, these three additional terms will be automatically moved to the right-hand side of the Einstein equations (3) and treated as some sort of energy source, thereby contributing an "effective" part to the energy-momentum tensor. In particular, $G^{(2)}_{\alpha\beta}$ is smoothly distributed in the space and hence contributes to dark energy, while $G^{(3)}_{\alpha\beta}$ and $G^{(4)}_{\alpha\beta}$ have the spatial dependence and contribute to dark matter. In the above discussion we have sketched the main idea. As a demonstration of this idea, we will show in the next section how homogeneous extra dimensions can lead to "effective" dark energy and consequently change the evolution pattern of a (nonrelativistic-)matter-dominated universe.[b]

3. Dark Energy from Homogeneous Extra Dimensions

We consider in this section the case of a (3+n+1)-dimensional space-time described by the unperturbed metric defined in Eq. (1), i.e., both the ordi-

[b]The part of "effective" dark matter originated from extra dimensions is currently under investigation, and will not be discussed in detail in the rest of this paper.

nary three-space and the extra space are homogeneous and isotropic. Assuming that the matter content in this higher-dimensional space is a perfect fluid with the energy-momentum tensor

$$T^{\alpha}{}_{\beta} = diag(\bar{\rho}, -\bar{p}_a, \ldots, -\bar{p}_b, \ldots), \tag{5}$$

we can write the Einstein equations as

$$3\left[\left(\frac{\dot{a}}{a}\right)^2 + \frac{k_a}{a^2}\right] + \frac{n(n-1)}{2}\left[\left(\frac{\dot{b}}{b}\right)^2 + \frac{k_b}{b^2}\right] + 3n\frac{\dot{a}}{a}\frac{\dot{b}}{b} = 8\pi\bar{G}\bar{\rho}\,, \tag{6}$$

$$2\frac{\ddot{a}}{a} + n\frac{\ddot{b}}{b} + \left[\left(\frac{\dot{a}}{a}\right)^2 + \frac{k_a}{a^2}\right] + \frac{n(n-1)}{2}\left[\left(\frac{\dot{b}}{b}\right)^2 + \frac{k_b}{b^2}\right] + 2n\left(\frac{\dot{a}}{a}\right)\left(\frac{\dot{b}}{b}\right) = -8\pi\bar{G}\bar{p}_a\,, \tag{7}$$

$$3\frac{\ddot{a}}{a} + (n-1)\frac{\ddot{b}}{b} + 3\left[\left(\frac{\dot{a}}{a}\right)^2 + \frac{k_a}{a^2}\right] + \frac{(n-1)(n-2)}{2}\left[\left(\frac{\dot{b}}{b}\right)^2 + \frac{k_b}{b^2}\right] + 3(n-1)\left(\frac{\dot{a}}{a}\right)\left(\frac{\dot{b}}{b}\right) = -8\pi\bar{G}\bar{p}_b\,, \tag{8}$$

where $\bar{\rho}$ is and the energy density in the higher-dimensional world, and $\bar{p}_a$ and $\bar{p}_b$ are the pressures in the ordinary three-space and the extra space, respectively.

In the previous work by Gu and Hwang,[9] the case with $k_a = k_b = 0$ was considered, in which the accelerating expansion of the present (nonrelativistic-)matter-dominated universe was proposed to be generated along with the evolution of extra dimensions. Here we also focus on a matter-dominated universe, setting $\bar{p}_a$ and $\bar{p}_b$ to zero accordingly, but consider a more general case in which only $k_b = 0$ is assumed while k_a is treated as a free parameter. In this case Eqs. (7) and (8) can be rearranged to become

$$(n+2)\frac{\ddot{a}}{a} + (2n+1)\left[\left(\frac{\dot{a}}{a}\right)^2 + \frac{k_a}{a^2}\right] + n(n-1)\frac{\dot{a}}{a}\frac{\dot{b}}{b} - \frac{n(n-1)}{2}\left(\frac{\dot{b}}{b}\right)^2 = 0, \tag{9}$$

$$(n+2)\frac{\ddot{b}}{b} - 3\left[\left(\frac{\dot{a}}{a}\right)^2 + \frac{k_a}{a^2}\right] + 6\frac{\dot{a}}{a}\frac{\dot{b}}{b} + \frac{(n-1)(n+4)}{2}\left(\frac{\dot{b}}{b}\right)^2 = 0\,, \tag{10}$$

and then we can rewrite the Einstein equations, using new variables $u(t) \equiv \dot{a}/a$ and $v(t) \equiv \dot{b}/b$, as

$$3\left(u^2 + \frac{k_a}{a^2}\right) + 3nuv + \frac{n(n-1)}{2}v^2 = 8\pi\bar{G}\bar{\rho}\,, \tag{11}$$

$$(n+2)\dot{u} + 3(n+1)u^2 + (2n+1)\frac{k_a}{a^2} + n(n-1)uv - \frac{n(n-1)}{2}v^2 = 0\,, \tag{12}$$

$$(n+2)\dot{v} - 3\left(u^2 + \frac{k_a}{a^2}\right) + 6uv + \frac{n(n+5)}{2}v^2 = 0\,. \tag{13}$$

Before getting numerical solutions, we use simple analytical operations to extract, from the above Einstein equations, essential features of these equations and the evolution patterns governed by them. We first obtain, from Eq. (9), conditions for the accelerating and the decelerating expansion:

$$\ddot{a} \begin{array}{lcl} > 0 & , & v/u < J_- \\ < 0 & , & J_- < v/u < J_+ \,, \\ > 0 & , & v/u > J_+ \end{array} \tag{14}$$

where

$$J_\pm \equiv 1 \pm \sqrt{\frac{(n+1)(n+2) + 2(2n+1)k_a/\left(a^2u^2\right)}{n(n-1)}}\,. \tag{15}$$

We then read off from Eq. (11) that the condition for positive energy density $\bar{\rho}$ is

$$v/u > K_+ \quad \text{or} \quad v/u < K_-\,, \tag{16}$$

where

$$K_\pm \equiv -\frac{3}{n-1} \pm \sqrt{\frac{3}{n(n-1)}\left(\frac{n+2}{n-1} - \frac{2k_a}{a^2u^2}\right)}\,. \tag{17}$$

Observing Eqs. (14)–(17), we notice that variables v/u and $k_a/(a^2u^2)$ play essential roles in the above expressions of these conditions. These two essential variables can also be recognized from Eqs. (11)–(13), which tell us that values of all the quantities in them are determined, up to an overall factor related to the initial value of u, once the values of v/u and $k_a/(a^2u^2)$ are given. It is therefore a good way to analyze the evolution of the universe governed by Eqs. (11)–(13) via a two-dimensional diagram described by v/u and $k_a/(a^2u^2)$.

Conditions in Eqs. (14)–(17) are summarized in Fig. 1, where the number of extra dimensions n is specified to be three as an example. The

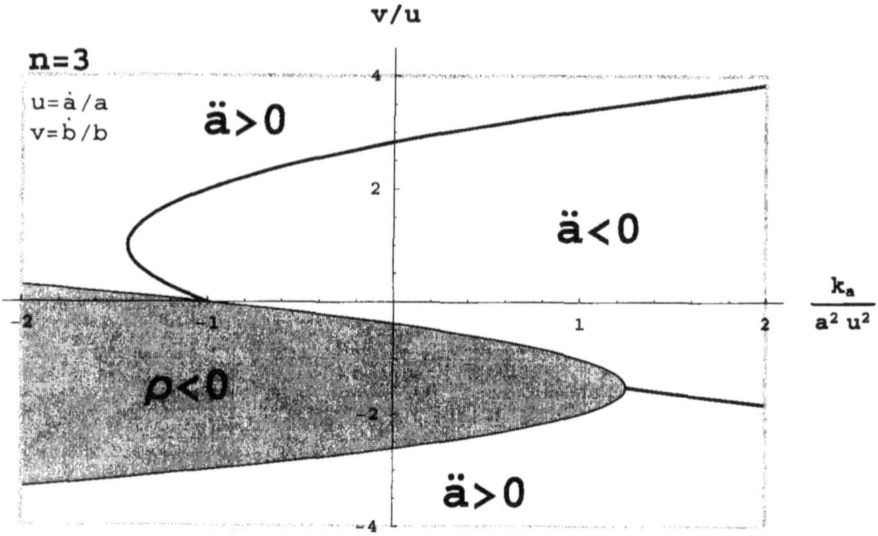

Figure 1. Conditions for various signs of energy density ρ and acceleration $\ddot{a}$ are illustrated, where the number of extra dimensions n is specified to be three.

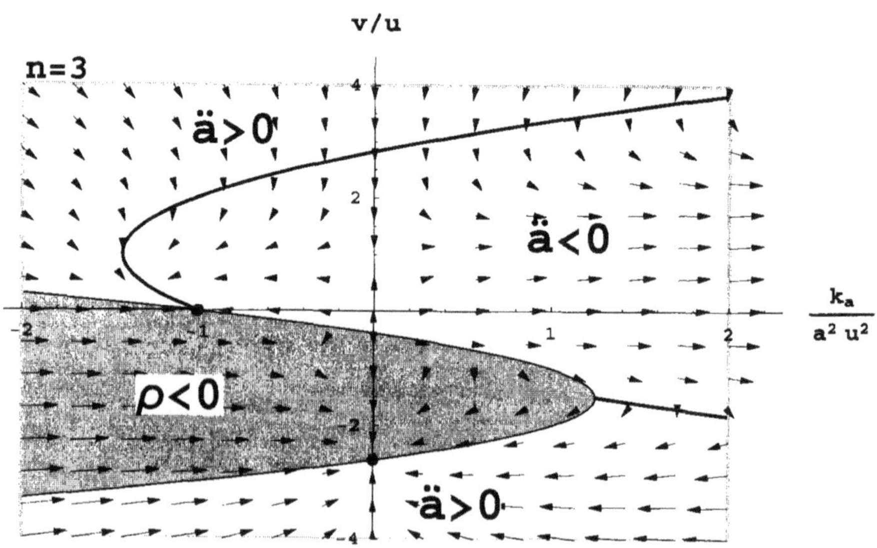

Figure 2. Flow vectors in $(k_a/(a^2u^2), v/u)$-diagram are plotted. Two grey dots denote two "attractors" at (-1,0) and $(0, -\left[3+\sqrt{3(n+2)/n}\right]/(n-1))$ (where $n = 3$), respectively.

grey area denoted by "$\rho < 0$" is a forbidden region if positive energy density is required. In addition, flow vectors in $(k_a/(a^2u^2), v/u)$-diagram, as determined by Eqs. (12) and (13), are plotted in Fig. 2 (where $n = 3$). There are two "attractors" denoted by grey dots in the flow diagram: one at $(k_a/(a^2u^2), v/u) = (-1, 0)$, and the other at $(k_a/(a^2u^2), v/u) = (0, -\left[3 + \sqrt{3(n+2)/n}\right]/(n-1))$.[c] The attractor at $(-1, 0)$ is on the margin of the forbidden region (i.e., indicating $\rho = 0$) and corresponding to a state of the higher-dimensional universe entailing stable extra dimensions and vanishing $\ddot{a}$. We note that the existence of solutions corresponding to stable extra dimensions is a good feature for building models in a higher-dimensional space-time. The other attractor is also on the margin, with zero energy density, of the forbidden region, entailing collapsing extra dimensions and positive acceleration.

For a concrete illustration, we now solve Eqs. (11)–(13) numerically for the case of $n = 3$. We plot in Fig. 3 four trajectories corresponding to four numerical solutions with respect to initial conditions, $(k_a/(a^2u^2), v/u) =$ (a) $(-0.0001, 4)$, (b) $(-0.001, 0)$, (c) $(0.0001, 4)$, and (d) $(1.3, -1.4)$. These four trajectories represent four different kinds of evolution path:

(a) **acceleration** $\rightarrow$ **deceleration** $\rightarrow$ **acceleration**, eventually approaching the attractor at $(-1, 0)$ with stable extra dimensions and zero acceleration, possessing negative spatial curvature.
(b) **deceleration** $\rightarrow$ **acceleration**, eventually merging to the trajectory (a) and approaching the attractor at $(-1, 0)$ with stable extra dimensions and zero acceleration, possessing negative spatial curvature.
(c) **eternal deceleration**, possessing increasing positive curvature contribution.
(d) **deceleration** $\rightarrow$ **acceleration**, eventually approaching the attractor at $(0, -\left[3 + \sqrt{3(n+2)/n}\right]/(n-1))$ with collapsing extra dimensions, possessing decreasing positive curvature contribution.

It is therefore indicated that there are many possibilities of evolution patterns in this higher-dimensional universe, in contrast to the unique manner of evolution, eternally decelerating expansion, for a matter-dominated universe in the standard cosmology without extra dimensions.

[c] Attractors are stable fixed points toward which the nearby points (or "state") tend to flow.

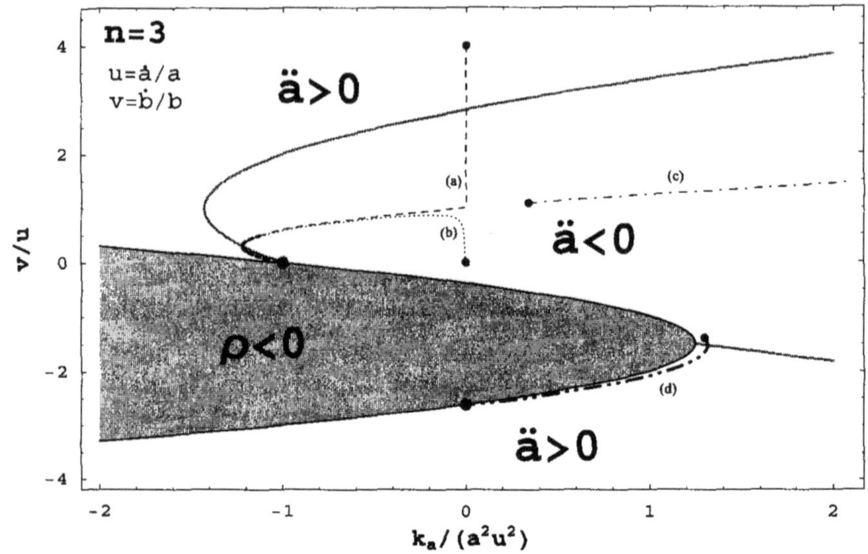

Figure 3. Four trajectories corresponding to four numerical solutions with respect to initial conditions, $(k_a/(a^2u^2), v/u)$ = (a) $(-0.0001, 4)$, (b) $(-0.001, 0)$, (c) $(0.0001, 4)$, and (d) $(1.3, -1.4)$, are plotted, where the black dot at one end of each trajectory denotes the initial position. (As in Fig. 2, two grey dots are "attractors" and $n = 3$.)

4. Discussion and summary

In this paper we make a point that there may be a deep relationship between "hidden" (or "dark") extra dimensions and the dark side of the universe, i.e., dark matter and dark energy. This conjecture is based on Einstein's general relativity, which indicates an important aspect that matter (with energy and momentum) and geometrical structures of a space-time are two faces of a single nature, to be called *matter face* and *geometry face*, respectively. In our point of view, if there exists a part of the *geometry face* which is beyond our poor vision, this missing part will be treated as a member of the *matter face*, and consequently provide mysterious, dark, "effective" energy sources. A possible missing part of the *geometry face* we consider in this paper is the existence of extra dimensions. This idea is sketched in Sec. 2 via analyzing the Einstein equations, including perturbations of both the metric tensor and the energy-momentum tensor, for a higher-dimensional world. We conclude that extra dimensions may manifest themselves as a source of energy in the ordinary three-space, such as "effective" dark energy, under the consideration of homogeneous extra dimensions, and "effective" dark matter, as contributed by inhomogeneities in the extra space or the ordinary three-space.

As a particular demonstration of the general idea, we consider in Sec. 3 a (nonrelativistic-)matter-dominated universe with homogeneous extra dimensions and show that the evolution of homogeneous extra dimensions can lead to "effective" dark energy and consequently change the evolution pattern of the universe. There are many possibilities of evolution patterns in this higher-dimensional universe, in contrast to the unique way of evolution, eternally decelerating expansion, for a matter-dominated universe in the standard cosmology without extra dimensions. It needs further detailed studies to determine which evolution pattern can appropriately describe our universe. In addition, there are various possible realizations of this idea worthy of further quests, and some are currently under our investigation.

As mentioned in Sec. 1, this work is motivated by a fundamental question: *Why are dark matter and dark energy so dark?* Through the preliminary studies of the general idea discussed in this paper, here comes up a possible answer: *Dark matter and dark energy are generated from the extra dimensions, a nature of geometry we are too blind to see.* This simple answer indicates an intriguing possibility of unifying these two kinds of dark entities, extra dimensions and dark energy sources, into one.

Acknowledgements

The author wishes to thank Professor W-Y. P. Hwang for helpful discussions. This work was supported by Taiwan CosPA project of the Ministry of Education (MOE 91-N-FA01-1-4-0).

References

1. M. S. Turner, arXiv:astro-ph/0207297.
2. S. Perlmutter *et al.* [Supernova Cosmology Project Collaboration], *Astrophys. J.* **517**, 565 (1999) [arXiv:astro-ph/9812133].
3. A. G. Riess *et al.* [Supernova Search Team Collaboration], *Astron. J.* **116**, 1009 (1998) [arXiv:astro-ph/9805201].
4. J. L. Sievers et al, arXiv:astro-ph/0205387.
5. N. Arkani-Hamed, S. Dimopoulos and G. Dvali, *Phys. Lett.* **B429**, 263 (1998) [arXiv:hep-ph/9803315]; I. Antoniadis, N. Arkani-Hamed, S. Dimopoulos and G. R. Dvali, *Phys. Lett.* **B436**, 257 (1998) [arXiv:hep-ph/9804398].
6. I. Antoniadis, Phys. Lett. **B246**, 377 (1990).
7. L. Randall and R. Sundrum, *Phys. Rev. Lett.* **83**, 3370 (1999) [arXiv:hep-ph/9905221]; **83**, 4690 (1999) [arXiv:hep-th/9906064].
8. T. Kaluza, *Sitzungsber. Preuss. Akad. Wiss. Berlin* (*Math. Phys.*) **K1**, 966 (1921); O. Klein, *Z. Phys.* **37**, 895 (1926) [*Surveys High Energ. Phys.* **5**, 241 (1926)].
9. Je-An Gu and W-Y. P. Hwang, *Phys. Rev.* **D66**, 024003 (2002) [arXiv:astro-ph/0112565].

X-ray Jets in Radio-loud Active Galaxies

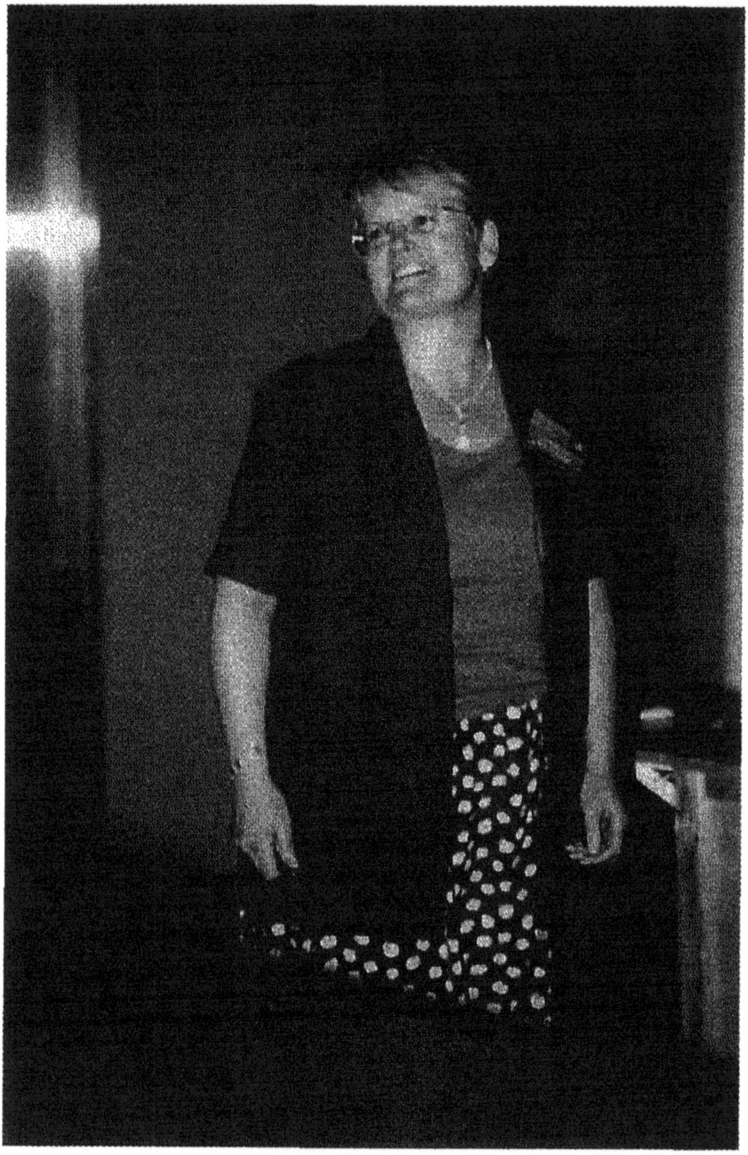

Diana Worrall

X-RAY JETS IN RADIO-LOUD ACTIVE GALAXIES

D.M. WORRALL

Department of Physics, University of Bristol,
Tyndall Avenue,
Bristol, BS8 1TL, UK
E-mail: d.worrall@bristol.ac.uk

Much can be learnt about the physics of the radio-emitting plasma in active galaxies through X-ray measurements. If inverse-Compton X-rays are seen from radio-emitting regions, the degeneracy between particle density and magnetic field strength can be broken, and if synchrotron X-rays are seen, key information is available concerning the highest-energy particles, those most subject to energy losses and thus most in need of re-acceleration. The *Chandra* X-ray Observatory routinely detects extended galaxy or cluster-scale X-ray emitting gas around radio-loud active galaxies, and this has a strong influence on overall source dynamics. In low-power (FRI) radio sources, resolved X-ray emission associated with the brighter, kpc-scale radio jet is commonly measured. There is good evidence that this X-ray jet emission is synchrotron in origin, and the short lifetimes of the TeV electrons involved means that considerable *in situ* particle acceleration is required. There are global similarities but small-scale differences in the radio and X-ray structures and spectra between sources, and this information can be used to probe details of the acceleration processes. The jets of high-power (FRII) radio sources are believed to slow down less, and propagate to kpc (and in some cases Mpc) distances at (at least mildly) relativistic speeds. *Chandra* results suggest that here inverse Compton processes are dominant in dictating the level of detected X-ray emission. In particular, in a number of hotspots and lobes the level of X-ray emission suggests that the structures are radiating at close to minimum energy, with similar total energy in relativistic particles and magnetic fields.

1. Introduction

Radio-loud active galaxies are believed to be powered by accretion onto super-massive black holes. The radio jets are formed of plasma which is highly relativistic, at least in the pc-scale inner regions probed by Very Long Baseline Interferometry (VLBI). Jets seen end on are thus strongly relativistically boosted and give rise to sources classed as BL Lac objects and quasars. Objects whose twin jets are close to the plane of the sky are classed as radio galaxies.

The large-scale radio structures are broadly of two types.[1] Fanaroff and Riley Type I (FRI) sources are of low total radio power. At kpc distances from the core, their jets are believed to have slowed down to less than about a tenth of the speed of light[2] such that the effects of Doppler boosting, important on the sub-kpc scales, can effectively be ignored. Unification models generally class FRI radio galaxies whose small-scale jets are pointed towards the observer as BL Lac objects. In contrast, Fanaroff and Riley Type II (FRII) sources are of high total radio power and their jets at termination blend into the lobe plasma that they supply. The jet to counter-jet asymmetries seen in FRII sources suggest that the radio-emitting plasma is still moving at speeds as large as 70 per cent of the speed of light at tens of kpc or more from the core.[3] Bright radio hotspots mark jet termination in these objects. In unification models, the FRII sources with their jets closest to the line of sight are generally radio-loud quasars.

The radio emission of FRI and FRII sources is known to be synchrotron radiation, and as such provides limited information on the physics of the sources. Radiation in the TeV band is the highest in energy to be observed. The current handful of detections at such very high energy are all of BL Lac objects, and the generally favored interpretation of their double-peaked spectral energy distributions (in $\log \nu S_\nu$ versus $\log \nu$) is that the low-energy peak is due to synchrotron radiation and the second (incorporating the TeV band) is Compton scattering, primarily synchrotron self-Compton (SSC) emission.[4]

The detection from a given relativistic electron population (in a known volume) of both synchrotron radio emission and higher-energy inverse-Compton emission (from scattering of a known photon field) provides a powerful probe of the source physics. Since the intensity of the inverse-Compton emission is proportional to the number of electrons, and the synchrotron luminosity is related to the number of electrons and the magnetic field strength, both the magnetic field and the density of electrons can be determined. This permits a test of the common assumption that radio sources have equal energy density in magnetic field and radiating particles (the 'equipartition' assumption, which roughly minimizes total source energy). Application to TeV sources is difficult as the emission volumes are not well known, and strong relativistic boosting effects are an added complication. However, the current generation of X-ray telescopes are sufficiently sensitive to measure even those components that are not strongly relativistically boosted, and *Chandra*[5] has a sufficiently narrow point spread function to resolve components in a number of active galaxies. At X-ray energies both synchrotron radiation and inverse Compton scattering can be important, and spectroscopy is usually required so that the shape of the X-ray spectrum (flat, similar to the radio slope in the case of inverse-Compton emission, and steep, due to spectral ageing from energy losses in the case of synchrotron radiation) can be used to determine the dominant emission process.

In either case of X-ray inverse Compton or synchrotron emission the coupling of radio and X-ray information provides powerful diagnostic tests. If inverse-Compton X-rays are seen from radio-emitting regions, the degeneracy between particle density and magnetic field strength can be broken, and if synchrotron X-rays are seen, key information is available concerning the highest-energy particles, those most subject to energy losses and thus most in need of re-acceleration. Furthermore, a gaseous medium is essential to the propagation (and probably the final disruption) of radio sources, and thus has a major influence on overall source dynamics: the galaxy and cluster potential wells of these active galaxies dictate that this medium is hot X-ray emitting gas.

Several leaps in understanding have occurred in the three years since the launch of the two powerful X-ray observatories *Chandra*[5] and *XMM-Newton*[6] in 1999. Previously there was a disappointingly low number of X-ray detections of resolved jets, lobes, and hotspots, but now the situation is changing rapidly, In what follows, the low-power FRIs and high-power FRIIs will be treated in turn.

2. FRIs

Before *Chandra*, although the type of large-scale X-ray-emitting atmosphere in which FRIs reside was reasonably well known,[7] uncertainties remained with regards the strength and nature of X-ray emission from the active nucleus. Even the best spatial resolution of *ROSAT* (with the High Resolution Imager) was unable to distinguish between point-like emission and the small-scale hot gas and cooling flows whose presence was inferred from the structure of the outer atmospheres. A trend for the compact X-ray emission to correlate with the radio core strength argued that at least part of the compact X-ray emission is associated with the radio jets of the active nucleus,[8,9] and this motivated us to initiate a *Chandra* program of imaging and spectroscopy (using the Advanced CCD Imaging Spectrometer) of FRIs to separate the physically-distinct emission components. $1''$ corresponds to about 1 kpc at the distances where one begins to find FRIs in large numbers, and so the arcsec spatial resolution of *Chandra* is crucial to resolving spatial structures on kpc scales. Prior to *Chandra*, kpc-scale jet emission had been detected only from the two closest FRIs:[10,11] Centaurus A at 3.4 Mpc and M 87 at ~ 17 Mpc.

We chose the sample of 50 B2 radio sources identified with bright elliptical galaxies as our primary FRI sample for *Chandra* observations, since these sources are well matched in isotropic optical and radio properties to BL Lac objects,[12] and thus can be used also to address unification models. A largely unbiased subsample of 40 of the galaxies was observed in *ROSAT* pointings, making this the best X-ray observed sample of FRI radio galaxies.[9] Our *Chandra* observations have generally been for modest exposure times of between 5 and 20 ks, and yet have revealed a considerable richness of structure. Most notably, four of the first five sources to be observed show not only unresolved cores but also one-sided extensions which are well aligned with the brighter of the twin radio jets (Fig. 1).

Our results for B2 FRIs have shown their X-ray emission to be complex. Such a radio galaxy generally contains galaxy-scale X-ray emit-

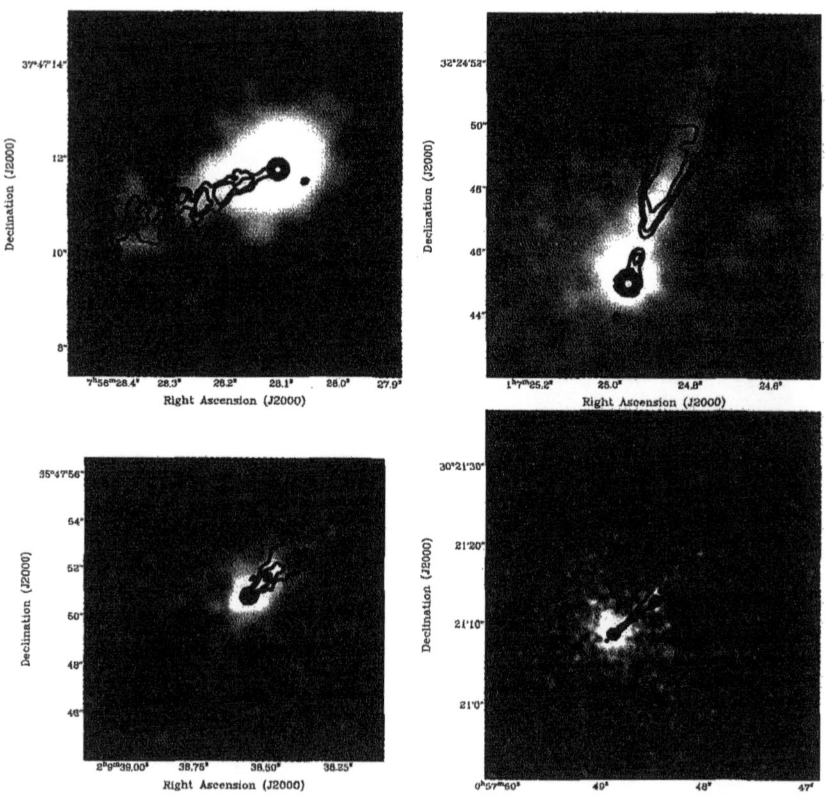

Figure 1. Radio contours on *Chandra* X-ray images of FRI radio galaxies B2 0206+35, B2 0104+32 (3C 31), B2 0055+30 (NGC 315), and B2 0755+37 (NGC 2484), clockwise from top left. The X-ray emission from the core and brighter radio jet are emphasized here, but all have galaxy-scale X-ray emitting gas also.

ting gas, a compact X-ray core, and X-ray jet emission associated with its brighter kpc-scale radio jet.[13] The presence also of cluster-scale gas is known from *ROSAT*, but this appears as structureless background in our high-resolution, reduced field-of-view *Chandra* images.

Evidence is mounting, based on X-ray spectra which are steeper than the radio in the kpc-scale jets of FRIs, and on the overall spectral distributions, that the X-ray emission mechanism in the jets is synchrotron radiation. Figure 2 illustrates this for two sources with some of the best available multiwavelength coverage, M 87[14] and 3C 66B[15]. The further examples of the B2 galaxy B2 0755+37 (NGC 2484)[13] and the BL Lac object PKS 0521-365[16] are shown in Figure 3. The integrated emission over the X-ray emitting jet region can be well fitted with a broken power law, and in each case the X-ray spectrum is consistent with the two-point optical to X-ray spectral index. The synchrotron radiation that is seen across nine decades in frequency requires an electron population spanning more than four decades in energy. A 'universal' integrated spectral distribution is emerging, and in an equipartition magnetic field, $B_{\rm eq}$, the position of the spectral break corresponds to electrons of energy roughly 300 GeV. The amount of the spectral break, $\Delta\alpha \approx 0.6$ to 0.9, is not so easily interpreted. Synchrotron energy losses predict $\Delta\alpha = 0.5$, and a high-energy cut-off in the electron spectrum would produce an exponential fall in the synchrotron spectrum. However, the synchrotron lifetimes of TeV electrons emitting the X-radiation are of the order of tens of years, and since the detected X-ray jets are thousands of light years in length, it is clear that *in situ* particle acceleration (or re-acceleration) is required.

Although the jet X-ray emission is strongest in the inner regions of the jet where the speed is likely still to be moderately fast ($\sim 0.7c$), before the deceleration which accompanies the larger jet opening angles, more diffuse X-ray emission can be traced further out in several sources. For FRIs with deep *Chandra* exposures, the morphological complexity of the jet emission becomes apparent. The phenomenon seen in 3C 66B[15] (Fig. 4), of X-ray bright regions offset upstream of (closer to the core than) radio-bright regions, is now reported in other sources. Figure 5 shows the examples of B2 0104+32 (3C 31)[17], where the galaxy-scale X-ray emitting gaseous atmosphere is also particularly prominent, and 3C 15[18]. The bright X-ray 'knots' lying upstream of radio 'knots' are perhaps best explained if the X-ray knots are sites of shocks and the X-ray/radio displacements are due to a combination of advection, particle diffusion and local ageing. The 'universal' overall spectral distribution must arise from a combination of electron acceleration and loss processes that are similarly balanced in all

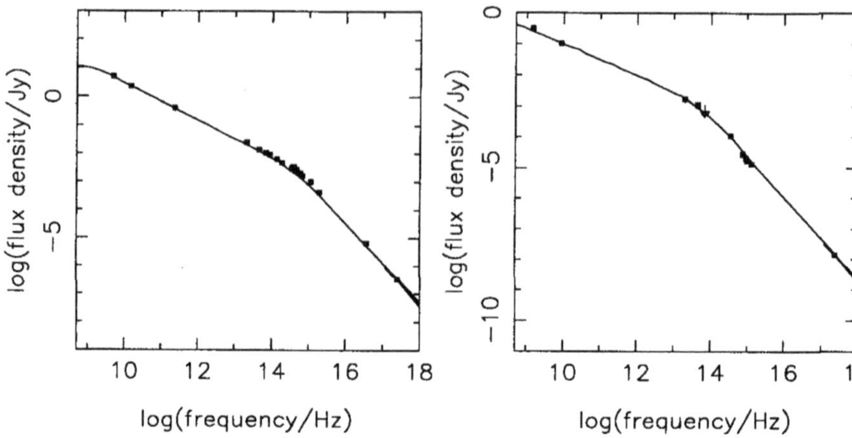

Figure 2. The overall spectra of kpc-scale jet emission from low-power radio galaxies and BL Lac objects (integrated over the region where jet X-rays are detected) are similar and can be fitted with single-component broken power laws as might be expected for synchrotron emission from ageing electron populations. Plot shows radio galaxies M 87 (left) and 3C 66B (right).

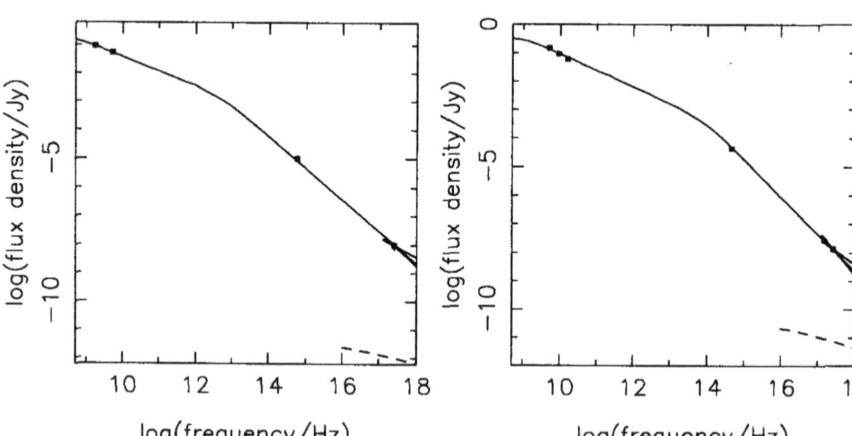

Figure 3. Radio to X-ray spectra of the low power radio galaxy B2 0755+37 (left) and the BL Lac object PKS 0521-365 (right). The predicted inverse-Compton emission (mostly SSC) is shown for $B = B_{\rm eq}$ as the dot-dashed line: it is a mis-match to the X-ray data both in intensity and X-ray spectral index.

the jets of Figures 2 and 3, perhaps because the populations of shocks capable of accelerating particles are similar. The granularity of the process will be studied with highest spatial resolution in the knotty X-ray jet of the closest radio galaxy, Centaurus A[19], which is being monitored in the radio

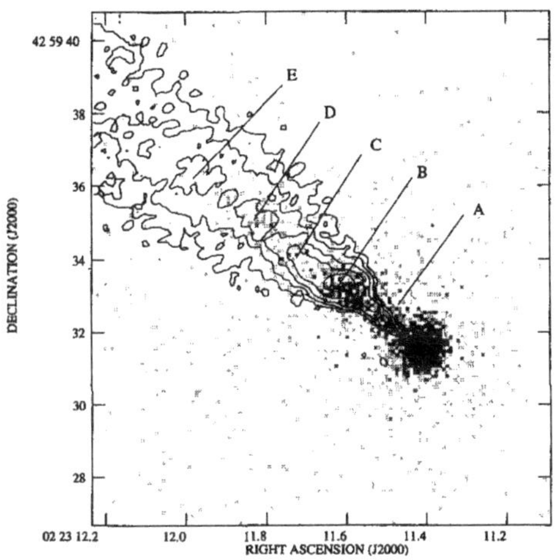

Figure 4. Radio contours on *Chandra* X-ray image of 3C 66B. The inner part of the radio jet is bright in X-rays, but the brightness distributions in X-ray and radio differ: the X-ray-bright 'knot' about 2″ from the core is slightly upstream of the bright radio emission in component B.

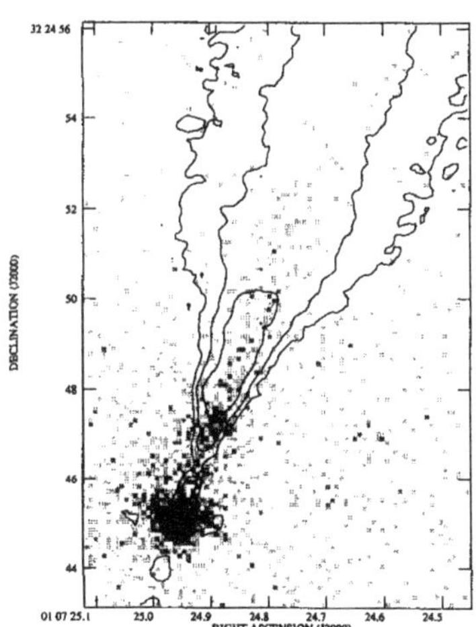

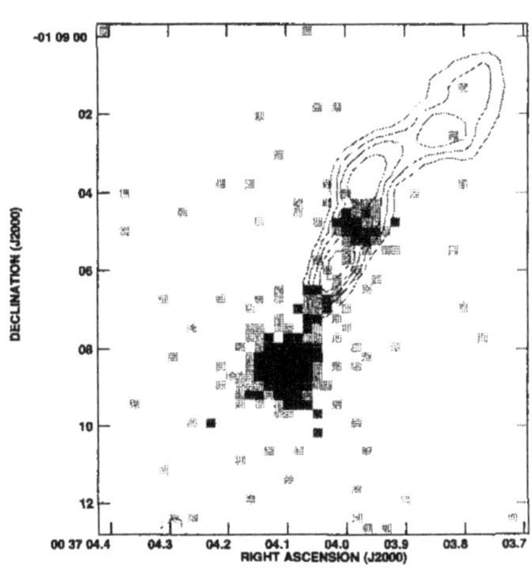

Figure 5. Radio contours on *Chandra* X-ray image of 3C 31 (left) and 3C 15 (right). These sources are similar to 3C 66B (Fig. 4) in that X-ray bright 'knots' lie upstream of radio 'knots'.

and X-ray.

Whilst the synchrotron origin of the X-ray emission from FRI jets now appears to be a solid result, and alternative emission mechanisms present problems, there is much still to be understood about the acceleration, diffusion, and ageing of the synchrotron-emitting relativistic electrons, and the magnetic-field structures. It is also vital to establish the speeds of the bulk flows. All these issues will require observations of more sample members and deeper observations of individual sources. Complexities deepen in that FRIs are in direct contact with the X-ray emitting atmospheres through which they propagate (as is obvious, for example, in 3C 84[20]). The deceleration of the radio jets from bulk relativistic speeds to less than about 0.1c is thought to be due to the entrainment of external material. A steep pressure gradient in the external X-ray-emitting medium is then required to prevent jet disruption. Galaxy atmospheres, such as those now routinely being detected, are thus expected. For 3C 31[17] is has been possible to use the measurements to measure the energy flux transported by the jet.[21]

3. FRIIs

A sketch of the standard model for one side of an FRII radio source is given in Figure 6. Simulations show that to reproduce the sharp-edged features seen in the radio morphology of high-power sources, the jets must be supersonic and lighter than the surrounding medium. Material is ejected from the core of the source in a supersonic beam which expands as it moves outwards, and terminates in a strong shock identified with a radio hotspot. The shocked beam fluid flows through this shock into the region associated with the radio lobes. As the head moves forwards through the X-ray-emitting cluster gas it is preceded by a strong bow shock. Heated X-ray emitting material must flow transversely out of the way of the jet, inflating a cocoon between the back-flow of old jet plasma and the forwards bow shock.

The X-ray-emitting cluster atmospheres of some FRIIs had been detected before *Chandra*, most notably at redshifts of about 0.6 [22,23] where these sources begin to be seen in large numbers and where 1″ corresponds to about 10 kpc. Although the only detected FRII X-ray jet was 3C 273 [24], X-ray emission associated with hotspots was known in a few sources.[25]

Chandra's spatial resolution has allowed the unambiguous separation of cluster emission from unresolved nuclear emission, as for example in the case of the $z = 0.62$ FRII radio galaxy 3C 220.1.[26] It has also detected X-ray emission from hotspots, jets, and lobes in several sources, but rarely are all these components detected in a single source. Figure 7 shows the $z = 0.55$ radio galaxy 3C 330, where X-rays are detected from the nucleus, hotspots, lobes and cluster.[27] The measurement of emission associated with the radio structures is particularly important when it can be shown to be inverse Compton scattering of a known photon field, since

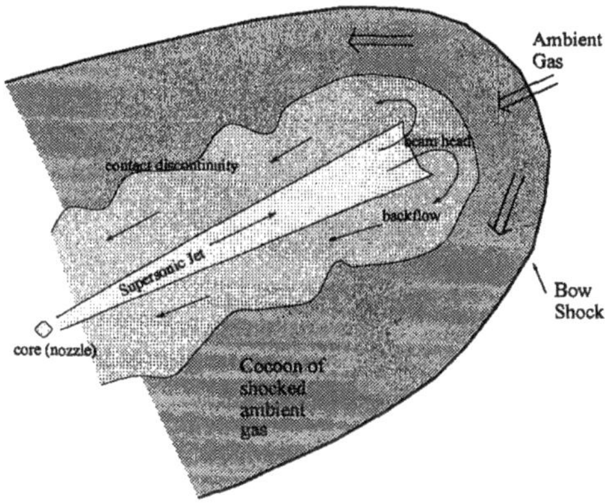

Figure 6. Sketch of the termination region of a powerful radio jet viewed in the rest frame of the bow shock. Radio lobe emission fills the region inside the contact discontinuity. Between the contact discontinuity and the bow shock we expect the ambient X-ray-emitting medium to be both compressed and heated with respect to the medium in front of the bow shock.

in combination with the radio measurements the magnetic field strength can be determined and compared with the equipartition value. In 3C 330, not only do the hotspots appear to be in broad agreement with a synchrotron self-Compton (SSC) model with $B \approx B_{\mathrm{eq}}$, but the emission from the lobes can be fitted with inverse-Compton scattering of the cosmic microwave background (CMB) radiation also with $B \approx B_{\mathrm{eq}}$ (Fig. 7, right). Hotspots agreeing with equipartition to within a factor of about two are now known in Cygnus A, 3C 295, 3C 123, 3C 207, 3C 263 and 3C 330.[27] Typical magnetic field strengths are between 10 and 30 nT. Although there are sources where the hotspot X-ray emission is too bright for $B \approx B_{\mathrm{eq}}$, it is interesting that none are under-bright, suggesting in general that magnetic fields are not a great deal larger than their equipartition values. Of course the situation concerning the magnetic fields in lobes is likely to be more complex, since these are large extended structures. Current work has shown the detectability of the features.[27] Future detailed work with *Chandra* and *XMM-Newton* is needed to trace magnetic-field substructure through detailed X-ray mapping.

Another remarkable achievement of *Chandra* has been to increase significantly (from the single case of 3C 273) the number of quasars for which resolved X-ray jet emission is detected. The first such *Chandra* detection was PKS 0637-752.[28,29] The broad-band spectral distribution and X-ray spectral index support an inverse Compton origin for the X-rays but, as illustrated in Figure 8, the component predicted to give the highest level of

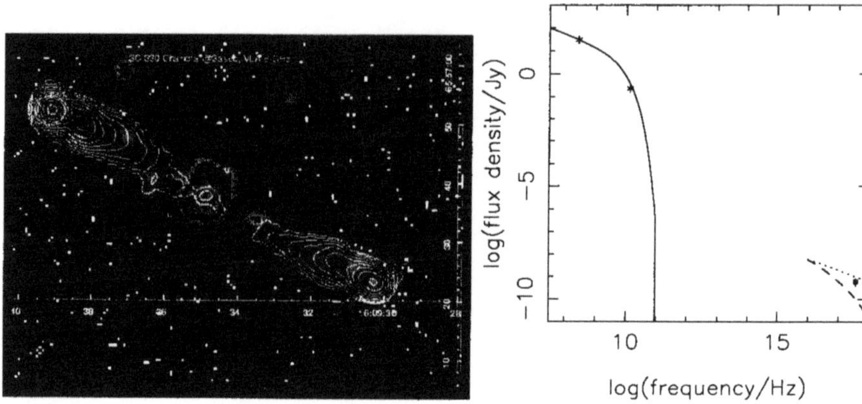

Figure 7. The left figure shows radio contours on a smoothed *Chandra* X-ray image of the powerful radio galaxy 3C 330. X-ray emission is detected from the hotspots, lobes, core, and cluster. From the figure on the right we infer that the X-ray emission from the lobes lies above a power-law extrapolation of the radio spectrum. A reasonable match to the lobe X-ray intensity is obtained with a simple model where the synchrotron emission cuts off at about 100 GHz, but inverse-Compton scattering of CMB photons (dotted line) is important in the X-ray. The model shown here assumes a uniform equipartition magnetic field over the lobe volume.

inverse-Compton X-ray emission for a jet moving at only modest bulk speed produces far too few X-rays if $B = B_{eq}$. A way to recover an equipartition magnetic field is to assume [30,31] that the highly relativistic jet ($\beta = 0.9987$) seen at a small angle to the line of sight in VLBI data does not measurably decelerate out to a distance of ~ 1 Mpc, in which case the X-rays can arise from up-scattering of the boosted (in the jet frame) CMB radiation field. If such high speeds are common (and the recent X-ray detection of several other quasar jets in surveys now underway suggests this may be the case [32,33]) then *Chandra* measurements suggest a departure from previously accepted bulk-flow speeds of $\beta \approx 0.7$, at least for the electrons responsible for the X-ray radiation.

The fact that FRIIs, unlike FRIs which are in close contact with the surrounding X-ray emitting medium, can be thought of as sealed boxes due to their supersonic jets and bow-shock structure (Figure 6) presents exciting possibilities for using X-ray observations to uncover some key physical parameters. It is not well understood whether radio sources are electron-positron or electron-proton, with arguments based on theory and observations on both sides. But the external gas pressure must not exceed the internal pressure, or radio lobes would collapse. X-ray instrumentation has now progressed to the stage where the various X-ray emitting components

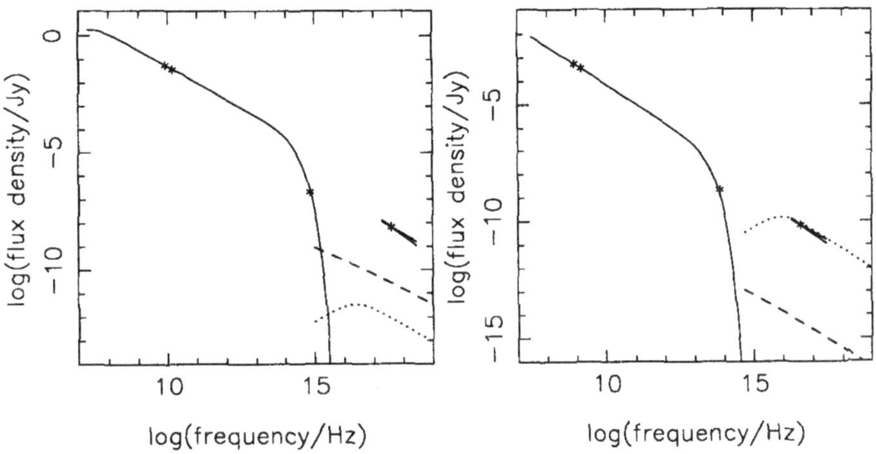

Figure 8. Rest-frame spectral distributions of the WK7.8 knot in the jet of the powerful quasar PKS 0637-752. Model fits show synchrotron emission as the solid line, SSC as dashed, and inverse Compton scattering of the CMB as dotted. Left: model assumes $B = B_{\rm eq}$ with negligible relativistic beaming in the jet, and SSC fails to predict the X-ray under this assumption. Right: Inverse-Compton scattering of the CMB matches the X-ray flux density in a model which assumes $B = B_{\rm eq}$ and a jet with $\beta = 0.9987$ at 5° to the line of sight

in and around an FRII can be identified, separated and measured. Thus the outer thermal emission measures the external pressure, while inverse-Compton lobe emission gives the total pressure in synchrotron-emitting relativistic particles and magnetic field. There is little gas in FRIIs that can supply supporting pressure, since the level of internal Faraday depolarization is low. Any difference in the sense of an apparently low internal pressure would therefore imply the presence of relativistic protons contributing to the internal energy. Thus there are good prospects for using X-ray observations to address the important issue of the composition of the plasma in radio jets and lobes.

The X-ray measurements of jets in FRII sources have brought jet speed into question. An exciting possibility is that of detecting and measuring the temperature and density of heated gas between the bow shock and radio lobe (Figure 6) as well as of the cluster medium into which the source is expanding. The shock jump conditions can then be used to measure the lobe advance speed. Further application of thrust and energy equations, requiring good radio information, could lead to independent measurements of jet speed and density. Encouragement has come from an unlikely source: the FRI galaxy Centaurus A. Although the jet is believed to be relatively sluggish, an interesting rim of heated X-ray gas is seen on the counter-jet side, closely resembling that expected from a cocoon.[34] In general, for FRIIs, the temperature of cocoon gas is predicted to be high, maybe twenty times hotter than the external cluster medium. Although we are attempting its detection using *XMM-Newton*, the measurement of jet speed may require the throughput of the more sensitive X-ray missions of the future, like *Constellation-X*[35].

Acknowledgments

I thank the many unsung heros of *Chandra* who have made the mission such an overwhelming success, and given us a new window to some of the mysteries of extragalactic radio jets. I also thank my collaborators on jet astrophysics, in particular Mark Birkinshaw, Martin Hardcastle, Ralph Kraft, Herman Marshall and Dan Schwartz. I am grateful to the Royal Society and the National Science Council for travel support.

References

1. B.L. Fanaroff and J.M. Riley, *MNRAS* **167**, 31P (1974).
2. R.A. Laing, P. Parma, H.R. de Ruiter and R. Fanti, *MNRAS* **306**, 513 (1999).
3. J.F.C. Wardle and S.E. Aaron, *MNRAS* **286**, 425 (1997).
4. L. Costamante and G. Ghisellini, *A&A* **384**, 56.
5. M.C. Weisskopf, H.D. Tananbaum, L.P. Van Speybroeck and S.L. O'Dell, in J.E. Trümper and B. Aschenbach, eds, *Proc. SPIE, X-Ray Optics, Instruments, and Missions III* **4012**, 2 (2000), and see http://asc.harvard.edu/.
6. F. Jansen et al., *A&A* **365**, L1 (2001), and see http://xmm.vilspa.esa.es/.
7. D.M. Worrall and M. Birkinshaw, *ApJ* **551**, 178 (2001).
8. D.M. Worrall and M. Birkinshaw, *ApJ* **427**, 134 (1994).
9. C.M. Canosa, D.M. Worrall, M.J. Hardcastle and M. Birkinshaw, *MNRAS* **310**, 30 (1999).
10. S. Döbereiner et al., *ApJ* **470**, L15 (1996).
11. J.A. Biretta, C.P. Stern and D.E. Harris, *AJ* **101**, 1632 (1991).
12. M.-H. Ulrich, in L. Maraschi et al., eds, *BL Lac Objects, Berlin: Springer-Verlag* 45 (1989).
13. D.M. Worrall, M. Birkinshaw and M.J. Hardcastle, *MNRAS* **326**, L7 (2001).
14. H. Böhringer et al., *A&A* **365**, L181 (2001).
15. M.J. Hardcastle, M. Birkinshaw and D.M. Worrall, *MNRAS* **326**, 1499 (2001).
16. M. Birkinshaw, D.M. Worrall and M.J. Hardcastle, *MNRAS* in press (2002).
17. M.J. Hardcastle, D.M. Worrall, M. Birkinshaw, R.A. Laing and A.H. Bridle, *MNRAS* in press (2002).
18. S. Nolan, M. Birkinshaw and D.M. Worrall, *MNRAS*, in preparation (2002).
19. R.P. Kraft, W.R. Forman, C. Jones, S.S. Murray, M.J. Hardcastle and D.M. Worrall, *ApJ* **569**, 54 (2002).
20. A.C. Fabian et al., *MNRAS* **318**, L65 (2000).

21. R.A. Laing and A.H. Bridle, *MNRAS* in press (2002).
22. M.J. Hardcastle and D.M. Worrall, *MNRAS* **309**, 969 (1999).
23. C.S. Crawford, I. Lehmann, A.C. Fabian, M.N. Bremer and G. Hasinger, *MNRAS* **308**, 1159 (1999).
24. D.E. Harris and C.P. Stern, *ApJ* **313**, 136 (1987).
25. D.E. Harris, K.M. Leighly and J.P. Leahy, *ApJ* **499**, L149 (1998).
26. D.M. Worrall, M. Birkinshaw, M.J. Hardcastle and C.R. Lawrence, *MNRAS* **326**, 1127 (2001).
27. M.J. Hardcastle, M. Birkinshaw, R.A. Cameron, D.E. Harris, L.W. Looney and D.M. Worrall, *ApJ*, in press (2002).
28. D.A. Schwartz et al., *ApJ* **540**, L69 (2000).
29. G. Chartas et al., *ApJ* **542**, 655 (2000).
30. F. Tavecchio, L. Maraschi, R.M. Sambruna, C.M. Urry, *ApJ* **544**, L23 (2000).
31. A. Celotti, G. Ghisellini and M. Chiaberge, *MNRAS* **321**, L1 (2000).
32. R.M. Sambruna, L. Maraschi, F. Tavecchio, C.M. Urry, C.C. Cheung, G. Chartas, R. Scarpa, J.K. Gambill, *ApJ* **571**, 206 (2002).
33. H.L. Marshall et al., *ApJ*, in preparation (2002).
34. R.P. Kraft, S. Vázquez, W.R. Forman, C. Jones, S.S. Murray, M.J. Hardcastle, D.M. Worrall and E. Churazov, *ApJ*, in preparation (2002).
35. N.E. White, J.A. Bookbinder and H. Tananbaum, in R. Giacconi, S. Serio and L. Stella, eds, *X-Ray Astronomy 2000, ASP Conference Series* **234**, 597 (2001), and see http://constellation.gsfc.nasa.gov/.

Neutrino Astrophysics at 10^{20} eV

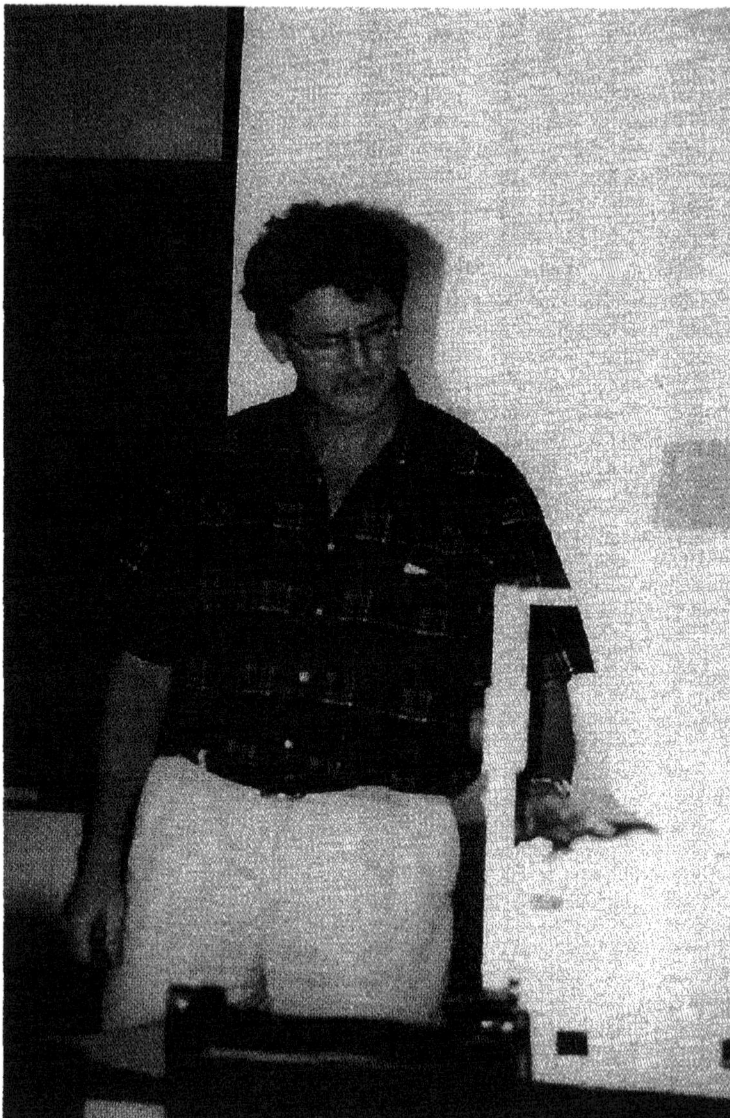

Thomas J. Weiler

NEUTRINO ASTROPHYSICS AT 10^{20} EV

THOMAS J. WEILER
Department of Physics & Astronomy
Vanderbilt University
Nashville TN, USA
E-mail: tom.weiler@vanderbilt.edu

Neutrinos offer a particularly promising view of the extreme Universe. Since neutrinos are not attenuated by the intervening CMB and other radiation fields, they are messengers from the very distant and very young universe. Since neutrinos are not degraded or absorbed by the source material at production, they carry information about central engine dynamics. Since neutrinos are not deflected by cosmic magnetic fields, they should point to their sources. This will allow astronomy to be performed. The neutrino cross-section at extreme-energy ($\gtrsim 10^{20}$ eV) may also offer a window to new particle physics above thresholds inaccessible to terrestrial accelerators. Measurement of an anomalously large neutrino cross-section would indicate new physics (e.g. low string-scale, extra dimensions, precocious unification), while a smaller than expected cross-section would reveal an aspect of QCD evolution. Here I focus on the significance of the neutrino cross-section at extreme-energy (EE), and how it may be determined; and on hints in the EE cosmic ray data which may already implicate $\gtrsim 10^{20}$ eV neutrinos.

1. Why Neutrinos at 10^{20} eV?

Detection of ultrahigh-energy neutrinos is one of the important challenges of the next generation of cosmic ray detectors. Their discovery will mark the advent of neutrino astronomy, allowing the mapping on the sky of the most energetic, and most distant, sources in the Universe. In addition, detection of extreme-energy (EE) neutrinos may help resolve the puzzle of cosmic rays (CRs) with energies beyond the Greisen-Zatsepin-Kuzmin cutoff by validating Z-bursts, topological defects, superheavy relic particles, neutrino strong-interactions, *etc.*

To mimic hadronically-induced air showers, the new neutrino cross section must be of hadronic strength, ~ 100 mb, above $E_{\rm GZK} \equiv 5 \times 10^{19}$ eV. Simple perturbative calculations of single scalar or vector exchange cannot provide an acceptably fast growth of the cross-section with energy.[1] However, the modern thoughts on large TeV-scale cross-sections are much more imaginative. A plethora of new states, possibly growing exponentially in s

or $\sqrt{s}$, is motivated by precocious unification, low-scale string theory, and modes from additional space-dimensions accessible at $\sqrt{s} \sim$ TeV.

Direct limits on the EE neutrino cross-section are quite weak. The vertical column density of our atmosphere is $X_v = 1033\text{g/cm}^2$. In terms of neutrino MFP λ_ν, this may be written $X_v/\lambda_\nu = \sigma_{\nu N}/1.6\,\text{mb}$. The horizontal slant depth X_h is 36 times larger, leading to $X_h/\lambda_\nu = \sigma_{\nu N}/44\mu\text{b}$. Since penetrating events are not observed above us or to our side, the neutrinos must be interacting high in the atmosphere (large cross-section) or interacting barely at all (small cross-section). Thus the cross-section range from $\sim 20\mu$b to ~ 1 mb must be excluded.

Several sources of 10^{20} eV neutrinos are possible, ranging from AGNs to exotic top-down production. A nice review of sources, classified according to their speculative nature, was given a few years ago by Protheroe.[2] It remains a useful source of possibilities to date. Of course, there is a reasonably guaranteed prediction for a flux F_ν of GZK neutrinos in the energy range 10^{15} to 10^{20} eV, based on the observed flux of UHECR protons at the GZK limit. This flux is expected to peak in the decade 10^{17} to 10^{18} eV for uniformly-distributed proton sources, and around 10^{19} eV for "local" sources within ~ 50 Mpc of earth.[3]

The growth of experiments continues, with Auger to follow HiRes and AGASA, and EUSO to follow Auger. Eventually, OWL may follow EUSO. A useful table of happening and proposed EECR experiments has been assembled by Peter Gorham. It is available at
http://astro.uchicago.edu/home/web/olinto/aspen/gorham_table.htm.
EUSO and OWL are proposed space-based observatories, triggering on the nitrogen fluorescence produced when an EECR traverses our atmosphere. In terms of U.S. states, the Auger field of view may equal the area of Rhode Island, while EUSO and OWL will equal Texas or more. For the space-based experiments, the 400-500 km height limits their sensitivity to events with energy above $\sim 10^{19}$ eV. Thus, the $1/r^2$ loss provides a natural filter to select only the most extreme CRs.

2. Dispersion Relations: the High-Low Energy Connection

Dispersion relations provide a rigorous, nonperturbative, model-independent calculation of the growth of the *elastic* neutrino amplitude at much lower energies due to any rising high-energy cross-section. If new physics dominates the neutrino total cross-section with a value σ^* above the lab energy E^*, then the dispersion relation determines the real part of the new strong-interaction elastic amplitude at lower energy E to be $\frac{1}{2\pi}\frac{E}{E^*}\sigma^*$.

Remarkably, significantly enhanced scattering rates may occur for elastic νN seven orders of magnitude lower in energy than the onset of a new total cross–section.[4] Such anomalous "low" energy scattering may be present in the neutrino data at Fermilab and CERN, and in $e^- p \to \nu_e n$ scattering at HERA.

How does this magic come about? Assuming only that the scattering amplitude is analytic, there results the following dispersion relation:[4]

$$\text{Re}\, A_\pm(E) - \text{Re}\, A_\pm(0) = \frac{E}{4\pi}\mathcal{P}\int_0^\infty dE' \left(\frac{\sigma_{tot}^{\nu N}(E',\pm)}{E'(E'-E)} + \frac{\sigma_{tot}^{\bar\nu N}(E',\pm)}{E'(E'+E)}\right) \quad (1)$$

where $A_\pm(E)$ are invariant ν-N amplitudes, labeled by the nucleon helicity, and $\mathcal{P}$ denotes the principle value of the integral. Next, suppose the new-physics cross-section dominates the dispersion integral (1) for $E' \geq E^*$; such has been hypothesized to explain the air showers observed above the GZK cutoff. Motivated by simplicity and the behavior of the Standard Model (SM) strong-interaction, let us assume that σ^* is independent of helicity and energy, and that the new component of the neutrino cross section obeys the Pomeranchuk theorem: $\sigma_{tot}^{\nu N}(E,\pm) - \sigma_{tot}^{\bar\nu N}(E,\pm) \stackrel{\nu\to\infty}{\longrightarrow} 0$. These assumptions and the dispersion relation lead directly a result for the real part of the amplitude at energy E:

$$\text{Re}\, A_\pm(E) \simeq \text{Re}\, A_\pm(0) + \frac{1}{2\pi}\frac{E}{E^*}\sigma^* \quad . \quad (2)$$

This result cannot be obtained in perturbation theory! The appearance of elastic amplitudes at $E = 0$ can be traced back to a subtraction required to arrive at the convergent integral in (1). They do not weaken the predictive power of the dispersion relation, for Re $A_\pm(0)$ is nothing but the low energy limit of the weak interaction, $\sim \frac{G_F}{2\sqrt{2}}$.

The most direct test of an anomalously large neutrino cross section would be a measurement of the refractive index

$$n_{\text{ref},\pm} - 1 = \frac{2\rho}{E}\text{Re}A_\pm(E)\,, \quad (3)$$

where ρ is the nucleon number density of the (possibly polarized) medium. The anomalous contribution to the right–hand side of (3) exceeds the SM contribution already at neutrino energies $E \gtrsim 100$ GeV. There could be sizeable new matter-effects on oscillations. However, if the anomalous reactions (if they exist) are flavor neutral, they produce a common phase and there will be no new matter effects associated with them.

A more promising observable consequence is available from the elastic cross section, obtained from the square of the elastic amplitude. The SM weak amplitude is energy-independent before renormalization, and weakly energy-dependent after renormalization. Therefore, we may immediately write down the ratio of the new amplitude to the SM amplitude:

$$\frac{\mathrm{Re}A(E)_{\mathrm{new}}}{\mathrm{Re}A(E)_{\mathrm{SM}}} \simeq \left(\frac{E/100\ \mathrm{GeV}}{E^{*}/10^{18}\ \mathrm{eV}}\right)\left(\frac{\sigma^{*}}{100\ \mathrm{mb}}\right) . \tag{4}$$

It is clear from (4), and striking, that order 100% effects in the real elastic amplitudes begin to appear already at energies seven orders of magnitude below the full realization of the strong cross section. 100 GeV neutrino data already exists at fermilab and CERN.

However, elastic neutrino scattering is challenging to measure. A low–energy recoil proton must be detected, with a veto on events with pions produced. Because the momentum transfer in elastic scattering is limited to $\lesssim 1\ \mathrm{GeV}^2$, the recoil nucleon has a kinetic energy of at most 0.5 GeV. Other related possibilities exist. Since the anomalous elastic cross-section grows quadratically with E, the anomalous event rate develops rapidly for $E > 100$ GeV. Thus, the event sample of a future underground/water/ice neutrino telescope optimized for TeV neutrinos could conceivably contain *1000* times more elastic neutrino events than predicted by the SM; and a telescope optimized for PeV neutrinos may contain 10^9 more elastic events. The dispersion result is robust enough to have ruled out some of the wilder brane-world cross-section's proposed for the EE neutrino.

There may be further tests of the strong–interaction hypothesis. If the neutrino develops a strong-interaction at high energy, do not the electron and the other charged-lepton $SU(2)$–doublet partners of the neutrinos also develop a similar strong-interaction? Is there new physics in the quasi-elastic $e^- p \to \nu_e n$ scattering channel at HERA energies? Although a possible enhancement in the quasi-elastic channel cannot be deduced from dispersion relations, a separate calculation can be made if certain aspects of the new high-energy strong-interaction are assumed. This is presently under investigation.[5]

3. Can't Lose Theorem for Smaller/larger Cross-Sections

Approved and proposed experiments plan to detect UHE neutrinos by observation of the nearly horizontal air showers (HAS) in the Earth's atmosphere resulting from ν-air interactions. The expected rates are proportional to $\sigma_{\nu N}$. Calculations of this cross section at 10^{20} eV necessarily use an extrapolation of parton distribution functions and SM parameters far beyond the reach of experimental data. The resulting cross section at 10^{20} eV is $\sim 10^{-31}\mathrm{cm}^2$. It has recently been argued that the extrapolation may overestimate the true neutrino cross section[6] at energies above about

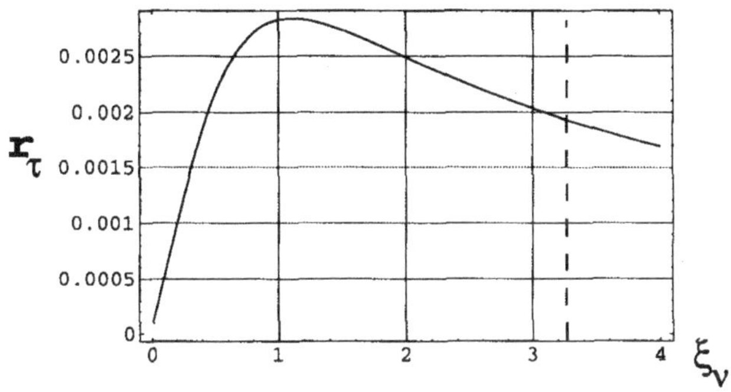

Figure 1. The ratio r_τ of the upward going τ flux to the incident tau neutrino flux F_{ν_τ} as a function of $\xi_\nu = \lambda_\nu/R_\oplus = 1/(\sigma_{\nu_N} n R_\oplus)$, with fixed $\lambda_\tau/R_\oplus = 3.5 \times 10^{-3}$, appropriate for events initiated by $\sim 10^{20}$ eV neutrinos. Here n is the mean nucleon number density. Assuming a monotonic cross section dependence on $\sqrt{s}$, the value of ξ is limited from above by the HERA measurements, as shown by a vertical dashed line. For neutrino trajectories through the Earth's mantle, a useful expression is $\xi_\nu = 0.66/\sigma_{33}$, where σ_{33} is the neutrino cross section in units of 10^{-33}cm^2.

$10^{17.5}$ eV. A smaller cross section would compromise the main detection signal proposed for UHE neutrino experiments. On the other hand, the extrapolated cross-section may be too low, for it ignores possible contributions from new physics that may enter in the ~TeV to PeV scale inaccessible to terrestrial accelerators.

It was recently shown [7] that the flux of up-going charged leptons (UCLs) per unit surface area produced by neutrino interactions *below* the surface is *inversely* proportional to $\sigma_{\nu N}$. This contrasts with the HAS rate which is proportional to $\sigma_{\nu N}$. As shown in Fig. 1, a lower cross section *increases* the UCL rate per surface area as $\sigma_{\nu N}^{-1}$ as long as the neutrino absorption mean free path (MFP) in Earth is small in comparison with the Earth's radius, $R_\oplus$. This relation holds for $\sigma_{\nu N} \gtrsim 2 \times 10^{-33}\text{cm}^2$. bogus

Let us now examine the physics of upward showers in some detail. UHE neutrinos are expected to arise from pion and subsequent muon decay. These flavors oscillate and eventually decohere during their Hubble-time journey. If $|U_{\tau 3}| \simeq |U_{\mu 3}|$ are near-maximal, as inferred from the Super-Kamiokande data, then $F_{\nu_\tau} \approx \frac{1}{3} F_\nu$ is expected. The energy-loss MFPs, λ_τ and λ_μ, for taus and muons to lose a decade in energy are 11 km and 1.5 km, respectively, in surface rock with density $\rho_{sr} = 2.65\,\text{g/cm}^3$. Tau and muon decay MFPs are long above 10^{18} eV: $c\tau_\tau = 490\,(E_\tau/10^{19}\text{eV})$ km, and $c\tau_\mu \sim 10^8\, c\tau_\tau$ for the same lepton energy. Because the energy-loss MFP for a τ produced in rock or water is much longer than that of a muon, the produced taus have a much higher probability to emerge from the Earth and to produce an atmospheric shower. Thus, the dominant primary for initiation of UAS events is the tau neutrino.

Consider an incident tau neutrino whose trajectory cuts a chord of length l in the Earth. The probability for this neutrino to reach a distance x is $P_\nu(x) = e^{-x/\lambda_\nu}$, where $\lambda_\nu^{-1} = \sigma_{\nu N}\,\rho$ (the conversion from matter density to number density via N_A/gm is implicit). The probability to produce a tau lepton in the interval dx is $\frac{dx}{\lambda_\nu}$. The produced τ carries typically 80% of the parent neutrino energy; we approximate this as 100%. The probability of a τ produced at point x to emerge with sufficient energy $E_{\rm th}$ to produce an observable shower can be approximated as $P_{\tau\to{\rm UAS}} = \Theta(\lambda_\tau + x - l)$, with $\lambda_\tau = \frac{1}{\beta_\tau\,\rho_{s\tau}}\ln(E_\nu/E_{\rm th})$; $\beta_\tau \approx 0.8\times10^{-6}{\rm cm}^2/g$ is the exponential energy-attenuation coefficient. For taus propagating through rock, one can take $\lambda_\tau \approx 22$ km for $E_\nu \sim 10^{20}$eV and $E_{\rm th} \sim 10^{18}$eV, while for taus propagating through ocean λ_τ is 2.65 times larger.

Taking the product of these conditional probabilities and integrating over the interaction site x we get the probability for a tau neutrino incident along a chord of length l to produce an UCL:

$$P_{\nu_\tau\to\tau}(l) = \int_{l-\lambda_\tau}^{l} \frac{dx}{\lambda_\nu}\, e^{-x/\lambda_\nu} = (e^{\lambda_\tau/\lambda_\nu} - 1)\, e^{-l/\lambda_\nu}\,. \tag{5}$$

The emerging tau decays in the atmosphere with probability $P_d = 1 - \exp(-2R_\oplus H/c\tau_\tau l)$, where $H \approx 10$ km parametrizes the height of the atmosphere. Thus, the probability for a tau-neutrino to produce an up-going air shower (UAS) is

$$P_{\nu_\tau\to{\rm UAS}}(l) = (1 - e^{-2R_\oplus H/c\tau_\tau l})\, P_{\nu_\tau\to\tau}(l)\,. \tag{6}$$

The fraction of neutrinos with chord lengths in the interval $\{l, l+dl\}$ is $\frac{l}{2R_\oplus^2}dl$. To get an event rate probability from the incident neutrino flux, there are two further geometric factors to be included: the solid angle factor π for a planar detector with hemispherical sky-coverage, and the tangential surface area A of the detector. Putting all probabilities together, we arrive at the rate of UCL and UAS events:

$$R_{\tau({\rm UAS})} = F_{\nu_\tau}\pi A \int_0^{2R_\oplus} \frac{l\,dl}{2\,R_\oplus^2}\, P_{\nu_\tau\to\tau({\rm UAS})}(l)\,. \tag{7}$$

The ratio $r_\tau = R_\tau/F_\nu \pi A$ is shown in Fig. 1, and the the number of expected UAS events per incoming neutrino is shown in Fig. 2, as a function of the neutrino cross section. In both figures, we have taken $\lambda_\tau \approx 22$ km, appropriate for "over-land" events. For comparison, we also show in Fig. 2 the number of expected HAS events per neutrino that crosses a 250 km field of view, up to an altitude of 15 km. It is clear that for the smaller

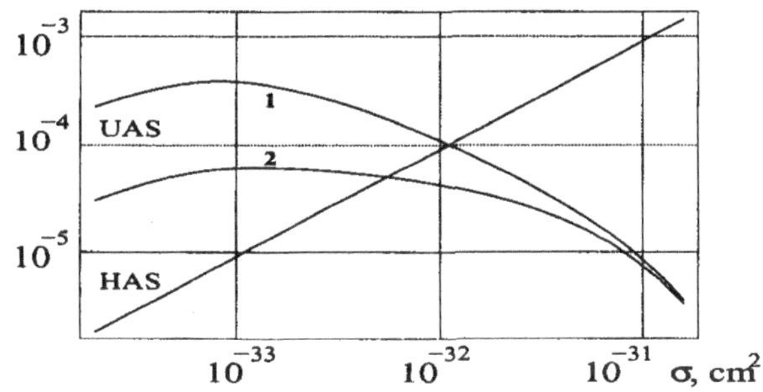

Figure 2. The air shower probability per incident tau neutrino $R_{\rm UAS}/F_{\nu_\tau}\pi A$ as a function of the neutrino cross section. The incident neutrino energy is 10^{20} eV and the assumed energy threshold for detection of UAS is $E_{\rm th} = 10^{18}$eV for curve 1 and 10^{19}eV for curve 2.

values of the cross section, UAS events will outnumber HAS events, and vice versa.

Taken together, up-going and horizontal rates ensure a healthy total event rate, regardless of the value of $\sigma_{\nu N}$. Moreover, by comparing the HAS and UAS rates, the neutrino-nucleon cross section can be inferred at energies as high as 10^{11} GeV or higher. This enables QCD studies at a minimum, and possibly discovery of a strong neutrino cross-section, at a cms energy three and two orders of magnitude beyond the reach of Fermilab's Tevatron and the LHC, respectively. $\sigma_{\nu N}$ may also be determinable from a measurement of the angular distribution of UCL/UAS events, in addition to the approach comparing UAS and HAS rates. One expects the angular distribution of UCL to peak near $\cos\theta_{\rm peak} \sim \lambda_\nu/2R_\oplus$, which implies $\sigma_{\nu N} \sim (2\,\langle\rho\rangle\, R_\oplus\, \cos\theta_{\rm peak})^{-1}$.

We give some examples of the UAS event rates expected from a smaller neutrino cross section at 10^{20} eV. Let us choose $\sigma_{\nu N} = 10^{-33}{\rm cm}^2$, for example. Taking the mantle density of $\rho_m = 4.0\,g/{\rm cm}^3$ and $R_\oplus = 6.37 \times 10^8$cm, one gets $\xi_\nu = 0.65$. Reference to Fig. 1 then shows that the $\nu_\tau \to \tau$ conversion probability is $r_\tau = 0.1\%$ for land events with $E_\nu \sim 10^{20}$ eV and $E_\tau \geq 10^{18}$ eV. Including the probability for a tau to decay in the atmosphere, the $\nu_\tau \to$ UAS probability is 4×10^{-4} (7×10^{-5}) for a shower-energy threshold $E_{\rm th} = 10^{18}$ eV (10^{19} eV), according to Fig. 2. EUSO and OWL have shower-energy thresholds $\sim 10^{19}$eV, corresponding to curve 2 in Fig. 2. They have apertures $\sim 6\times10^4$ km^2 and 3×10^5 km^2, respectively, for a wide angular-range of UAS. These detectors should observe F_{20} and $7F_{20}$ UAS events per year, respectively (not including duty cycle); here F_{20} is the incident neutrino flux at and above 10^{20} eV in units of km^{-2}sr^{-1}yr^{-1}, one-third of which are ν_τ's. Including showers from taus originating outside the field of view, and direct tau events, increases these rates. The rates may be further increased in space-based detectors by tilting toward the horizon so as to maximize the acceptance for events with smaller chord lengths (where the neutrino attenuation is less and the field of view is greater, but the energy threshold is higher) and to allow more atmospheric path length for tau decay. The rates will also increase if $E_{\rm th}$ can be reduced.

Recently, it was pointed out that production of the lightest supersymmetric particle (LSP) from annihilating dark matter may present a detectable flux of LSP-CRs.[8] The LSP cross-section may be as small as 10^{-2} times the SM neutrino cross-section, making the LSP cross-section a perfect prototype for measurement via either the UAS/HAS method or the angular-dependence method.

4. Puzzles in the Extreme-Energy Cosmic Rays (EECRs)

The discoveries by the AGASA, Fly's Eye, Haverah Park, and Yakutsk collaborations of air shower events with energies above the Greisen-Zatsepin–Kuzmin (GZK) cutoff of $\sim 5 \times 10^{19}$ eV challenge the SM of particle physics and the hot big-bang model of cosmology. Not only is the mechanism for particle acceleration to such extreme energies controversial, but also the propagation of EECRs over cosmic distances is problematic. It must be mentioned, however, that the HiRes experiment does not confirm the prior rate for events $\sim 10^{20}$ eV. The situation will remain murky until the Auger hybrid detector provides clear guidance, expected in about two years. Above the resonant threshold for Δ^* production, $\sim 5 \times 10^{19}$ eV, protons lose energy by the scattering process $p + \gamma_{2.7K} \to \Delta^* \to N + \pi$; $\gamma_{2.7K}$ denotes a photon in the $2.7K$ cosmic background radiation. This is the mechanism for the much–heralded GZK cutoff. For every mean free path ~ 6 Mpc of travel, the proton loses 20% of its energy on average. A proton produced at its cosmic source a distance D away will on average arrive at earth with only a fraction $\sim (0.8)^{D/6\,\rm Mpc}$ of its original energy. Of course, proton energy is not lost significantly if the highest energy protons come from a rather nearby source, $\lesssim 50$ to 100 Mpc. However, acceleration of protons to $\sim 10^{20}$ eV, if possible at all, is generally believed to require the most extremely–energetic compact sources, such as active galactic nuclei (AGNs) or gamma–ray bursts (GRBs). Since AGNs and GRBs are hundreds of megaparsecs away, the energy requirement at an AGN or GRB for a proton which arrives at earth with a super-GZK energy is unrealistically high. A primary nucleus mitigates the cutoff problem (energy per nucleon is reduced by 1/A), but above $\sim 10^{19}$ eV nuclei should be photo–dissociated by the 2.7K background, and possibly disintegrated by the particle density ambient at the astrophysical source. Gamma–rays and neutrinos are other possible primary candidates for the highest energy events. The mean free path, however, for a $\sim 10^{20}$ eV photon to annihilate on the radio background to e^+e^- is believed to be only 10 to 40 Mpc.

Concerning the neutrino hypothesis, the Fly's Eye event occurred high in the atmosphere, whereas the event rate expected in the SM for early development of a neutrino–induced air shower is down from that of an electromagnetic or hadronic interaction by six orders of magnitude. On the other hand, there is evidence that the arrival directions of some of the highest–energy primaries are paired in directions with events having an order of magnitude lower energy, and displaced in time by about a year.[9] Such event-pairing may argue for stable neutral primaries coming from a source of considerable duration. Neutrino primaries do satisfy this criterion. Furthermore, a recent analysis of the arrival directions of the super–GZK events offers a tentative claim of a correlation with the directions of BL-Lac quasars.[10] If this correlation is validated with future data, then the propagating cosmic particle must be charge neutral and have a negligible magnetic moment. The neutrino again emerges as the only candidate among the known particles.

A rather conservative and economical scenario for understanding the super-GZK events involves cosmic ray neutrinos scattering resonantly on the cosmic neutrino background (CNB) predicted by Standard Cosmology, to produce boosted ($\gamma \sim 10^9$) Z-bosons known as "Z-bursts".[12] A more radical proposal is that neutrinos at EE acquire a large non-SM cross section that allows them to initiate air showers high in the atmosphere. This strong neutrino cross-section has been addressed in sections 2 and 3. The Z-burst hypothesis is discussed in section 4.2.

Several non-neutrino solutions for the origin of the exceptional EECRs have been proposed, ranging from unseen Zevatron accelerators (1 ZeV = 10^{21} eV) and decaying supermassive particles and topological defects in the Galactic vicinity, to exotic primaries, exotic new interactions, and even exotic breakdown of conventional physical laws.[11] Generally, these models are distinguished from the neutrino scenarios by the lack of pairing and pointing. However, magnetic caustics have been proposed as effective lenses which focus charged particles into pairs.

4.1. *Directional pairs and a triplet?*

To assess the significance of the small-scale clustering observed in the highest-energy AGASA cosmic ray data, analytic formulae for the probability of random cluster configurations have been derived.[13] These formulae also offer a quick study of the strong potential of future experiments for deciding whether any observed clustering is most likely due to non-random sources. For detailed comparison to data, this analytical approach cannot compete with Monte Carlo simulations including experimental systematics. However, the derived formulae do offer two advantages: (i) easy assessment of the significance of any observed clustering, and most importantly, (ii) an explicit dependence of cluster probabilities on the chosen angular bin-size.

To derive the combinatoric formula for the probability of various event distributions in angle, imagine that the sky coverage consists of a solid angle Ω divided into N equal angular bins, each with solid angle $\omega \simeq \pi\theta^2$ steradian; the number of bins with cone half-angle θ is

$$N \simeq \frac{\Omega}{\pi\theta^2} = 1045\,\frac{(\Omega/1.0\text{ sr})}{(\theta/1.0^\circ)^2}, \tag{8}$$

where Ω is the solid angle (sidereal or galactic) on the celestial sphere covered by the experiment. Toss n events at random into these bins. (Such a chance distribution of events is just what is expected in some models for the EECRs, e.g. randomly situated decaying SMPs, or charged-particle or monopole primaries traversing incoherent magnetic fields.) Define each event distribution by specifying the partition of the n total events into a number m_0 of empty bins, a number m_1 of single hits,a number m_2 double hits, etc., among the N angular bins. The probability to obtain a given event topology is:

$$\begin{aligned} &P(\{m_j\}, n, N) \\ &= \frac{1}{N^n}\,\frac{N!}{m_0!\; m_1!\; m_2!\; m_3!\ldots}\,\frac{n!}{(0!)^{m_0}\;(1!)^{m_1}\;(2!)^{m_2}\;(3!)^{m_3}\ldots}. \end{aligned} \tag{9}$$

The $N!$ and $n!$ factors in the numerator count the permutations of the bins and the events, respectively. The $m_j!$ and $j!$ factors in the denominator remove the overcounting of those bins containing j events, and the events within those bins, respectively. The normalization factor N^n is total number of ways to partition n events among N bins.

The variables in the probability are not all independent. The partitioning of events is related to the total number of events by $\sum_{j=1} j \times m_j = n$, and to the total number of bins by $\sum_{j=0} m_j = N$. Because of these constraints, one infers that the process is not described by a simple multinomial or Poisson probability distribution. It is useful to use these constraints to rewrite the exact probability (9) as

$$P(\{m_j\}, n, N) = \frac{N!}{N^N} \frac{n!}{n^n} \prod_{j=0} \frac{(\overline{m_j})^{m_j}}{m_j!}\,, \tag{10}$$

where

$$\overline{m_j} \equiv N \left(\frac{n}{N}\right)^j \frac{1}{j!}\,. \tag{11}$$

In the $n \ll N$ limit, $\overline{m_j}$ is expected to approximate the mean number of j-plets, and eq. (10) becomes roughly Poissonian. As an approximate mean, $\overline{m_j}$ defined in eq. (11) provides a simple estimate of cluster probabilities due to chance for the $n \ll N$ case.

Two large-number limits of interest are $N \gg n \gg 1$, and $n > N \gg 1$. With bin numbers typically $\sim 10^3$, the first limit applies to the AGASA, HiRes, Auger and Telescope Array experiments; the second limit becomes relevant for the EUSO/OWL/AW experiment after a year or more of running.

When $N \gg n$, the number m_0 of empty bins is of order N, and the number of bins m_1 with single events (singlets) is order n; the number of clusters (doublets, triplets, etc.) is small. It is sensible to explicitly evaluate the not-so-interesting $j = 0$ and 1 terms in eqs. (10) and (11). With the use of Stirling's approximation for the factorials, one arrives at a simple form for the probability, valid when $N \gg n \gg 1$:

$$P(\{m_i\}, n, N) \approx \mathcal{P} \left[\prod_{j=2} \frac{(\overline{m_j})^{m_j}}{m_j!} e^{-\overline{m_j}\, r^j (j-2)!}\right], \tag{12}$$

where $r \equiv (N - m_0)/n \approx 1$, and the prefactor $\mathcal{P}$ is

$$\mathcal{P} = e^{-(n-m_1)} \left(\frac{n}{m_1}\right)^{m_1+\frac{1}{2}}\,. \tag{13}$$

In the "sparse events" case here, where $N \gg n$, one expects the number of singlets m_1 to approximate the number of events n. In this case the prefactor is near unity. The non-Poisson nature of Eq. (12) is reflected in the factorials and powers of r in the exponents, and the deviation of the prefactor from unity.

In the case where $n > N \gg 1$, higher j-plets are common and the distribution of clusters can be rather broad in j. Already at $j = 1\,(2)$, Stirling's approximation to $j!$ is good to 8% (4%), and so we may write $\overline{m_j}$ in the approximate form for $j \geq 1$:

$$\overline{m_j} \approx \sqrt{\frac{N^3}{2\pi e n}} \left(\frac{en}{jN}\right)^{j+\frac{1}{2}}\,. \tag{14}$$

Extremizing this expression with respect to j, one learns that the most populated j-plet occurs near $j \sim n/N$. Combining this result with the broad distribution expected for large n/N, one expects clusters with j up to several $\times\ \frac{n}{N}$ to be common in the EUSO/OWL/AW experiment.

Shown in Fig. 3 is an updated assessment[14] of the AGASA-plets, five doublets plus a triplet, obtained using formula (12). Two features of the

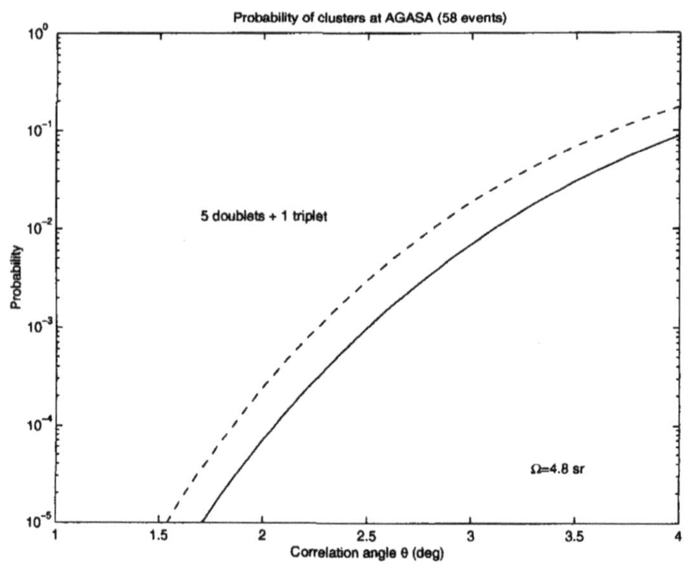

Figure 3. Exact (solid) and Poissonian (dashed) inclusive probabilities for five doublets and one triplet in the 58-event AGASA sample (from Ref. 14).

figure are noteworthy. The first is the rather extreme sensitivity of the statistical significance to the angular binning size. AGASA claims a 2.5° resolution, which puts the significance of their clusters at 10^{-3}. If the resolution were 3° (2.0°), the significance would be a factor of six weaker (fifteen stronger). The second feature is the error made when Poisson statistics are blindly applied. For the 2.5° resolution, Poisson statistics underestimate the significance by a factor of three.

The optimal bin-size for elucidating the physics (if any) underlying clustering is an open question. If clustered events originated from a common source and traveled without bending, then the experimental angular resolution is the optimal bin-size. On the other hand, if primary trajectories are somewhat bent by cosmic magnetic fields, then the optimal bin-size may exceed the experimental angular resolution. If clustering results from magnetic focusing, then the angular size of magnetic caustics may be the relevant bin-size. If clustering results from density fluctuations in the Galactic halo, then the angular size of the fluctuations on the sky may be the optimal bin-size. In fact, since photons are not bent by magnetic fields whereas protons are bent, it is likely that the optimal bin-size for photon-initiated events is smaller than that for proton-initiated events. The analytic formulae provide the random background for any chosen angular bin-size, which should prove quite useful in the future. the formulae are surprisingly robust, in that it has been shown that seasonal variation of sky-coverage with time (right ascension) inherent in a fluorescence detector (like HiRes) does not invalidate the analytic approach.

4.2. *Z-bursts*

In the Z-burst mechanism, EECR neutrinos scatter resonantly on the cosmic neutrino background (CNB) predicted by Standard Cosmology to produce Z-bosons.[12] These Z-bosons in turn decay to produce a highly boosted "Z-burst", containing on average twenty photons and two nucleons above $E_{\rm GZK}$ (see Fig. 4). The photons and nucleons from Z-bursts produced within 50 to 100 Mpc of earth can reach earth with enough energy to initiate the air-showers observed at $\sim 10^{20}$ eV.

The energy of the neutrino annihilating at the peak of the Z-pole is

$$E_\nu^R = \frac{M_Z^2}{2m_\nu} = 4\,(eV/m_\nu)\,\mathrm{ZeV}\,. \tag{15}$$

Accordingly, the boost factor is $\gamma_Z = E_\nu^R/M_Z = M_Z/2m_\nu = 4.5 \times 10^{10}(\mathrm{eV}/m_\nu)$. The resonant-energy width is narrow, reflecting the narrow width of the Z-boson: at FWHM $\Delta E_R/E_R \sim \Gamma_Z/M_Z = 3\%$. The mean energies of the ~ 2 baryons and ~ 20 photons produced in the Z decay are easily estimated. When the Z-burst energy is averaged over the mean multiplicity of 30 secondaries in Z-decay, one has

$$\langle E_p \rangle \sim \frac{E_R}{30} \sim 1.3 \left(\frac{\mathrm{eV}}{m_\nu}\right) \times 10^{20}\mathrm{eV}\,. \tag{16}$$

The photon energy is reduced by an additional factor of 2 to account for their origin in two-body π^0 decay:

$$\langle E_\gamma \rangle \sim \frac{E_R}{60} \sim 0.7 \left(\frac{\mathrm{eV}}{m_\nu}\right) \times 10^{20}\mathrm{eV}\,. \tag{17}$$

Even allowing for energy fluctuations about mean values, it is clear that in the Z-burst model the relevant neutrino mass cannot exceed ~ 1 eV. On the other hand, the neutrino mass cannot be too light or the predicted

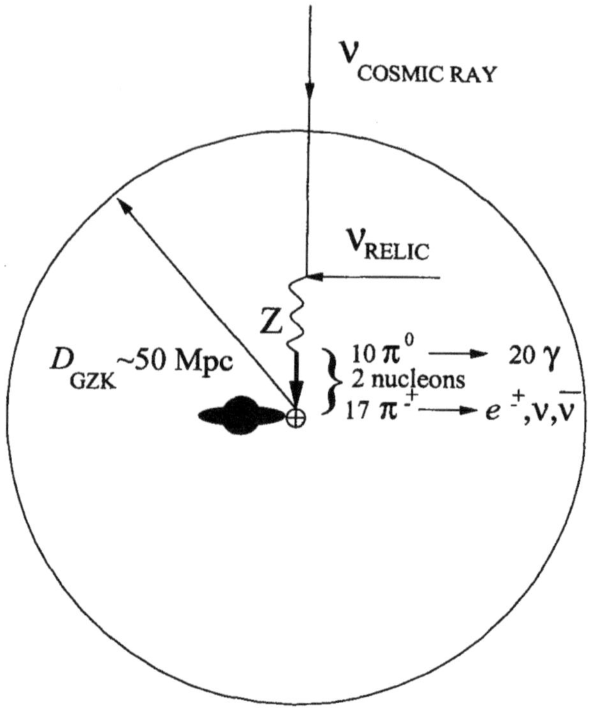

Figure 4. Schematic diagram showing the production of a Z-burst resulting from the resonant annihilation of a cosmic ray neutrino on a relic (anti)neutrino. If the Z-burst occurs within the GZK zone ($\sim$ 50 to 100 Mpc) and is directed towards the earth, then photons and nucleons with energy above the GZK cutoff may arrive at earth and initiate super-GZK air-showers

primary energies will exceed the observed event energies.[a] In this way, one obtains a rough lower limit on the neutrino mass of ~ 0.1 eV for the Z-burst model, when allowance is made for an order of magnitude energy-loss for those secondaries traversing 50 to 100 Mpc.

It is worth emphasizing that the physics of the Z-burst mechanism is entirely SM physics; there are no add-ons. There are, however, two necessary conditions[12] to be provided by nature if the mechanism is to be measurable: a sufficient flux of neutrinos at $\gtrsim 10^{21}$ eV, and a neutrino mass scale of the order $0.1 - 1$ eV. The first condition seems challenging, while the second is quite natural in view of the recent oscillation data. Labelling the mass

[a]Also, the neutrino mass cannot be too small without pushing the primary neutrino flux to unattractively higher energies.

eigenstates as $m_3 > m_2 > m_1$, one has for the mass-squared differences

$$m_3^2 = m_2^2 + \delta m_{\rm atm}^2\,, \quad \text{and} \quad m_2^2 = m_1^2 + \delta m_{\rm sun}^2\,. \tag{18}$$

(The alternative "inverted hierarchy" splitting is disfavored.) Oscillations are directly sensitive to these nonzero neutrino mass-squared differences. Fits to data yield $\delta m_{\rm sun}^2 \sim 10^{-5}$ to 10^{-4} eV^2, and $\delta m_{\rm atm}^2 \sim (1.6-5)\times 10^{-3}$ eV^2. These mass-squared differences imply lower bounds on the masses m_3 and m_2. The atmospheric bound is $m_3 \geq \sqrt{\delta m_{\rm atm}^2} \sim$ 0.05 eV, which is encouraging for mass-sensitive experiments. A very recent estimate[15] of the total neutrino mass in the Universe, based on the distribution of large-scale structures, is $\sum_j m_{\nu_j} \sim 2$ eV. It appears the neutrino mass is squeezed to lie within just the 0.1 to 1.0 eV range most beneficial to the Z-burst model! Several successful fits to the EECR data are available in Refs. 16 and 17. The most recent fit to the EECR data in the Z-burst paradigm[17] has provided a candidate neutrino mass, $0.08\,\text{eV} \leq m_\nu \leq 1.3\,\text{eV}$ at 68% CL.

Acknowledgments

I thank the CosPA faculty at National Taiwan University for a highly enjoyable meeting and a very hospitable environment graced by excellent food. Support of the U.S. Department of Energy grant no. DE-FG05-85ER40226 is gratefully acknowledged.

References

1. G. Burdman, F. Halzen, and R. Ghandi, *Phys. Lett.* **B417**, 107 (1998).
2. R. Protheroe, astro-ph/9809144 (1998).
3. R. Engel, D. Seckel, T. Stanev, *Phys. Rev.* **D64**, 093010 (2001).
4. H. Goldberg and T. J. Weiler, *Phys. Rev.* **D59**, 113005 (1999).
5. H. Goldberg, L. G. Song, and T. J. Weiler, in preparation.
6. D. Dicus, S. Kretzer, W. Repko, and C. Schmidt, *Phys. Lett.* **B514**, 103 (2001).
7. A. Kusenko and T. J. Weiler, *Phys. Rev. Lett.* **88**, 161101 (2002).
8. C. Barbot, M. Drees, F. Halzen, and D. Hooper, hep-ph/0207133 (2002).
9. M. Takeda et al. (AGASA Collaboration), Proc. ICRC 2001, Hamburg, Germany, updates Y. Uchihori et al., *Astropart.Phys.* **13**, 151 (2000).
10. P. Tinyakov and I. Tkachev, astro-ph/0204360 (2002); ibid. *JETP Lett.* **74**, 445 (2001).
11. Recent reviews of EECR data, puzzles, and models include: P. Biermann, *J. Phys.* **G23**, 1 (1997); P. Bhattacharjee and G. Sigl, *Phys. Rept.* **327**, 109 (2000) [astro-ph/9811011]; A.V. Olinto, astro-ph/0102077; X. Bertou, M. Boratov, and A. Letessier-Selvon, *Int. J. Mod. Phys.* **A15**, 2181 (2000); M. Nagano and A.A. Watson, *Rev. Mod. Phys.* **72**, 689 (2000); G. Sigl,

Science **291**, 73 (2001); T. J. Weiler, *Proc. RADHEP2000*, Nov. 16-18, 2000, UCLA, ed. P. Gorham and D. Saltzberg, [hep-ph/0103023]; F.W. Stecker, astro-ph/0207629.

12. T.J. Weiler, *Phys. Rev. Lett.* **49**, 234 (1982); ibid. *Astrophys. J.* **285**, 495 (1984); ibid. *Astropart. Phys.* **11**, 303 (1999); D. Fargion, B. Mele and A. Salis, *Astrophys. J.* **517**, 725 (1999).
13. H. Goldberg and T. J. Weiler, *Phys. Rev.* **D64**, 056008 (2001).
14. L, Anchordochi et al., *Mod. Phys. Lett.* **16**, 2033 (2001) [arXiv:astro-ph/0106501].
15. O. Elgaroy et al., 2dFGRS team, arXiv:astro-ph/0204152 and *Phys. Rev. Lett.* **89** (2002), to appear.
16. S. Yoshida, G. Sigl, and S. Lee, *Phys. Rev. Lett.* **81**, 5505 (1998); O. Kalashev, V. Kuzmin, D. Semikoz, and G. Sigl, *Phys. Rev.* **D65** (2002) 103003; G. Gelmini and G. Varieschi, hep-ph/0201273.
17. Z. Fodor, S. Katz, and A. Ringwald, *JHEP* **0206**, 046 (2002); ibid. *Phys. Rev. Lett.* **88**, 171101 (2002).

New Window for Observing Cosmic Neutrinos at $10^{15} - 10^{18}$ Electron Volts

George Wei-Shu Hou

NEW WINDOW FOR OBSERVING COSMIC NEUTRINOS AT 10^{15}–10^{18} ELECTRON VOLTS

GEORGE W.S. HOU

Department of Physics, National Taiwan University, Taipei, Taiwan 10764, R.O.C.
E-mail: wshou@phys.ntu.edu.tw

The detection of very energetic cosmic neutrinos demands "km^3" size detectors, such as IceCube, Auger, or the future EUSO space telescope. Here we explore an alternative, novel *Neutrino Telescope* (NuTel) with a mountain as target and a valley as shower volume. Converting $\nu_\tau \to \tau$ in the mountain, one picks up the direct Cherenkov pulse from τ decay shower in the valley, via a wide field, fast electronics telescope placed on a second mountain. Thus, "Seeing an AGN (or GC) from Behind a Mountain" also constitutes a $(\nu_\mu \to)\nu_\tau$ appearance experiment! The focus of our discussion is the efficiency, acceptance and rate of such a construction. We then take the Mauna Loa – Mt. Hualalai combination on Hawaii Big Island as potential site to study the sensitivity and sky coverage. The detector concept will emulate EUSO. Collaboration formation is briefly discussed.

1. Backdrop

The High Energy Physics group at NTU, NTUHEP, joined the Ministry of Education "Cosmology and Particle Astrophysics" (CosPA) 4-year "Academic Excellence" project, being in charge of the CosPA-2 subproject. We are the "PA" arm to complement the other subprojects involving radio, infrared and optical telescopes. Written in 1999, the objectives of CosPA-2 were 1) to gain strength in mainstream HEP, and 2) to venture into genuine particle astrophysics. The first objective is on track. In fact, the NTUHEP group has come of age in the last two years, gushing forth a lot of results in Belle physics analysis. However, the original plan for the second two years was to construct a small subdetector prototype for the JLC, which has to be postponed because of delays in the Linear Collider project in Japan.

The original plan for venturing into particle astrophysics, the R&D and feasibility study on direct cold dark matter (CDM) search, was conducted successfully, concluding that we should not continue. Together with the attained strength of NTUHEP, we were therefore in the blessed/cursed position of having some funding under CosPA-2, with the conviction that it was really time to enter *genuine* Particle Astrophysics.

2. Cosmic/Astro Neutrinos and Their Detection

The measured cosmic ray (CR) spectrum ranges over 30 orders of magnitude. At around 10^{16} eV, or the Knee, one has a change in power. Beyond which, there is another change in slope at $\sim 10^{18}$ eV, or the Ankle. Further beyond lies the "GZK" cutoff at $\sim 10^{19}$ eV due to UHECR absorption by CMB photons. In general, cosmic rays hitting matter would produce pions. Neutrinos are produced by $\pi^\pm$ decay, which would in general lead to 2 ν_μs and 1 ν_e. Thus, one expects "GZK neutrinos" with flux peaking around GZK cutoff. In similar vein, although the acceleration mechanism in Active Galactic Nuclei (AGN) is probably not directly responsible for CR in the Knee region, it should have no trouble generating energetic particles at such energies or higher. Given that, unlike all other particles, neutrinos can easily reach the Earth, the "AGN neutrino" flux is an important means to check its acceleration mechanism.

The closest abundant sources of neutrinos, of course, are the core of our Sun and the Earth's atmosphere when hit by CRs. Neutrino astrophysics has come of age by great advances in these two areas: the detection of deficit in solar neutrinos and upward going atmospheric neutrinos. The two leading detectors, SuperK[1] and SNO,[2] are both huge in volume and deep underground. An important discovery implied by atmospheric neutrino deficit is maximal mixing in $\nu_\mu \to \nu_\tau$ oscillations. This means the $\nu_e : \nu_\mu : \nu_\tau \cong 1 : 2 : 0$ at production could become $1 : 1 : 1$ when reaching Earth!

For VHE neutrino detection, one has even larger detector volumes, such as[3] the IceCube which is literally km^3 in size, and *1.4 km below Antarctica ice!* Because of the large volume and difficult working conditions, deployment will last until 2009. There is special interest in detecting ν_τ, where one relies on "double bang" topology, one from $\nu_\tau \to \tau$ reaction and one from τ decay. Because of boost, beyond $\sim 10^{15}$ eV, one bang occurs outside of volume ("lollipop") and detection becomes more difficult. Thus, the maximum energy is limited to 1 PeV or so.

Another type of neutrino telescope is adapted from UHECR detectors aimed at detecting extensive air showers. These, such as Auger,[4] are limited by small neutrino conversion efficiency in atmosphere. Even for horizontal air showers, the use of fluorescence technique taps only a tiny fraction of shower energy hence implying a very high energy threshold, of order 10^{18} eV or higher. In general, one aims for GZK neutrinos with these detectors.

Our survey thus shows a window of opportunity for AGN neutrino detection, i.e. 10^{15} to 10^{18} eV.

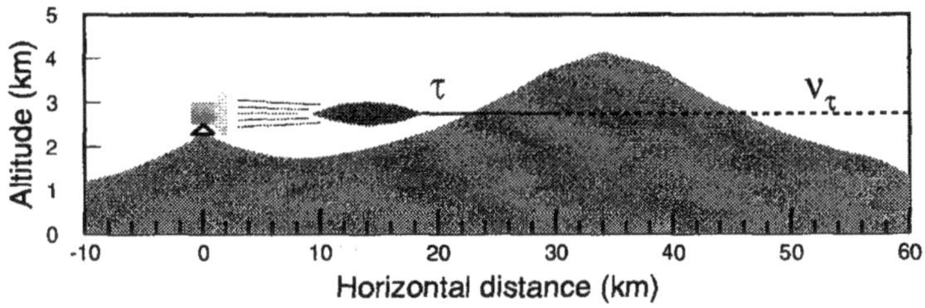

Figure 1. Principle of neutrino telescope: $\nu_\tau \to \tau$ conversion in mountain, τ exits into valley, decays and showers. A telescope on second mountain picks up Cherenkov pulse.

3. Alternative Approach: Mountain/Valley ν-Tel

The alternative technique (Fig. 1) to study HE cosmic νs is as follows: [5]

- HE ν_τ presumed, with $\nu_\tau \to \tau$ in mountain: ν_τ *appearance!*
- τ exits and decays in valley, generates shower.
  N.B. $\nu_e \to e$ — Shower in mountain.
  $\nu_\mu \to \mu$ — Pass through valley without interaction.
- Telescope (second mountain) picks up Cherenkov pulse.
  — Fast electronics, similar to γ ray Cherenkov telescopes.

We now briefly account for each step.

3.1. *Source: AGN Jets, CRs and $\nu_\mu \to \nu_\tau$ Oscillations*

AGNs such as good old M87 show clear "jet" phenomena. Though not yet understood, a typical model involves a supermassive black hole at the center with an accretion disc around it. The jet gets ejected perpendicular to the disc by some yet uncertain acceleration mechanism. The latter could be the source of CRs in the knee region. The proton content of jets would generate pions, leading to 2 ν_μs (and 1 ν_e) which, by current understanding, will evolve into 1 ν_μ and 1 ν_τ when impinging on Earth.

Thus, the "mountain watch" approach is exploring both the proton content of AGN jets as well as constituting a ν_τ appearance experiment.

3.2. *Conversion Efficiency for $\nu_\tau \to \tau$ from Mountain*

Figure 1, in simplified form, leads to an overall efficiency for ν_τ conversion, τ survival through mountain then decay in valley, and final detection,

$$\varepsilon = P_1 \left(\int_0^L P_2(x) P_3(L-x) \frac{dx}{\lambda_\nu} \right) P_d. \qquad (1)$$

In Eq. (1), P_1 is the ν_τ survival probability in atmosphere, which will be taken as 1; $P_2(x) = \exp(-x/\lambda_\nu)$ is ν_τ survival probability at distance x in rock, when $\nu_\tau \to \tau$ conversion occurs; $P_3(L-x) = \exp(-(L-x)/\lambda_\tau)$ is for the τ to survive the rest of rock; and P_d summarizes detection efficiency after τ exits the mountain.

The neutrino interaction length λ_ν is inversely proportional to the rock density $\rho \cong 2.65$ g/cm^3 and νN cross section $\sigma_{\nu N}$, i.e.

$$\lambda_\nu^{-1} = N_A\, \rho\, \sigma_{\nu N} \;\propto\; E_\nu^{0.4}, \tag{2}$$

since[6] $\sigma_{\nu N} \propto E_\nu^{0.4}$ for $E_\nu > 10^{14}$ eV. For charged current $\nu_\tau + N \to \tau + X$ reaction, the τ energy is parameterized as $E_\tau = (1-y)E_\nu$, where y is the energy fraction carried by the nucleon fragment. Taking the mean of $\langle y \rangle \sim 1/4$, we have $E_\tau \simeq 3/4\, E_\nu$. In turn, ignoring energy loss, λ_τ is the τ decay length, $\lambda_\tau = \gamma c \tau_\tau = (E_\tau/10^{15}\text{ eV}) \times 48.92$ m.

Since we have ignored the τ energy loss, simple integration gives the conversion efficiency $P_{\nu\to\tau}$ for τ exiting mountain,

$$P_{\nu\to\tau} = \int_0^L e^{-x/\lambda_\nu} e^{-(L-x)/\lambda_\tau} \frac{dx}{\lambda_\nu} = \frac{\lambda_\tau}{\lambda_\nu - \lambda_\tau}\left(e^{-L/\lambda_\nu} - e^{-L/\lambda_\tau}\right). \tag{3}$$

Knowing that $\lambda_\nu \gg L,\ \lambda_\tau$, we find $P_{\nu\to\tau} \propto E^{1.4}$ for $L \gg \lambda_\tau$ to good approximation, until $L \sim \lambda_\tau$. Differentiating with respect to L, the optimal thickness is $L_m/\lambda_\tau \simeq \log \lambda_\nu/\lambda_\tau$, i.e. the effective interaction occurs several decay length prior to exiting the mountain. $P_{\nu\to\tau}$ is plotted vs. L in Fig. 2, with optimal thickness indicated by an arrow.

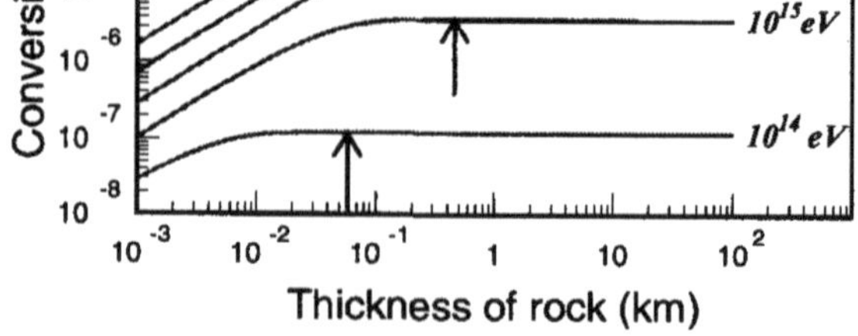

Figure 2. Efficiency vs. mountain thickness at various energies for converting $\nu_\tau \to \tau$ with τ exiting mountain. The arrows give optimal thickness without energy loss. With τ energy loss, the curves saturate earlier.

3.3. *Effect of τ Energy Loss*

The τ loses energy in the form of dE/dx as it propagates through the mountain. At the simplest level, this is approximated by a "β" term, i.e. $E(x) = E_0 e^{-\beta x}$. This leads to a modulation of τ survival probability by

$$P(x) = e^{-\frac{1}{\kappa\beta}\left(\frac{1}{E(x)} - \frac{1}{E_0}\right)}. \tag{4}$$

This is not all bad. Despite the degraded energy resolution, and the reduced τ range, it turns out that this increases the acceptance for higher energy τs. The conversion efficiency saturates earlier with consideration of energy loss, which is shown in Fig. 2 for 10^{17} and 10^{18} eV.

3.4. *Detection Probability*

The τ exits the mountain and decays. Only fraction $R \sim 0.83$ of τ decays generate showers ($\tau \to \mu\nu\nu$ will not). Besides the τ decay length, the critical parameter is the depth of shower maximum. The equivalent distance is defined as $X_{\max}$. To simplify, we require that shower maximum is reached before the detector. Denoting the distance between mountain and detector as D, the detection probability is

$$P_d \simeq R\left(1 - e^{-(D-X_{\max})/\lambda_\tau(E')}\right) P_{\text{abs}}, \tag{5}$$

where one is modulated by atmospheric absorption factor

$$P_{\text{abs}} = e^{-(D-X_{\max})/d_s}, \tag{6}$$

which is due to Mie (aerosol/cloud) and Rayleigh (atmospheric molecular) scattering. We take the scale distance d_s to be 20 km.

4. Acceptance and Event Rate

The detection probability defined so far is detector independent, and gauges only the amount of light that can reach the detector. The actual acceptance would depend on detector design, where we assume device based on photomultiplier tubes (PMT), with an individual PMT reading a particular "pixel". The event rate is

$$R(E) = a(E) \times \varepsilon(E) \times \Phi(E), \tag{7}$$

where $a(E)$ is the acceptance = area × solid angle [$\text{cm}^2\,\text{sr}^1$], $\varepsilon(E)$ is the neutrino conversion efficiency discussed above, and $\Phi(E)$ is the unknown cosmic neutrino flux [$\text{cm}^{-2}\,\text{s}^{-1}\,\text{sr}^{-1}$].

The effective solid angle is the Cherenkov light cone. The lateral development of the air shower itself leads to a light cone that extends to $\theta_c \sim 5°$.

This results in a solid angle $\Omega = 2\pi(1-\cos\theta_c) \cong 0.0239$ sr. The differential effective area is the area where τ decays and initiates shower as seen by a particular pixel. The τ decays at a couple λ_τs from mountain, and recalling D as the distance between mountain surface and detector, we have

$$da(E) = (D(\omega) - \lambda_\tau(E))^2 \, d\omega, \tag{8}$$

where $d\omega$ is the solid angle of each pixel. Integrating over the field of view (FOV), one can get the effective area $a(E)$.

5. Site Selection: Possible Sensitivity for Hawaii Big Island

The field of view, differential acceptance, and the final sensitivity would be site specific, so we now choose a specific site as illustration to make further studies.

From our discussions so far, we can already see that a good site for this "mountain watch" approach requires:

- Cross-section of target mountain as large as possible.
- Valley as wide as few tens km:
  - Shower max. $\sim$ 500–700 g/cm^2 $\Longrightarrow$ $X_{\max} \sim$ 4.5–7.8 km for atmosphere at 1–3 km.
  - τ decay distance.
- Optical (UV) detection, so atmosphere dry and not cloudy.
- Night sky dark and free from artificial light.
- Preferred if Galactic Center, or GC visible, since Sag A* may be lair of our "local" massive black hole.
- Usual logistics considerations.

Clearly, a large fraction of the criteria above are the same as usual astronomical telescopes. With the existence of two large volcanic mountains rising from the ocean floor, we tentatively fix our mind on the Hawaii Big Island, which, after all, is the astronomer's dream site.

At first sight, the peak-to-peak separation of $\sim$ 40 km between Mauna Loa and Mauna Kea seems tailor made for the purpose. On further thought, we find that it may be better to sit on top of the smaller Mt. Hualalai, at 2500 m, situated on the dryer west side, which has a good view of the *broad* Mauna Loa. Indeed, Mauna Loa provides long baseline at $\sim$ 90 km wide and 4 km high. Figure 1 in fact is a cross sectional view with Mt. Hualalai to the left and Mauna Loa to the right!

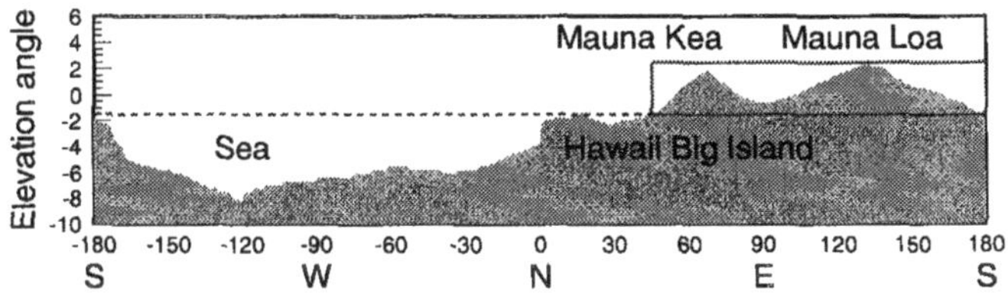

Figure 3. Panoramic view from Mt. Hualalai, with dashed line as the horizon and the terrain of Hawaii Big Island shaded, where the sea is below the horizon.

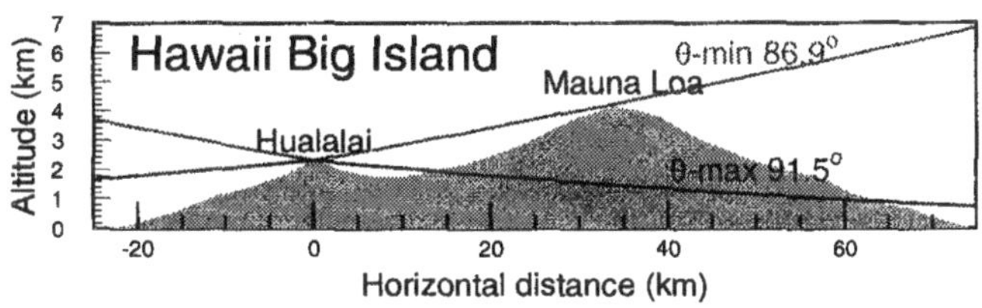

Figure 4. Cross-sectional view of Big Island along the line from Mt. Hualalai to Mauna Loa, where effect of Earth's curvature can be seen.

5.1. *Field of View and Differential Rate*

Fig. 3 shows a schematic panorama from the top of Mt. Hualalai. The field of view of Mauna Loa and Mauna Kea of the detector is the shaded mountain region inside the box. The azimuth angle extends from south to east. The minimum zenith angle of 86.9° is set by the line from the summit of Mt. Hualalai to that of Mauna Loa. The maximum zenith angle of 91.5° is set by the line from the summit of Mt. Hualalai to the horizon at the base of Mauna Loa. A cross-section of the Big Island along the line from Mt. Hualalai to Mauna Loa is shown in Fig. 4.

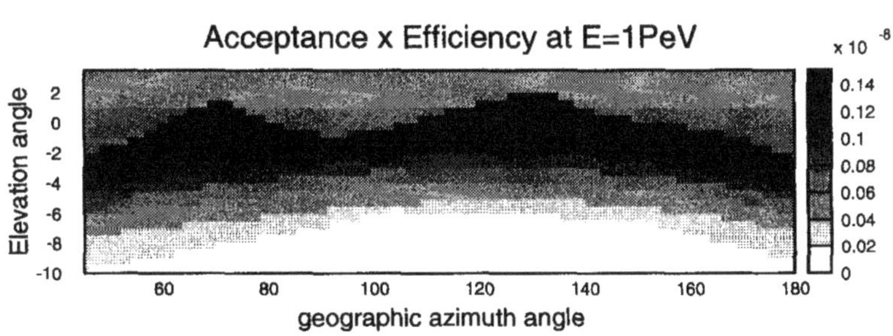

Figure 5. Differential acceptance × efficiency for Mauna Loa and Kea at $E = 1$ PeV.

Assuming each pixel covers 0.5° × 0.5°, the differential acceptance $da(\theta, \phi, E)$ is calculated using Eq. (8). Together with the conversion efficiency $\varepsilon(\theta, \phi, E)$ of Eq. (1), we show the differential acceptance × efficiency in Fig. 5. The differential rate is obtained by $dR(E) = da(E) \times \varepsilon(E) \times \Phi(E)$ if the flux $\Phi(E)$ is known.

5.2. *Mean Acceptance and Sensitivity*

The cosmic neutrino flux $\Phi(E)$ is actually not known, and is in fact the target for measurement. It is useful then to get the mean, or effective, acceptance, by averaging the differential acceptance × efficiency over the full target solid angles.

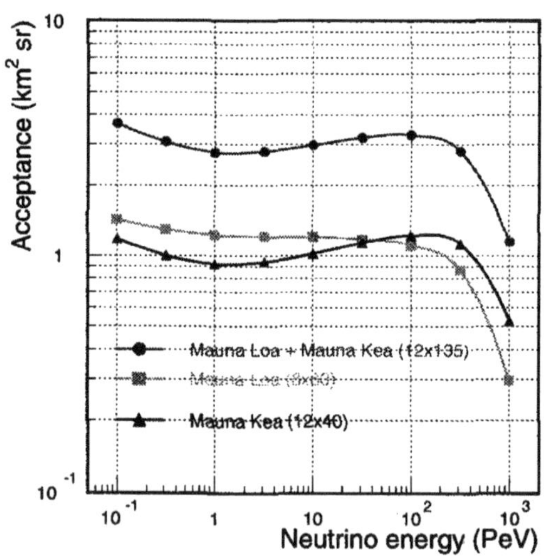

Figure 6. Effective acceptance of Mauna Loa and Kea vs. energy.

Taking Mauna Loa as extending over $\theta = 88°$–$96°$, $\phi = 110°$–$170°$, and Mauna Kea as extending over $\theta = 88°$–$100°$, $\phi = 51°$–$91°$, and the combined coverage of $\theta = 88°$–$100°$, $\phi = 45°$–$180°$, the effective acceptance is plotted in Fig. 6. It is interesting that the sensitivity of Mauna Kea at higher energy fares better than Mauna Loa. This is because Mauna Kea is a steeper volcano, and good conditions extend to larger zenith angles, while for Mauna Loa, at large zenith angles, there maybe insufficient distance for shower to develop.

We define the sensitivity as the flux that produces 0.3 events per year

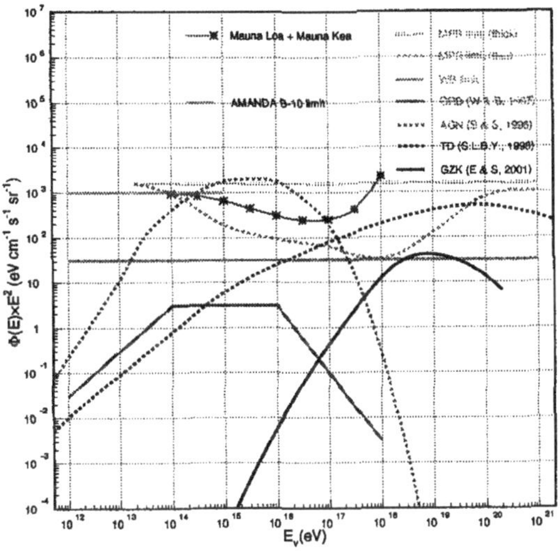

Figure 7. Sensitivity of Mauna Loa and Kea combined vs. energy, together with various sources and bounds.

per 1/2 decade of energy. From the mean acceptance, the (flux) sensitivity defined this way is plotted in Fig. 7. We see that one might explore MPR limits. Also, although the limit of sensitivity is not much better than AMANDA-B10,[7] the mountain watch approach is rather complementary in that it covers higher energy range. In fact, it seems to fill nicely the niche of 10^{15}–10^{18} eV. Conventional km^3 neutrino detectors are limited by target volume, while UHECR detectors are limited by useful energy fraction, and in part because they aim at GZK neutrinos at order 10^{19} eV.

We also note that we have used a diffuse source approach here. The sensitivity to a few nearby point sources remains to be explored.

5.3. *Run Time and Sky Coverage*

To get some sense of actual operation, we consider the actual run time for period of 12/2003 to 12/2006. Optical detectors operate in moonless/cloudless nights. There are $\sim$ 5000 hours of moonless nights in the period. Deducting fraction of cloudy or foggy nights, one probably has 10%-15% actual duty time, or of order 3000 hours. But because of nano second UV pulse, and the fact that one is not observing the sky itself, the operating conditions of "cloudlessness" should be different from usual optical telescopes.

It is very important to see how much of the sky is covered and for how

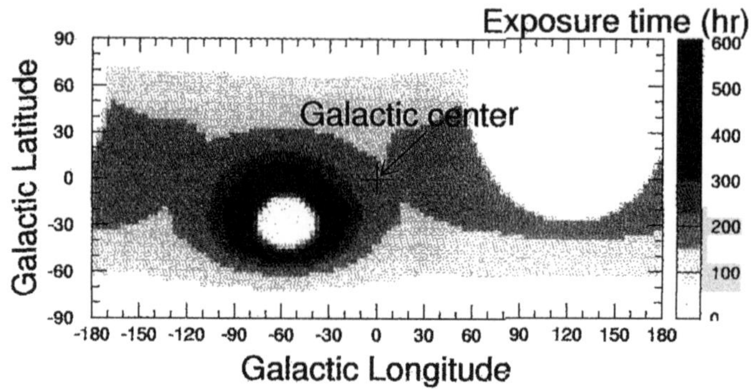

Figure 8. Exposure time for viewing Mauna Loa and Mauna Kea from Mt. Hualalai.

long. Considering the field of view of Mt. Hualalai site, and viewing both Mauna Loa and Mauna Kea, for the period of 12/2003 – 12/2006 with *20%* duty time, the sky coverage is plotted in Fig. 8. It is important to note that the GC is visible, with an observing time of order 170 hours. Although the Hawaii site may not be ideal for observing GC, one is *able* to see it.

6. Detector Concept and Collaboration Formation

From our discussions, it is clear that we need very wide field of view, PMT level fast electronics, and UV sensitivity.

6.1. *EUSO-type Detector Concept*

The EUSO (Extreme Universe Space Observatory) space telescope,[8] to be deployed on the International Space Station, is a mission supported by the European Space Agency for the study of extreme energy cosmic rays (EECR), $E > 10^{20}$ eV. Perched in orbit at 400 km, it has a 60° FOV and watches an area of 150000 km^2 (about 4× size of Taiwan), much larger than the area covered by Auger. It picks out extensive air showers (EAS) either by fluorescence, or by reflected Cherenkov light. Thus, the detector has wide field optics and MAPMT (multi-anode PMT) electronics, and seems ideal as starting place for us to adapt to our "mountain watch" needs.

To be a little more specific, the EUSO telescope is a compact, monocular instrument, with custom designed double Fresnel lens to achieve very wide field of view, and has already settled on MAPMTs with 8×8 pixels, fitted with optical adaptors. The MAPMT modules sit on the focal surface, achieving very fine spot size of 2mm × 2mm. Our adaptation probably does not need such fine resolution, and will likely involve a much smaller telescope, since we detect direct Cherenkov pulse, rather than fluorescent or reflected Cherenkov light. Instead of looking downwards, we shall turn sideways to watch the mountain.

6.2. *Early Stage Collaboration Formation*

Born out of above considerations, a "Very High Energy Neutrino Telescope Workshop" was held on NTU campus during March 21–23, 2002, to discuss the feasibility of NuTel, as well as explore the possible formation of a collaboration. The conclusion was positive, and an *Executive Summary* was written in the names of

George W.S. Hou, Taiwan	(Belle and CMS)
Francois Vannucci, Paris	(NOMAD and EUSO)
Osvaldo Catalano, Palermo	(EUSO and GAW)
John G. Learned, Hawaii	(SuperK and other neutrino expts.)

which is posted on the web page http://hep1.phys.ntu.edu.tw/VHENTW/. Useful advice was also given by many theorists and Auger and HiRes[9] groups. There is also strong domestic theory support from NCTU/NCTS (Guey-Lin Lin *et al.*) in Taiwan. Participation from RIKEN (Hiro Shimizu), responsible for Fresnel lens for EUSO, is also sought, but at present funding needs to be resolved.

It is clear that EUSO members constitute a sizable component of the formative collaboration. Indeed, Palermo is one of the leading institutions in EUSO. They are also developing the GAW (Gamma Air Watch) project,[10] again EUSO-like, that aims at studying gamma ray generated air showers. It is not yet conclusive[11] whether a small size telescope with coarse-grained resolution, or a GAW like finer-grained telescope, or a combination would be needed for the "mountain watch". What seems clear is that, although one has the advantage of a short signal pulse at ns level, because of the extremely low expected signal event rate of ~ 1/year, there are potentially many backgrounds. One therefore needs multiple coincidence trigger, hence at least two telescopes.

It should be clear that the acceptance is quite site-limited. Thus, a serious site search should be conducted, but Hawaii Big Island is a good start. Whatever the site, for sake of mobility, one needs the design of a compact detector with low noise and high gain.

The participation of RIKEN is still being pursued. It was decided at the NTU workshop that a second workshop would be held on Hawaii Big Island, at or around the time of the big SPIE astronomical instrument meeting. The date has now been fixed to August 24–25. The goal of the second workshop is to discuss various preliminary studies towards a prototype, including simulations and methods, and make a site visit up Mt. Hualalai. We also intend to face more seriously the collaboration formation issues of responsibility and resource sharing.

7. Conclusion and Outlook

The optimal energy range for detecting ν_τ by conversion in mountain or Earth appears to be 10^{15}–10^{18} eV. The conversion efficiency is high, and the energy resolution is reasonable. The energy range fits in the niche between conventional ν detectors such as IceCube, and UHECR ν detectors such as Auger (and EUSO). Just by seeing a physical event that is correlated in direction with known cosmic or astronomical sources, one can pin down things like the AGN jet mechanism, while constituting a $\nu_\mu \to \nu_\tau$ appearance experiment. As such, this is a great combination of particle and astrophysics, and a good chance to initiate a first experiment. The CosPA-2 subproject has been approved by the MOE, mid-course, for pursuing such a direction for the next two years, which would likely continue afterwards.

The tentative site of Mt. Hualalai on Hawaii Big Island, viewing the twin peaks of Mauna Loa and Kea, is a good one. It has good weather, large acceptance of ~ 1 km^2 sr, and has similar sensitivity as AMANDA-B10. Since it has potential sensitivity at higher energy, it may even complement the formidable IceCube experiment at the South Pole. It is important that the Galactic Center is visible. One should in any case seek out alternative sites, but one major consideration may be logistics.

So far our considerations have been grossly simplistic. On one hand we have ignored issues like detector efficiency. On the other hand, there are also potential means for increasing acceptance (probably at extra cost!). One can add *earth skimming* below the horizon ($\theta > 91.5°$), even including ocean-skimming events (question of reflection from waves?). One could also consider adding fluorescent mode of operation.

Acknowledgments

I am indebted to Alfred M.H. Huang for much of the results presented in this report.

References

1. See http://www-sk.icrr.u-tokyo.ac.jp/doc/sk/.
2. See http://www.sno.phy.queensu.ca/.
3. See http://icecube.wisc.edu/.
4. See http://auger.cnrs.fr/pages.html.
5. George W.S. Hou and Alfred M. Huang, astro-ph/0204145.
6. R. Gandhi *et al.*, *Phys. Rev.* **D58**, 093009 (1999).
7. G.C. Hill, AMANDA Collaboration, Proceedings of the XXXVIth Recontres de Moriond, *Electroweak Interactions and Unified Theories*, March 2001, astro-ph/0106064.
8. See http://www.ifcai.pa.cnr.it/ EUSO/.
9. See http://hires.physics.utah.edu/.
10. See http://www.ifcai.pa.cnr.it/Ifcai/gaw.html.
11. See http://hep1.phys.ntu.edu.tw/VHENTW/

Comparison of High-Energy Galactic and Atmospheric Tau Neutrino Flux

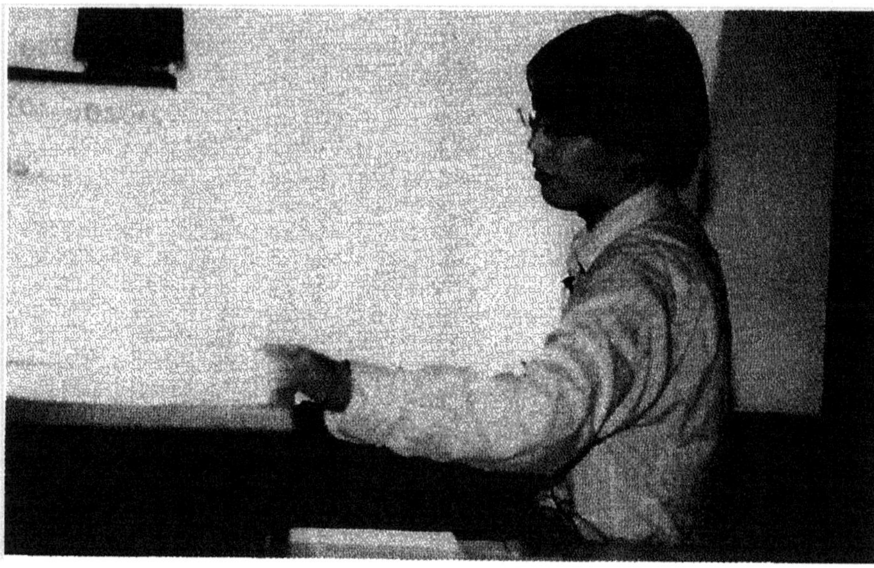

Jie-Jun Tseng

COMPARISON OF HIGH-ENERGY GALACTIC AND ATMOSPHERIC TAU NEUTRINO FLUX

H. ATHAR

Institute of Physics, National Chiao Tung University, Hsinchu 300, Taiwan
E-mail: athar@phys.cts.nthu.edu.tw

KINGMAN CHEUNG

Physics Division, National Center for Theoretical Sciences, Hsinchu 300, Taiwan
E-mail: cheung@phys.cts.nthu.edu.tw

GUEY-LIN LIN

Institute of Physics, National Chiao Tung University, Hsinchu 300, Taiwan
E-mail: glin@cc.nctu.edu.tw

JIE-JUN TSENG

Institute of Physics, National Chiao Tung University, Hsinchu 300, Taiwan
E-mail: geny.py86g@nctu.edu.tw

We compare the tau neutrino flux arising from the galaxy and the earth atmosphere for $10^3 \leq E/\text{GeV} \leq 10^{11}$. The intrinsic and oscillated tau neutrino fluxes from both sources are calculated. The intrinsic galactic ν_τ flux ($E \geq 10^3$ GeV) is calculated by considering the interactions of high-energy cosmic-rays with the matter present in our galaxy, whereas the oscillated galactic ν_τ flux is coming from the oscillation of the galactic ν_μ flux. For the intrinsic atmospheric ν_τ flux, we extend the validity of a previous calculation from $E \leq 10^6$ GeV up to $E \leq 10^{11}$ GeV. The oscillated atmospheric ν_τ flux is, on the other hand, rather suppressed. We find that, for $10^3 \leq E/\text{GeV} \leq 5 \cdot 10^7$, the oscillated ν_τ flux along the galactic plane dominates over the maximal intrinsic atmospheric ν_τ flux, i.e., the flux along the horizontal direction.

1. Introduction

Searching for high-energy tau neutrinos ($E \geq 10^3$ GeV) will yield quite useful information about the highest energy phenomenon occurring in the universe.[1] The same search may also provide evidence for physics beyond the standard model.[2]

The high-energy tau neutrinos can be produced in pp interactions taking place in cosmos. There can be several astrophysical sites where the pp interactions may occur. Examples of these include the relatively nearby and better known astrophysical sites such as our galaxy and the earth atmosphere, where the basic pp interactions occur as pA interactions. The pp interactions in these sites form a rather certain background to the extragalactic high-energy tau neutrino searches. It is possible that such interactions are the only sources of high-energy tau neutrinos should the search for high-energy tau neutrinos originating from several proposed distant sites such as AGNs, GRBs, as well as groups and clusters of galaxies, turns out to be negative.

Both the galactic and atmospheric tau neutrinos can be categorized into intrinsic and oscillated ones. Here, intrinsic ν_τ flux refers to the ν_τ produced directly by an interaction while oscillated ν_τ refers to the ν_τ resulted from the $\nu_\mu \to \nu_\tau$ oscillation. In this work, we calculate the intrinsic ν_τ flux from our galaxy by using the perturbative and nonperturbative QCD approaches to model the pp interactions, and taking into account all major tau neutrino production channels up to $E \leq 10^{11}$ GeV. To calculate the oscillated galactic and atmospheric ν_τ fluxes, we apply the two-flavor neutrino oscillation analysis.[3] The intrinsic atmospheric tau neutrino flux has been calculated for $E \leq 10^6$ GeV.[5] In this work, we extend the calculation up to $E \leq 10^{11}$ GeV. Such an extension requires the input of cosmic-ray flux spectrum for an energy range beyond that considered in Ref. 5. Furthermore, for a greater neutrino energy, the solutions of cascade equations relevant to the neutrino production behave differently. For the oscillated ν_τ flux, it is interesting to note that the oscillation length for $\nu_\mu \to \nu_\tau$ for the energy range $10^3 \leq E/\text{GeV} \leq 10^{11}$ is much greater than the thickness of the earth atmosphere. Hence the oscillated atmospheric ν_τ flux in this case is highly suppressed.

The organization of the paper is as follows. In Section II, we discuss the calculation of intrinsic high-energy tau neutrino flux from our galaxy. In Section III, we present our result on the intrinsic atmospheric ν_τ flux and compare it with the galactic one. In Section IV, we discuss the effects of neutrino flavor mixing. The total galactic ν_τ flux (the sum of intrinsic and oscillated fluxes) is compared to its atmospheric counterpart, and the dominant energy range for the former flux is identified. Finally, we summarize in Section V.

2. The intrinsic galactic tau neutrino flux

2.1. *The tau-neutrino flux formula and the galaxy model*

We use the following formula for computing the ν_τ flux:

$$\frac{\mathrm{d}N_{\nu_\tau}}{\mathrm{d}E} = Rn_p \int_E^\infty \mathrm{d}E_p\, \phi_p(E_p)\, \frac{\mathrm{d}\sigma_{pp\to\nu_\tau+Y}}{\mathrm{d}E}\,. \tag{1}$$

In the above equation, E is the tau neutrino energy and the cosmic-ray flux spectrum, $\phi_p(E_p)$, is given by [7]

$$\phi_p(E_p) = \begin{cases} 1.7\,(E_p/\mathrm{GeV})^{-2.7} & \text{for } E_p < E_0, \\ 174\,(E_p/\mathrm{GeV})^{-3} & \text{for } E_p \geq E_0, \end{cases} \tag{2}$$

where $E_0 = 5 \cdot 10^6$ GeV and $\phi_p(E_p)$ is in units of $\mathrm{cm}^{-2}\ \mathrm{s}^{-1}\ \mathrm{sr}^{-1}\ \mathrm{GeV}^{-1}$. The $\lambda_{pp}(E_p) = (\sigma_{pp}^{\mathrm{incl}} n_p)^{-1}$ is the pp interaction length and R is a representative distance in the galaxy along the galactic plane. The target particles are taken to be protons with a constant number density of 1 cm^{-3} and R is taken to be $\sim$ 10 kpc. Furthermore, we calculate the intrinsic tau neutrino flux along the galactic plane only to obtain the maximal expected tau neutrino flux. In this work, we include all major production channels of tau neutrinos, namely, via the D_s meson, b-hadron, $t\bar{t}$, W^*, and Z^*.

2.2. *Tau neutrino production*

2.2.1. *Via D_s mesons*

The lightest meson that can decay into a τ-ν_τ pair is the D_s meson. Therefore, we first look at the production of D_s mesons. Here, we employ two approaches to calculate the production of D_s mesons: (i) the perturbative QCD (PQCD) and (ii) the quark-gluon string model (QGSM).

In the PQCD approach, we use the leading-order result for $pp \to c\bar{c}$, while the QGSM approach is nonperturbative and is based on the string fragmentation.[8] The production cross section of the D_s meson is given by the sum of n-pomeron terms. A comparison of these two approaches for D_s meson is shown in Fig. 1. The dashed line in the figure is the spectrum of the injected proton flux given by Eq. (2). The ν_τ spectra calculated by these two approaches agree well with each other for $E \leq 10^6$ GeV. Beyond this energy, the QGSM approach gives a relatively harder spectrum. Nevertheless, in the region where the two approaches differ, the tau neutrino flux is already small. The current highest energy collider experiment for D_s meson production is at the FERMILAB TEVATRON with $\sqrt{s} = 1.8 \cdot 10^3$ GeV, which corresponds to an $E_p \sim 1.7 \cdot 10^6$ GeV, as $s \sim 2m_pE_p$ in our setting. Note that up to this $\sqrt{s}$, the agreement between

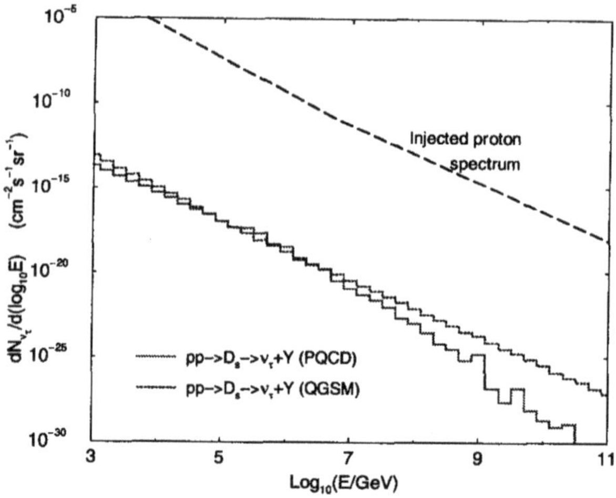

Figure 1. A comparison between the PQCD and QGSM approach to the energy spectrum of the intrinsic galactic ν_τ flux coming from the D_s meson. The thick dashed curve is the injected proton flux spectrum given by Eq. (2).

the two approaches is quite good, according to Fig. 1. We have used a factorization scale $Q^2 = \hat{s}/4$ and the one-loop running strong coupling constant α_s with the value $\alpha_s(Q^2 = M_Z^2) = 0.118$, and the $m_c = 1.35$ GeV.

2.2.2. *Via $b\bar{b}$, $t\bar{t}$, W^* and Z^**

The production of $b\bar{b}$ and $t\bar{t}$ in pp interactions can be calculated quite reliably by the PQCD approach, similar to the calculation of $c\bar{c}$. The relevant matrix elements, including the ones for W^* and Z^*, are listed in Appendix A. The results are shown in Fig. 2. A few observations can be drawn from the figure.[9] (i) The production via D_s mesons dominates for $E \leq 10^9$ GeV, followed by b-hadrons, W^*, Z^*, and $t\bar{t}$ respectively. (ii) For $E \geq 10^9$ GeV all these production channels become *comparable*. (iii) The intrinsic tau neutrino flux is about $10-12$ orders of magnitude smaller than the injected proton flux.

3. The intrinsic atmospheric tau neutrino flux

The earth atmosphere is an interesting extra-terrestrial site where the basic pp interaction occurs in the form of a pA collision with A the nuclei present in the earth atmosphere. Incidently, it is the only known nearby extra-terrestrial site from where the intrinsic neutrinos are observed as a result of high-energy cosmic-ray interactions.

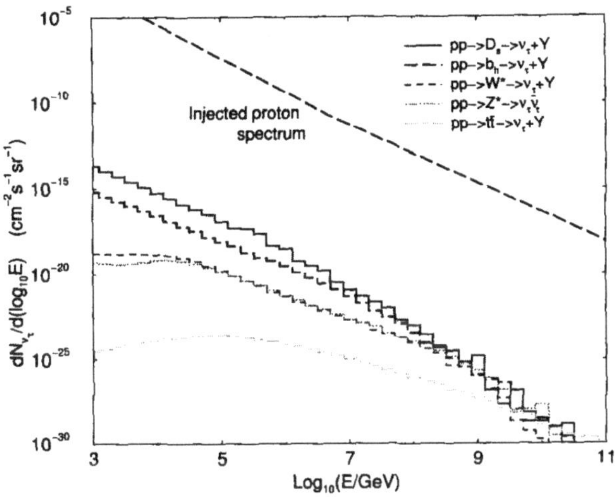

Figure 2. Intrinsic galactic tau neutrino flux calculated via various intermediate states and channels: D_s, b-hadron, W^*, Z^*, and $t\bar{t}$. The injected proton flux spectrum is also shown.

We have calculated the downward and horizontal intrinsic atmospheric ν_τ flux for the energy range $10^3 \leq E/\text{GeV} \leq 10^{11}$. We used the nonperturbative QCD approach mentioned in the last section to model the production of D_s mesons in pA interactions. Since the tau neutrino flux is determined by the flux of D_s meson, we briefly discuss the cascade equation for the D_s flux. In general, we have [10]

$$\frac{\mathrm{d}\phi_{D_s}}{\mathrm{d}X} = -\frac{\phi_{D_s}}{\lambda_{D_s}} - \frac{\phi_{D_s}}{\rho d_{D_s}} + Z_{D_s D_s}\frac{\phi_{D_s}}{\lambda_{D_s}} + Z_{pD_s}\frac{\phi_p}{\lambda_p}, \tag{3}$$

where X is the slant depth, i.e., the amount of atmosphere (in g/cm^2) traversed by the D_s meson ($X = 0$ at the top of the atmosphere), $\phi_p(E, X)$ and $\phi_{D_s}(E, X)$ are the fluxes of protons and D_s mesons respectively. The λ_{D_s} and d_{D_s} are the interaction thickness (in g/cm^2) and the decay length of the D_s meson respectively. Finally, the Z-moments Z_{pD_s} and $Z_{D_s D_s}$ describe the effectiveness of generating D_s meson from the higher-energy protons and D_s mesons respectively.

We note that Eq. (3) should be solved together with the cascade equation governing the propagation of high-energy cosmic-ray protons. In fact, the proton flux equation can be easily solved such that

$$\phi_p(E, X) \approx \exp(-\frac{X}{\Lambda_p})\, \phi_p(E, 0), \tag{4}$$

where $\Lambda_p \equiv \lambda_p/(1 - Z_{pp})$ is the proton attenuation length with λ_p the proton interaction thickness (in g/cm^2) and the Z_{pp} given by

$$Z_{pp}(E) = \int_E^\infty \mathrm{d}E' \frac{\phi_p(E',0)}{\phi_p(E,0)} \frac{\lambda_p(E)}{\lambda_p(E')} \frac{\mathrm{d}n_{pA\to p+Y}(E,E')}{\mathrm{d}E}. \tag{5}$$

An analytic solution of Eq. (3) can be obtained for either the low or the high energy limit. Such limits are characterized by whether the D_s decays before it interacts with the medium or vice versa. The critical energy separating the two limits is approximately $\sim 8.5 \cdot 10^7$ GeV. In the low energy limit, we disregard the first and third terms in the R.H.S. of Eq. (3). On the other hand, one can drop the second term in the high energy limit. We first obtain two ν_τ fluxes, valid for low and high energy limits respectively, in terms of Z-moments and then interpolate the two fluxes.

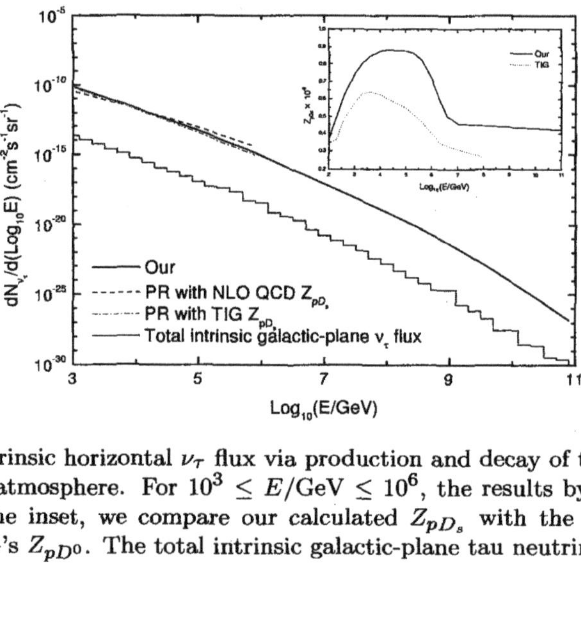

Figure 3. Intrinsic horizontal ν_T flux via production and decay of the D_s meson in the earth atmosphere. For $10^3 \leq E/\text{GeV} \leq 10^6$, the results by PR are also shown. In the inset, we compare our calculated Z_{pD_s} with the one given by rescaling TIG's Z_{pD^0}. The total intrinsic galactic-plane tau neutrino flux is also shown.

In Fig. 3, we show our result for the intrinsic atmospheric ν_τ flux along the horizontal direction. For comparison, the results by Pasquali and Reno (PR),[5] valid for $10^3 \leq E/\text{GeV} \leq 10^6$, are also shown. The ν_τ flux along all the other direction is small. We remark that the major uncertainty for determining the above ν_τ flux is the Z-moment Z_{pD_s}. In Ref. 5, the authors calculate Z_{pD_s} using two different approaches, which then give rise to different results for the ν_τ flux. The first approach is based upon next-to-leading order (NLO) perturbative QCD[11], while the second approach rescales the Z_{pD^0} given by the PYTHIA[12] calculation of Thunman, Ingelman, and Gondolo (TIG)[13]. In the inset of Fig. 3 , we also show our calculated Z_{pD_s} in comparison with the one given by rescaling TGI's result for Z_{pD^0}. We do not show the Z_{pD_s} obtained by NLO perturbative QCD since it is not explicitly given in Ref. 5.

4. Effects of oscillations

In the context of two neutrino flavors, ν_μ and ν_τ, the total ν_τ flux, $\mathrm{d}N_{\nu_\tau}^{\mathrm{tot}}/\mathrm{d}(\log_{10} E)$, is given by [14]

$$\mathrm{d}N_{\nu_\tau}^{\mathrm{tot}}/\mathrm{d}(\log_{10} E) = P \cdot \mathrm{d}N_{\nu_\mu}/\mathrm{d}(\log_{10} E) + (1-P) \cdot \mathrm{d}N_{\nu_\tau}/\mathrm{d}(\log_{10} E). \quad (6)$$

Here $P \equiv P(\nu_\mu \to \nu_\tau) = \sin^2 2\theta \cdot \sin^2(l/l_{\mathrm{osc}})$. The neutrino flavor oscillation length for $\nu_\mu \to \nu_\tau$ is $l_{\mathrm{osc}} \sim (E/\delta m^2)$. For $10^3 \leq E/\mathrm{GeV} \leq 10^{11}$ and with $\delta m^2 \sim 10^{-3}\,\mathrm{eV}^2$, we obtain $10^{-8} \leq l_{\mathrm{osc}}/\mathrm{pc} \leq 1$. We assume maximal flavor mixing between ν_μ and ν_τ.

For intrinsic neutrinos produced along the galactic plane, we take $\mathrm{d}N_{\nu_\mu}/\mathrm{d}(\log_{10} E)$ given by Ingelman and Thunman in Ref. 4 by extrapolating it up to $E \leq 10^{11}$ GeV, whereas for $\mathrm{d}N_{\nu_\tau}/\mathrm{d}(\log_{10} E)$, we use our results obtained in Section II. For galactic-plane neutrinos, we note that $l_{\mathrm{osc}} \ll l$, where $l \sim 5$ kpc is the typical average distance the intrinsic high-energy muon neutrinos traverse after being produced in our galaxy. Eq. (6) then implies that, on the average, half of the muon neutrino flux will be oscillated into tau neutrino flux, reducing its intrinsic level to one half. For downward going neutrinos produced in the earth atmosphere, we take $l \simeq 20$ km as an example. We use the (prompt) $\mathrm{d}N_{\nu_\mu}/\mathrm{d}(\log_{10} E)$ given in Ref. 15, whereas for $\mathrm{d}N_{\nu_\tau}/\mathrm{d}(\log_{10} E)$, we use our results obtained in Section III. For horizontal and upward going atmospheric neutrinos, $l \sim 10^3 - 10^4$ km. Here, $l_{\mathrm{osc}} \gg l$, and so the intrinsic atmospheric tau neutrino flux dominates over the oscillated one for $E \geq 10^3$ GeV, essentially irrespective of the incident direction. We present these results in Fig. 4, along with the GZK oscillated tau neutrino flux briefly mentioned in Section I. For the GZK neutrinos, $l \geq$ Mpc and $\mathrm{d}N_{\nu_\mu}/\mathrm{d}(\log_{10} E)$ is taken from Ref. 6. From the figure, we note that the galactic-plane oscillated ν_τ flux *dominates* over the intrinsic atmospheric ν_τ flux for $E \leq 5 \cdot 10^7$ GeV, whereas the GZK oscillated tau neutrino flux dominates for $E \geq 5 \cdot 10^7$ GeV.

5. Discussion and Conclusions

We have calculated the ν_τ flux due to *pp* interactions in our galaxy. This flux consists of intrinsic tau neutrino flux and that arising from the oscillations of muon neutrinos. We note that the latter flux is dominant over the former

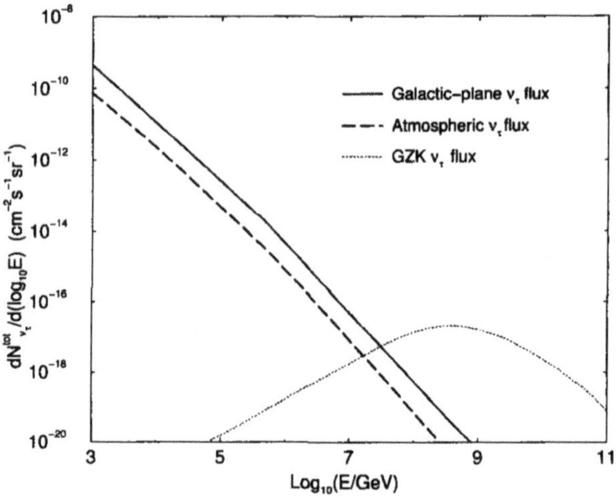

Figure 4. Galactic-plane, horizontal atmospheric and GZK tau neutrino fluxes under the assumption of neutrino flavor oscillations.

by four to five orders of magnitude for the considered neutrino energy range. From Fig. 4, one can see that the main background for the search of high-energy extra-galactic tau neutrino is due to the muon neutrinos produced in the galactic-plane, which then oscillate into tau neutrinos. Therefore it is clear that searching for extra-galactic tau neutrinos orthogonal to the galactic-plane is more prospective.

In the calculation of galactic tau neutrino flux, we have used a simplified model of matter distribution along our galactic plane to obtain the maximal intrinsic tau neutrino flux. We have explicitly calculated the contribution of heavier states such as $b\bar{b}$, $t\bar{t}$ as well as W^* and Z^* in addition to the more conventional D_s channel to the intrinsic tau neutrino flux. For D_s channel, we have used both perturbative and nonperturbative QCD approaches. We have also extended a previous calculation of intrinsic atmospheric ν_τ flux from $E \leq 10^6$ GeV up to $E \leq 10^{11}$ GeV. Here, we used the nonperturbative QCD approach to calculate the production of D_s mesons in pA interactions. In comparison with the intrinsic galactic-plane ν_τ flux, it is large. However, since the distance between the detector and the neutrino source in the galactic plane is sufficiently large, the neutrino flavor oscillations of non-tau neutrinos into tau neutrinos makes the eventual tau neutrino flux along the galactic plane greater than the atmospheric tau neutrino flux for $10^3 \leq E/\text{GeV} \leq 5 \cdot 10^7$.

In summary, we have completed the compilation of all definite sources of tau neutrino flux, i.e., those from our galaxy and from the earth atmosphere. Such a compilation is needed before one conducts the search for tau neutrinos from extra-galactic sources.

Acknowledgments

H.A. and K.C. are supported in part by the Physics Division of National Center for Theoretical Sciences under a grant from the National Science Council of Taiwan. G.L.L. and J.J.T. are supported by the National Science Council of R.O.C. under the grant number NSC90-2112-M009-023.

References

1. For a recent review article, see, for instance, F. Halzen, arXiv:astro-ph/0111059 and references therein.
2. Y. Fukuda *et al.* [Super-Kamiokande Collaboration], *Phys. Rev. Lett.* **81**, 1562 (1998).
3. S. Fukuda *et al.* [Super-Kamiokande Collaboration], *Phys. Rev. Lett.* **85**, 3999 (2000).
4. F. W. Stecker, *Astrophys. J.* **228**, 919 (1979); G. Domokos, B. Elliott and S. Kovesi-Domokos, *J. Phys.* **G19**, 899 (1993); V. S. Berezinsky, T. K. Gaisser, F. Halzen and T. Stanev, *Astropart. Phys.* **1**, 281 (1993); G. Ingelman and M. Thunman, arXiv:hep-ph/9604286. *Phys. Rev.* **D55**, 1297 (1997).
5. L. Pasquali and M. H. Reno, *Phys. Rev.* **D59**, 093003 (1999).
6. R. Engel, D. Seckel and T. Stanev, *Phys. Rev.* **D64**, 093010 (2001).
7. T. H. Burnett *et al.* [JACEE Collaboration], *Astrophys. J. Lett.* **349**, L25 (1990).
8. A. B. Kaidalov, *Phys. Lett.* **B116**, 459 (1982); A. B. Kaidalov and O. I. Piskunova, *Sov. J. Nucl. Phys.* **43**, 994 (1986) [*Yad. Fiz.* **43**, 1545 (1986)]; G. G. Arakelian and P. E. Volkovitsky, *Z. Phys.* **A353**, 87 (1995); G. H. Arakelian, *Phys. Atom. Nucl.* **61**, 1570 (1998) [*Yad. Fiz.* **61**, 1682 (1998)]; G. H. Arakelian and S. S. Eremian, *Phys. Atom. Nucl.* **62**, 1724 (199) [*Yad. Fiz.* **62**, 1851 (1999)].
9. For a more detailed discussion, see K. Cheung, in: H. Athar, G.-L. Lin and K.-W. Ng (Eds.), *Proceedings of the First NCTS Workshop on Astroparticle Physics*, 2001, Kenting (Taiwan) [to be published].
10. T. K. Gaisser, *Cosmic Rays And Particle Physics*, Cambridge University Press, NewYork, 1990.
11. S. Frixione, M. L. Mangano, P. Nason, and G. Ridolfi, *Nucl. Phys.* **B431**, 453 (1994).
12. T. Sjostrand, *Comput. Phys. Commun.* **82**, 74 (1994).
13. M. Thunman, G. Ingelman and P. Gondolo, *Astropart. Phys.* **5**, 309 (1996).
14. See, for instance, H. Athar, M. Jezabek and O. Yasuda, *Phys. Rev.* **D62**, 103007 (2000) and references therein.
15. L. V. Volkova and G. T. Zatsepin, *Phys. Atom. Nucl.* **64**, 266 (2001) [*Yad. Fiz.* **64**, 313 (2001)].

PQCD Analysis of Atmospheric Tau Neutrino

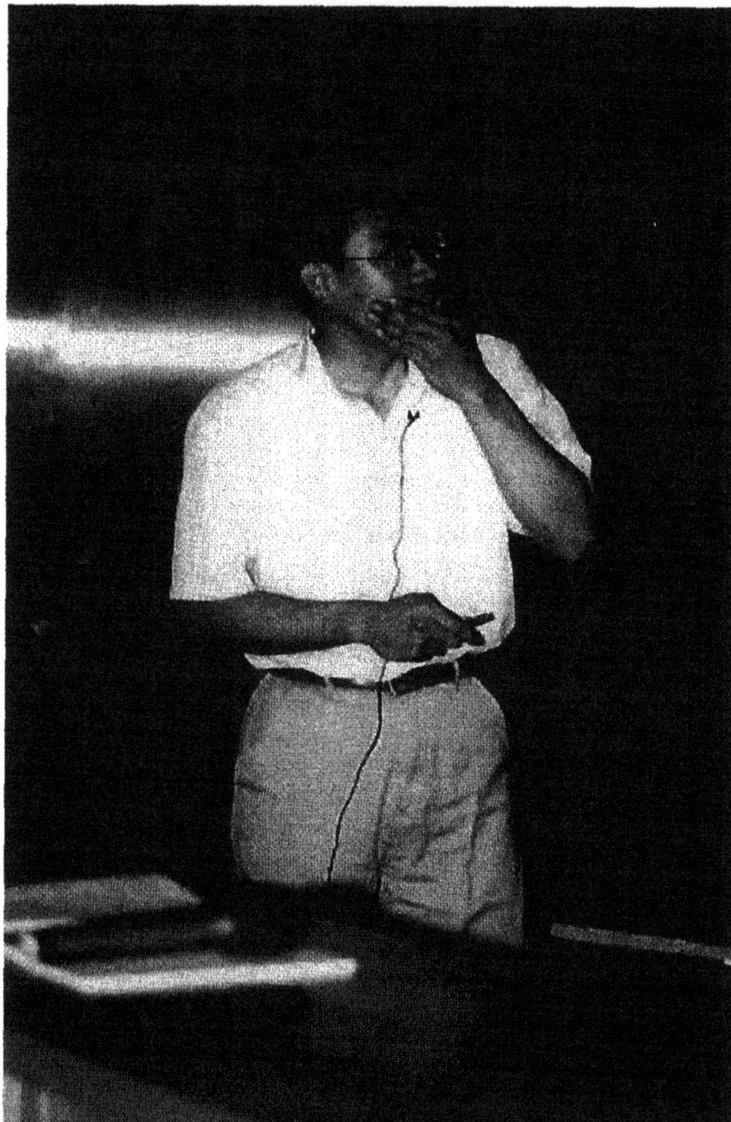

Tsung-Wen Yeh

PQCD ANALYSIS OF ATMOSPHERIC TAU NEUTRINO

TSUNG-WEN YEH

Institute of Physics, National Chiao-Tung University, Hsinchu 300, Taiwan

We calculate the flux of ν_τ produced in the atmosphere by cosmic ray-air collisions. The production channels of ν_τ from the process $pA \to X \to \nu_\tau(\bar{\nu}_\tau)Y$ with $X = D_s, W^\pm, Z, h_b, t\bar{t}$, are considered. The calculations are performed by employing perturbative QCD (PQCD). The predicted ν_τ flux $\phi_{\nu_\tau}(E)$ is given for energy range $10^3\text{GeV} \leq E \leq 10^8\text{GeV}$.

1. Introduction

The observation of high energy cosmic rays (CRs) with energies greater than 10^3 GeV has stimulated considerable experimental and theoretical investigations. Since the flux of the primary CRs is very rare in high energies, they can then be better observed indirectly through their secondary products from the interactions between the rays and the atmospheric nuclei. The primary CRs are mainly composed of charged nucleon, the proton. The secondary products from the proton air interactions could be hadrons, such as pion, kaons, and charm mesons and baryons, and, also the leptons, such as electrons, muons, taus, and their corresponding neutrinos. The leptons and neutrinos produced from the pion and kaon decays are called the conventional rays, and those from the charm hadron decays the prompt rays. Except of the τ lepton flux, the observation of the other lepton and neutrino fluxes would be the central issues. Unlike μ, e, ν_μ, ν_e could be conventional and prompt ones, the atmospheric ν_τ flux needs to be produced from the D_s meson decays. The ν_τ flux becomes equally important as the ν_μ flux for energies greater than 10^6 GeV. This is because the life time of μ lepton being much longer than that of τ lepton. The τ decays into ν_τ becomes dominant channel. Observation of high energy atmospheric ν_τ flux then appears as another ways for revealing the properties of the high energy primary cosmic rays.

Besides the above interests related to the cosmic ray physics, the high energy ν_τ can also entangle some problems for small x physics arising from the accelerator experiments. It is already known that the estimations of prompt fluxes could vary within a few orders of magnitude, since different models are required for calculations of the charm hadron cross section and energy spectra. The huge model dependence is due to the need of extrapolating charm production data obtained at accelerator energies to the orders of magnitude higher energies of the relevant cosmic ray collisions. The higher energy charm production cross section from cosmic ray air interactions could touch the x range even smaller that can be very hardly reached by the accelerator experiments in the near future. Thus, although the available data may not be enough to draw definitive conclusions on the flux of ν_τ, but the present and planed building large neutrino telescopes, such as AMANDA, BAIKAL, DUMAND, ICECUBE, EURO, and NESTOR could set a useful limits on the issues we shall discuss in this paper.

We shall investigate many theoretical uncertainties relating to the extrapolation of the parton distribution functions (PDFs) to cosmic ray relevant energies, where the present forms of the PDFs may not be applicable. The different sets of PDFs are the main uncertainty for the prediction of the ν_τ spectra. The concerned PDFs are restricted to the CTEQ ones, i.e., the CTEQ3, CTEQ4, CTEQ5 and CTEQ6 sets. There are also other sets of PDF's, such as MRS, GRV, etc, but they will not be investigated in this paper. Another kind of uncertainty come from the small x physics about which we still have little knowledge. We shall give a quantitative estimation for the uncertainties from different scenarios of the small x predictions.

2. The tau neutrino flux

The primary cosmic rays are mainly composed of protons with energy spectrum

$$\phi_p(E)[\frac{\text{nucleons}}{\text{cm}^2\ \text{s sr GeV}/A}] = \begin{cases} 1.7(E^{-2.7}) & \text{for} E < 5\cdot 10^6\text{GeV} \\ 164(E^{-3}) & \text{for} E > 5\cdot 10^6\text{GeV}\ . \end{cases} \quad (1)$$

The normalization factors are determined from experiments. Only primary protons are considered here, and the cosmic ray flux is assumed to be isotropic.

As produced and traveling through the atmosphere, the flux of the atmospheric ν_τ can be described by the transport equation

$$\frac{d\phi_{\nu_\tau}(X,E)}{dX} = -\frac{\phi_{\nu_\tau}}{\lambda_{\nu_\tau}} + \sum_{i=1}^{4} S_i(pA \to \nu_\tau) \quad (2)$$

where the first term is the sink term and the second the source term. The λ_{ν_τ} represents the attenuation length of the ν_τ in the atmosphere. At the concerned energies, the small cross section for the ν_τ air interactions implies that the sink term is much smaller than the source term and can be ignored within our estimation. The source terms $S_i(pA \to \nu_\tau)$ describe the production channels for the ν_τ produced from the proton-air collisions. The subscript i of the source terms S_i denotes a specific production channel. For a specific i, the source term has expression

$$S_i(pA \to \nu_\tau) = \int_E^\infty dE_p \frac{\phi_p(E_p, X)}{\lambda_p(E_p)} \frac{dn_i(E; E_p)}{dE} , \quad (3)$$

where $dn_i(E; E_p)/dE$ represents the normalized differential distribution for the channel i. We shall consider the following production channels for ν_τ: the charm meson channel $S(pA \to D_s \to \nu_\tau)$ with $i = 1$, the bottom hadron channel $S(pA \to h_b \to D_s \to \nu_\tau)$ with $i = 2$, the top quark channel $pA \to t\bar{t} \to \nu_\tau$ with $i = 3$, the W boson channel $pA \to W \to \nu_\tau$ with $i = 4$ and the Z boson channel $pA \to Z \to \nu_\tau$ with $i = 5$. The neutrino and antineutrino are considered together. Thus, the ν_τ flux ϕ_{ν_τ} describes $\nu_\tau + \bar{\nu}_\tau$. The normalized differential distribution $\frac{dn^i(E;E_p)}{dE}$ depends on the production channel. For the charm meson channel $pA \to D_s \to \nu_\tau$, the $\phi_(\nu_\tau)$ can be approximately expressed as

$$\phi_{\nu_\tau}^{high}(E) = \frac{Z_{pD_s}(E) Z_{D_s\nu_\tau}(E)}{1 - Z_{pp}(E)} \frac{\ln(\Lambda_{D_s}/\Lambda_p)}{1 - \Lambda_{D_s}/\Lambda_p} \frac{m_{D_s} c h_0}{E \tau_{D_s}} f(\theta) \phi_p(E, 0) , \quad (4)$$

and

$$\phi_{\nu_\tau}^{low}(E) = \frac{Z_{pD_s}(E) Z_{D_s\nu_\tau}(E)}{1 - Z_{pp}(E)} \phi_p(E, 0) , \quad (5)$$

where we have separated the solutions into the high and low energy limit with respect to the ν_τ energy $E = 10^8$ GeV. The Z-moment Z_{pD_s} is expressed as

$$Z_{pD_s}(E) = 2 f_{D_s} \int_0^1 \frac{dx_E}{x_E} \frac{\phi_p(E/x_E, X = 0)}{\phi_p(E, X = 0)} \frac{1}{\sigma_{pA}(E)} \frac{d\sigma_{pA \to c\bar{c}}(E/x_E)}{dx_E} , \quad (6)$$

where $x_E = E/E_p$ with E the energy of charm meson D_s, E_p is the proton energy and $f_{D_s} = 0.13$ for D_s fraction in the hadronization $c \to H_c$ and a factor 2 to count for two charm quarks in the final states. The differential cross section for proton air collision is modeled as

$$\frac{d\sigma_{pA \to c\bar{c}}(E/x_E)}{dx_E} = A^\gamma K \frac{d\sigma_{pp \to c\bar{c}}(E/x_E)}{dx_E} . \quad (7)$$

The σ_{pA} means the total inelastic cross section for proton-air collisions. The values of parameters are chosen as: $A = 14.5, \gamma = 1, K \approx 2.2, m_c = 1.35$ GeV. The Z-moments for the ν_τ from the D_s decay processes compose of two parts, the two body decay ($D_s \to \tau\nu_\tau$) and the decay chain ($D_s \to \tau \to \nu_\tau Y$)

$$Z_{D_s\nu_\tau}(E) = Z^{2body}_{D_s\nu_\tau}(E) + Z^{Chain}_{D_s\nu_\tau}(E) \tag{8}$$

and

$$Z^{2body}_{D_s\nu_\tau}(E) = \int_0^1 \frac{dx_E}{x_E} \frac{\phi_{D_s}(E/x_E)}{\phi_{D_s}(E)} \frac{d_{D_s}(E)}{d_{D_s}(E/x_E)} \frac{dn_{D_s\to\nu_\tau\tau}(E/x_E, E)}{dx_E}, \tag{9}$$

$$Z^{chain}_{D_s\nu_\tau}(E) = \int_0^1 \frac{dx_E}{x_E} \int_0^1 \frac{dx'_E}{x'_E} \frac{\phi_{D_s}(E/(x_E x'_E))}{\phi_{D_s}(E)} \frac{d_{D_s}(E)}{d_{D_s}(E/(x_E x'_E))} \times \frac{dn_{D_s\to\tau\nu_\tau}(E/(x_E x'_E), E/x_E)}{dx'_E} \frac{dn_{\tau\to\nu_\tau Y}(E/x_E, E)}{dx_E}. \tag{10}$$

The ν_τ flux produced from the $W^\pm$ and Z intermediate states are expressed as

$$\phi^{high}_{\nu_\tau}(E) = \frac{Z^W_{p\tau}(E) Z^W_{\tau\nu_\tau}(E)}{1 - Z_{pp}(E)} \frac{\ln(\Lambda_\tau/\Lambda_p)}{1 - \Lambda_\tau/\Lambda_p} \frac{m_\tau c h_0}{E\tau_\tau} f(\theta)\phi_p(E,0) \tag{11}$$

and

$$\phi^{low}_{\nu_\tau}(E) = \frac{Z^W_{p\nu_\tau}(E) + Z^W_{p\tau}(E) Z^W_{\tau\nu_\tau}(E)}{1 - Z_{pp}(E)} \phi_p(E,0) . \tag{12}$$

From the Z boson production $pA \to Z \to \nu_\tau\overline{\nu}_\tau$, the ν_τ flux has the form

$$\phi_{\nu_\tau}(E) = \frac{Z^Z_{p\nu_\tau}(E)}{1 - Z_{pp}(E)} \phi_p(E,0) . \tag{13}$$

For the b hadron production $pA \to h_b \to \nu_\tau$, the ν_τ flux takes similar form to that form from the D_s production. We assume that the probability of b hadronization is 10% and $Br(H_b \to \tau\nu_\tau) = 0.026$. The t quark production is similar to the W boson production by process $pA \to t\bar{t} \to b\bar{b}W^+W^-$ with the probability of W boson decay into $\tau\nu_\tau$ as 1/9. The Z moments $Z^W_{p\nu_\tau}$, $Z^Z_{p\nu_\tau}$ and $Z^W_{\tau\nu_\tau}$ are expressed as

$$\begin{aligned} Z^W_{p\nu_\tau}(E) &= Z^W_{p\tau}(E) \\ &= \int_0^1 \frac{dx_E}{x_E} \frac{\phi_p(E/x_E)}{\phi_p(E)} \frac{\lambda_p(E)}{\lambda_p(E/x_E)} \frac{dn_{pA\to W^\pm\to\tau\nu_\tau X}(E/x_E)}{dx_E}, \end{aligned} \tag{14}$$

$$Z^Z_{p\nu_\tau}(E) = \int_0^1 \frac{dx_E}{x_E} \frac{\phi_p(E/x_E)}{\phi_p(E)} \frac{\lambda_p(E)}{\lambda_p(E/x_E)} \frac{dn_{pA\to Z\to\bar{\nu}_\tau\nu_\tau X}(E/x_E)}{dx_E}, \tag{15}$$

$$Z^W_{\tau\nu_\tau}(E) = \int_0^1 \frac{dx_E}{x_E} \frac{\phi_\tau(E/x_E)}{\phi_\tau(E)} \frac{d_\tau(E)}{d_\tau(E/x_E)} \frac{dn_{\tau\to\nu_\tau X}(E/x_E)}{dx_E} . \tag{16}$$

In practical calculations, the dominant part of ν_τ flux is from the charm meson channel and other parts are suppressed due to weak interaction constant or the threshold effects. However, these suppressed channels would become important as the ν_τ energy is higher than 10^{11} GeV.

3. Perturbative QCD

One central quantity in the above normalized differential distribution is the differential charm production cross section

$$\frac{d\sigma_{pp\to c\bar{c}}(E/x_E)}{dx_E} = \sum_{a,b=u,d,g} \int_0^{\frac{x_E^2}{4}} \frac{d\tau}{x_L} \sigma_{ab\to c\bar{c}}(E/x_E) c_{ab} \times (f_a(x_a,Q^2)\bar{f}_b(x_b,Q^2) + f_b(x_b,Q^2)\bar{f}_a(x_a,Q^2)) \quad (17)$$

where $x_b = \tau/x_a$ and $f_a(x_a, Q^2)$ represents the parton distribution function and $c_{ab} = 1$ for $a, b = u, d, s$ and $c_{ab} = 1/2$ for $a, b = g$. In order to explore the uncertainty arising from the parton distribution function, we then employ the CTEQ3, CTEQ4, CTEQ5 and CTEQ6 sets of PDFs for calculation of the ν_τ flux from the production channels we have described in last section. The evolution of the parton distribution functions in the time can represent the progress in the understanding of the QCD dynamics. The gluon contributions dominate over the light quark contributions by about an order of magnitude. Therefore, measuring the ν_τ flux can indirectly determine the gluon distribution function. As shown in Fig. 1 the predic-

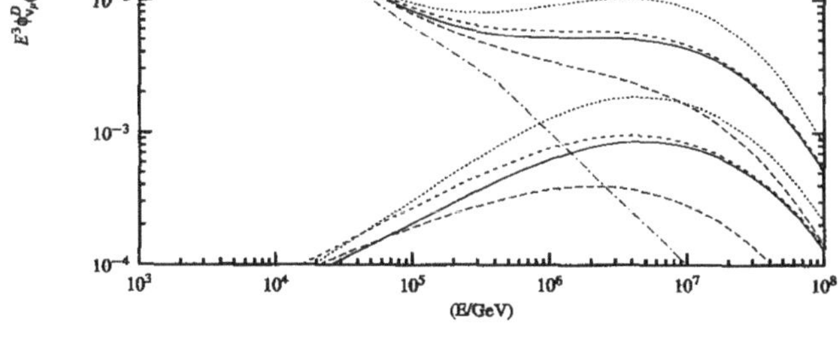

Figure 1. The ν_τ spectra are predicted from the CTEQ sets of PDFs. The ν_μ spectra are shown for comparison.

tions of the ν_τ spectra from four CTEQ sets of PDFs, we observe that the CTEQ3 spectrum dominates the other CTEQ predictions over the entire energy range. The ν_τ spectra from the CTEQ4 and CTEQ6 sets are very close to each other. The CTEQ5 predicted spectrum is smaller than the others. The largest difference between the CTEQ predictions may reach about one order of magnitude. For comparison, we also show our calculation for the ν_μ spectra predicted from the same CTEQ sets of PDFs. The ν_μ and ν_τ fluxes are equally important for the neutrino energy greater than 10^6 GeV.

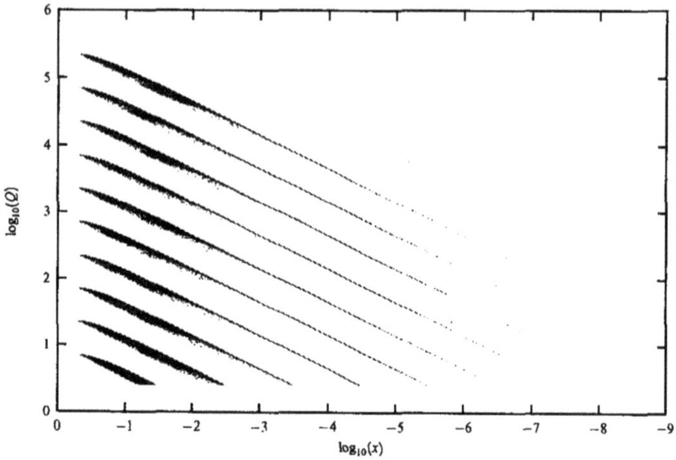

Figure 2. The kinematic map for charm production in the cosmic ray air collisions. The Q represents the scale involved in the process and the x the momentum fraction carried by the partons inside the proton and air nuclei. The lines order from bottom to top represent the ν_τ energy $E = 10^2, 10^3, 10^4, 10^5, 10^6, 10^7, 10^8, 10^9, 10^{10}, 10^{11}$ GeV.

From the kinematic map for the ν_τ flux as depicted in Fig. 2, the ν_τ with higher energies can have larger probabilities to reach the small x region ($x < 10^{-4}$). The present and planned accelerator experiments can populate at the region with $0 \leq \ln Q \leq 1.2$ GeV and $0 \leq -\ln x \leq 4$. The evolution of the parton distribution functions in the different regions in the above kinematic map is determined by different evolution equations. For large Q and moderate x, the DGLAP equation is most important. On the other hand, for moderate Q and small x, the BFKL equation becomes relevant. There are also the CCFM equation, which combines the DGLAP and BFKL equations and governs both large and small x evolution, and the GLR equation for the large Q and small x. The CTEQ sets of PDFs are only determined by employing the DGLAP equation. The BFKL equation predicts that the PDFs at small x can take a power law $xf(x) \sim x^{-\lambda}$ with $\lambda \approx 0.5$.[1,2] However, it is known that the parameter λ should depend on Q. It implies that the DGLAP equation should be introduced to give the Q dependence of the λ. For the gluon distribution function $G(x)$, the DGLAP equation leads to

$$xG(x, Q^2) \sim \exp[2\sqrt{\frac{N_c}{\beta_0} \ln\frac{1}{x} \ln\frac{t}{t_0}}] \tag{18}$$

with $t_{(0)} = \ln(Q^2_{(0)}/\Lambda^2_{\rm QCD})$. By comparing the above equation with $x^{-\lambda}$ for different Q^2 and a fix Q_0^2, we may observe that different value of Q^2 can lead to different λ. Without going into deeper investigation of these complicate small x physics, we try to give a simple estimation for the small x and Q evolution behavior of parton distribution functions. We extrapolate the CTEQ3 PDFs into the following form

$$xf(x, Q) = \begin{cases} xf(x, Q) \ , \text{ for } x > 10^{-4} \\ x^{-\lambda}c(Q) \ , \text{ for } x < 10^{-4} \end{cases} \tag{19}$$

where $c(Q) = f(x, Q)/x^{1+\lambda}$ at $x = 10^{-4}$. The values of λ are run through $0, 0.08, 0.2, 0.3, 0.4, 0.5$. The value of $\lambda = 0.08$ is understood from the prediction of the BFKL equation with low $\ln Q$.[3,4] For $\lambda = 0.3$, the extrapolated distribution is matched with the CTEQ PDFs. The result is shown in Fig. 3. To measure the ν_τ events in large neutrino telescopes, the neutrino

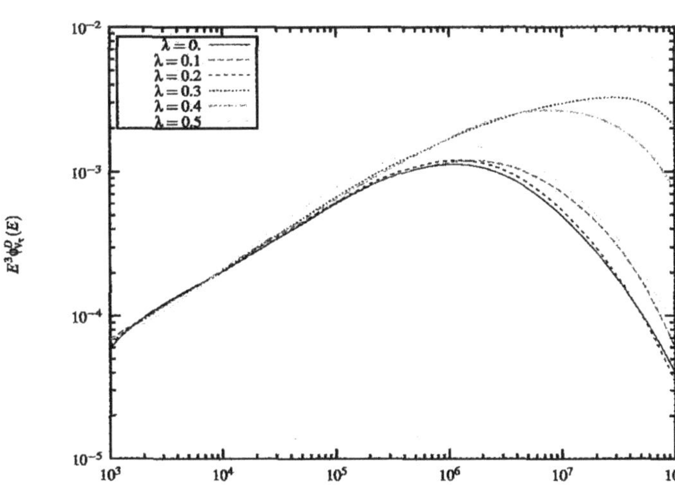

Figure 3. The ν_τ spectra are evaluated from the small x extrapolation of CTEQ3M PDFs. The PDFs are assumed to take the form as $xf(x, Q) \sim x^{-\lambda}c(Q)$.

nucleon interactions are exploited by tagging the τ lepton signals from the two bang events[5,6,7], or the Lollypop events[8]. The two bang events contain two hadronic showers. The first shower comes from the process $\nu_\tau N \to \tau X$ and the second shower arises from the τ lepton decays. The Lollypop events track the second shower and the tau lepton tracks. The atmospheric ν_τ's at high energies are the background for the search of the cosmic neutrinos. The investigation of the ν_τ flux given here can help the experiments to distinguish the events from different sources.

4. Conclusion

In this paper, we have shown our preliminary result for the atmospheric ν_τ. The uncertainties from different CTEQ sets of PDFs and different extrapolation forms for small x of PDFs are considered. There remained many theoretical uncertainties for future experiments. The theoretical predictions have an order of magnitude difference; but, the results can be taken as a background for experimentalists to design more efficient experiments. The nonperturbative QCD dynamics will be explored at many order of higher energies than the accelerator experiments.

5. Acknowledgement

The author appreciates many helpful discussions with K. Chung, H. Athar, G.L. Lin and the other members of the working group at NCTU. This work was supported in part by the National Science Council of R.O.C. under Grant No. NSC89-2811-M-009-0024.

References

1. E. A. Kuraev, L.N. Lipatov and V.S. Fadin, *Sov. Phys. JETP* **45**, 199(1977).
2. Ya. Ya. Balitsky, L.N. Lipatov, *Sov. J. Nucl. Phys.* **28**, 822(1978).
3. A.D. Martin, *Acta Physica Pol.* **B25**, 265 (1994).
4. J. Kwiecinski, *Phys. Rev.* **D52**, 1445(1995).
5. J. Alvarez-Muniz and F. Halzen, Detection of Tau Neutrinos in IceCube(2001).
6. H. Athar, G. Parente and E. Zas, *Phys. Rev.* **D62**, 093010 (2000) [hep-ph/0006123].
7. J. G. Learned and Sandip Pakavasa, *Astropart. Phys.* **3**, 267(1995).
8. IceCube Design Document, www.icecube.wisc.edu.

Ultra High Energy Cosmic Rays from Supermassive Objects with Magnetic Monopoles

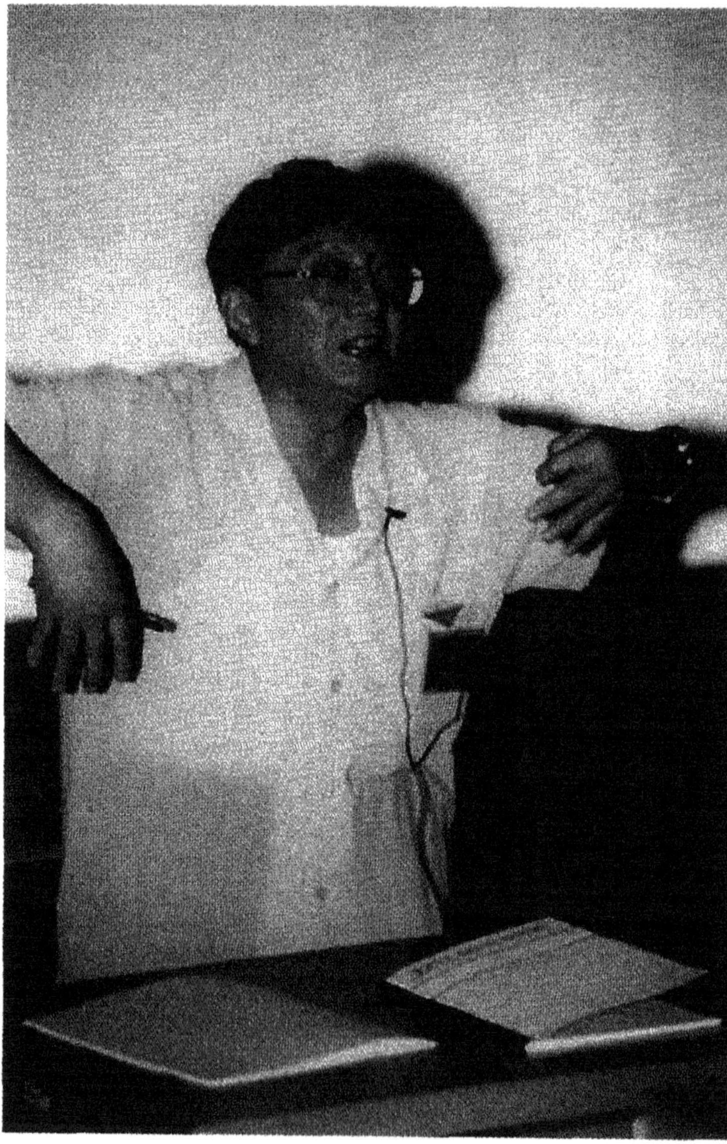

Qiu-He Peng

ULTRA HIGH ENERGY COSMIC RAYS FROM SUPERMASSIVE OBJECTS WITH MAGNETIC MONOPOLES

QIU-HE PENG[1,2,3,4], CHIH-KANG CHOU[2,5], AND MIN LONG[1]

1 Department of Astronomy, Nanjing University, Nanjing China 210093, qhpeng@nju.edu.cn

2 Institute of Astronomy and Department of Physics, National Central University, Chung-Li, 32054, Taiwan, China Taipei, chou@astro.ncu.edu.tw

3 Joint Astrophysics Center of Chinese Academy of Science-Beijing University

4 The Open Laboratory for Cosmic Ray and High Energy Astrophysics, Chinese Academy of Sciences

5 National Astronomical Observatories, Chinese Academy of Sciences, Beijing 100012, China

Although many ultra high energy (UHE) events with individual particle energy is beyond 4×10^{19} eV have been observed, the origin and signature of such ultra high energy cosmic rays (UHECRs) are still somewhat mysterious. How particles could obtain such enormous energies? In this paper, we present an effective acceleration mechanism for these UHE events based on super-massive stellar objects (SMSOs) with saturation of magnetic monopoles of t'Hooft-Polyakov type in the universe. These SMSOs have remarkable characteristics. Catalyzed by magnetic monopoles, nucleons may decay to energetic charged particles and radiation, and can travel to great distance from their source because of the absence of horizon and central singularity of SMSOs, due to the Rubakov-Callan effect. These rapidly rotating collapsed objects have radial magnetic fields in the local co-rotating frame. Moreover, the induced electromagnetic field in the rest frame is rather strong and quite different from the usual dipole field. As a result, the energetic charged particles may be further accelerated to 10^{21} eV. The denser the SMSO is, the higher the particle energy obtainable from our accelerator model becomes and the maximum particle energy may reach 10^{23} eV. The flux of UHECRs is briefly estimated by considering the production of protons and antiprotons via the interaction of the synchrotron photons with the thermal photons. The estimated flux of UHECRs from our theoretical model compares very favorably with the observed results.

1. Observation of Ultra High Energy Cosmic Ray

The origin and signature of cosmic rays continues to be an "unsolved problem" even after its 90 years discovery by Hess in 1912. The unexpected detection of Ultra High Energy Cosmic Rays (UHECRs) with energies in excess of 50EeV (1EeV=10^{18}eV) is much more mysterious than those low energy cosmic rays. Hundreds of such events have been observed by different kinds of detectors, such as the old Volcano Ranch[1] and Haverah Park[2] experiments, the Yakutsk experiment in Russia[3], the Akeno Giant Air Shower Array (AGASA)[4-6], the Sydney University Giant Air Shower Recorder (SUGAR)[7], the Fly's Eye detector[8] and high Resolution Fly's Eye (HiRes)[9]. Especially thirties of events with energy beyond 10^{20} eV are recorded. The top two highest energy events are 3.2×10^{20} eV observed by the Fly's Eye, and 2.1×10^{20} eV seen by AGASA. These pose severe challenge to the usual explanation for the origin of such UHECRs.

First, the greatest challenge to the theorists is the observed extreme high energy that is difficult to explain by most conventional acceleration mechanisms in the usual range of physics. It is almost impossible to accelerate particles to such high energy even with the most powerful astrophysical sources[10-12], such as radio galaxies and active galactic nuclei (AGN).

Second, hypothetic high energy particles interacting with Cosmic Microwave Background (CMB) photons will lose energy seriously, if the energy of the particles exceeds the Greisen-Zatsepin -Kuz'min cutoff[13,14]. For a CMB photon with mean energy 7.0×10^{-4} eV (T_b = 3.0 K), the threshold value of a nucleon $E_{th} = 10^{20}$ eV . The free propagation length is less than about tens of Mpc. Moreover, ultra high energy electrons lose most of their energies via synchrotron radiation and even originate much closer to us. The other hypothetic cosmic ray primaries, such as electrons, nuclei, photons and neutrinos (which are disputed[15,16]) also interact with the CMB photons, the inferred background photons and the geomagnetic fields and cascade well before reaching the Earth.[17]

Third, the UHECRs is basically isotropic. The locus of the UHECRs from its origin to the Earth is almost a straight line due to the weakness of galactic and extragalactic magnetic fields, but no counterpart object along UHECR directions has ever been found.

In order to explain these difficulties, all major scenarios contemplated by astrophysicists so far may be classified as "Bottom-Up", "Top-Down" and hybrid models.

The purpose of the "Bottom-Up" scenarios involves seeking for plausible accelerators named Zevatrons[18] that may accelerate cosmic particles to ZeV (10^{21} eV). In such scenario, it is still a challenge to explain the extremely high energy of the UHECRs observed in terms of the convectional models. Even if it is possible, it is extremely difficult in most cases to draw the particles out of the dense region.[19] And the values of the relevant properties required for acceleration are somewhat extreme.[12]

In the alternative "Top-Down" scenario, new exotic particles beyond the Standard Model are introduced, for example, the Grand Unified Theory (GUT) scale (supermassive) particles (named X-particles, typical X-particles masses vary between $10^{22} - 10^{25}$ eV/c^2). Thus, ZeV energies are not difficult for these models and no acceleration mechanism is needed. Some relics originated from GUT symmetry breaking phase transition during the inflation of early universe, such as topological defects, superheavy relic particles also may be invoked to generate the UHECRs. When topological defects are destroyed or superheavy relic particles decay, they can produce cascades of high energy nucleons, γ -ray and neutrinos. For these "Top-Down" scenarios, the real difficulty is the fact that the predicated flux is much lower than the observed flux,[20] since the typical distances between topological defects is the Hubble Scale $H^{-1} \approx 3h^{-1}$ Gpc.[21] Moreover, this scenario is model dependent.

On the other hand, the hybrid models make use of the physics beyond the standard model in a bottom-up way to generate ultra high energy particles. To evade the GZK constraint, some Zevatrons have been introduced to produce exotic particles such as neutrinos with small mass[22], topological defects, and vortons[23] instead of protons and photons. More detailed discussions for the theoretical aspects and the experimental background for the UHECRs may be obtained in recent reviews.[21,24]

In this paper, we present a new mechanism for the UHECRs on the basis of the previous model of AGN[25-29] given by Peng and his collaborators. More specifically, we propose that the supermassive objects (SMOs) with magnetic monopoles of the t' Hooft-Polyakov[30,31] type may be the reasonable sources for the UHECRs.

2. An AGN Model with the Magnetic Monopoles of t'Hoft-Polyakov Type

2.1. *Magnetic Monopoles and the Rubakov-Callen Effect*

Tiny amount of magnetic monopoles may be produced, according to the grand unified theories, due to violent oscillation and thermal fluctuation of higgs field during the phase transition in the primordial universe which is very hot ($kT > 10$ GeV) and in a highly chaotic state (Guth, 1982).[30] It is 't Hooft-Polyakov type (super-heavy) MM[31,32] with mass $m_m = 10^{16}$ GeV, magnetic charge $g_m = 3(g_m)_{Dirac} = 3\hbar c/2e = 9.88\times10^{-8}$G, as stable topological entities in any grand unified theory that breaks down to electromagnetism.[33] Although there are many aspects that still imperfectly understood, nevertheless, the prediction from the grand unification theory for magnetic monopole catalysis of baryon decay is solidly founded[34]

$$\begin{aligned} P + M &\rightarrow M + e^{+} + \text{mesons} \\ P + M &\rightarrow M + e^{+} + \text{hadrons} \end{aligned} \tag{1}$$

These are strong interaction processes and they must have a cross section of typical strong interaction $\sigma \approx (10^{-27} \sim 10^{-37})cm^2$. The reaction rate of the Rubakov-Callen (RC) effect for a celestial object is

$$r_{RC} = N_B N_m < \sigma v >\approx 2\times10^6 \rho^2 \frac{\zeta}{\zeta_n}\xi \tag{2}$$

$$\xi = \frac{<\sigma v/c>}{10^{-27}} \tag{3}$$

where ζ denotes the content of magnetic monopoles for the SMO, $\zeta \equiv N_m / N_B$, and ζ_n is the Newtonian saturation content while the Coulomb magnetic repulsive force is balanced by the Newtonian gravity of the SMO. The value of ζ_n is given by[25]

$$\zeta_n = Gm_B m_m / g_m^2 \approx 1.9\times10^{-25}(m_m/10^{16}m_B), \tag{4}$$

which is much lower than the Paker's upper limit[35,36] for the number of monopoles in the universe $\varsigma \le \varsigma^0 = 10^{-20\pm1}$. The maximum saturation content of magnetic monopoles for a relativistic SMSO is[26]

$$\zeta_s^{(\max)} \approx 214\zeta_n \tag{5}$$

The monopoles in some SMOs are primarily produced in a dense plasma state with high temperature in the primordial epoch of galaxy formation. The content of monopoles in SMOs such as quasars and AGN may be rather high[25] due to strong monopole-plasma interaction[37], although the content of monopoles for stars and planets is very tiny.[25]

2.2. *A Model of AGN with the Magnetic Monopoles*[25-29]

By invoking the RC effect as the energy source, an AGN model with saturation of magnetic monopoles has been proposed. Its gravitational properties is similar to the usual black hole model and an accretion disk around the central compact object may still exist, but their main energy source is due to the induced nucleon decay catalyzed by monopoles rather than due to the mass accretion. The main features of AGN with magnetic monopoles are[26-29]

1) Quasars and AGN should be strong infrared sources with a Planck spectrum, if they contain monopoles with saturation content. For the Galactic Center, the surface temperature of the SMO is 121K.

2) A strong flux of 0.511Mev γ-ray line radiation and higher energy γ-ray can be produced from AGN by the RC effect. For the Galactic Center, $S_{e^+} \approx 6.5\times10^{42} e^+ / \sec$, $S_\gamma > 10^{37} erg/s$.
3) The radial magnetic field the co-rotating frame of the SMSO[26,28] is

$$H(R) = Q_m / R^2 = 2.55\times10^6 \frac{\zeta}{\zeta_n}\left(\frac{M}{10^8 M_{Sun}}\right)^{-1}\left(\frac{R}{R_g}\right)^{-2} \qquad (6)$$

where $R_g = 2GM/c^2$ is the Schwarzschild radius. The corresponding radial magnetic field for the Galactic Center is approximately $H(R) \approx (20 \sim 100)$ *gauss*.
4) The SMOs may rotate very fast with the spinning angular velocity

$$\Omega \approx \frac{\alpha}{2}\frac{R_g}{R}\frac{c}{R},\ V^{(rot)}(R) \approx \frac{\alpha}{2}\frac{R_g}{R}c,\ \alpha \approx (0.5-0.99). \qquad (7)$$

5) Neither horizon nor central singularity can exist in these rapidly rotating objects with the critical content of magnetic monopoles due to the RC effect, even though their radii may be smaller than the Schwarzschild radius. The rate of the catalytic reaction due to the RC effect is proportional to the square of the density.[29] Hence it increases rapidly towards the center. The intense radiation and the weakening of the central gravitational field due to the lose of mass resulting from the RC effect will prevent the objects from collapsing indefinitely, the more violent the collapse, the faster the process proceeds, though the objects are still very dense. Moreover, the strong thermal pressure resulting from the rapidly increasing temperature in the central region of the object would enhance the process. Therefore there is no singularity at the center of the object, so long as the RC effect is effective.

The critical content of Magnetic monopoles for the SMO may be found via general relativity with the result[26]

$$\varsigma_c = \varsigma_{c0}(1 - 4a^2 / R_g^2)^{1/2} \qquad (8)$$

where $\varsigma_{c0} = \sqrt{G} m_B / g_m$, $a \equiv J/Mc = \alpha(GM/c^2)$, J is the angular momentum of the SMO.

For the Galactic Center, $\xi \approx 10^{-7}$. Based on the near infrared observation, the parameter of the SMO for the Galactic Center may be estimated:[29] $R \approx 8.1\times 10^{15}\ cm$, $R/Rg = 1.1 \times 10^4$.

3. UHECR Production and Acceleration from SMOs with Magnetic Monopoles

3.1. *The Induced electric field of the SMOs with Magnetic Monopoles*

In the rest frame of the observer, there is a radial electric field around the SMOs with the magnetic monopoles. This radial electric field is induced by the radial magnetic field of the SMOs via the Lorentz transform. The induced electric field may be calculated through the Kerr-Newman-Kasaya metric of a dyonic object as follows.

$$\begin{aligned} ds^2 &= g_{\mu\nu}dx^\mu dx^\nu \\ &= \frac{a^2\sin\theta-\Delta}{\rho^2}dt^2+\frac{(r^2+a^2)^2-\Delta a^2\sin^2\theta}{\rho^2}\sin^2\theta d\phi \\ &+2\frac{\Delta-(r^2+a^2)}{\rho^2}a\sin^2\theta dtd\phi+\frac{\rho^2}{\Delta}dr^2+\rho^2 d\theta^2 \end{aligned} \tag{9}$$

$$\rho^2 = r^2+a^2\cos^2\theta$$

$$\Delta = r^2-R_g r+a^2+\kappa(Q_e^2+Q_m^2)$$

where Q_m, Q_e are the content of magnetic monopoles and electric charge, respectively, $\kappa \equiv G/c^4$, while the electromagnetic vector potential is[26]

$$A_t = -Q_e\frac{r}{\rho^2}+Q_m\frac{a\cos\theta}{\rho^2}$$

$$A_r = A_\theta = 0$$

$$A_\phi = Q_e\frac{ar\sin^2\theta}{\rho^2}+Q_m[\pm 1-\cos\theta\frac{r^2+a^2}{\rho^2}] \tag{10}$$

The induced electric field may then be calculated easily.

$$\begin{aligned} F_{\mu\nu} &= A_{\nu;\mu}-A_{\mu;\nu} = \frac{\partial A_\nu}{\partial x^\mu}-\frac{\partial A_\mu}{\partial x^\nu}, \\ E_r &= F_{r0} = \frac{\partial A_0}{\partial r}-\frac{\partial A_r}{\partial t} = -\frac{2Q_m ar\cos\theta}{\rho^4}, \\ E_\theta &= F_{\theta 0} = \frac{1}{r}\frac{\partial A_0}{\partial\theta}-\frac{\partial A_\theta}{\partial t} = \frac{Q_m a}{r}[-\frac{\sin\theta}{\rho^2}+\frac{2a^2\cos^2\theta\sin\theta}{\rho^4}], \\ E_\phi &= F_{\phi 0} = 0. \end{aligned} \tag{11}$$

The induced electromagnetic field in the rest frame is rather strong. Charged particles catalyzed by monopoles via the RC effect will be further accelerated in such electric field and further escape from such SMOs. The maximum energy of the accelerated protons may be estimated as follows

$$E_p^{(\max)} \approx eA_0 \approx eQ_m \frac{a}{\rho^2} \approx \frac{eQ_m}{R_g}\frac{a}{R_g}(\frac{R}{R_g})^{-2} = \frac{eg_m c^2}{2Gm_B}\zeta_n \frac{\zeta}{\zeta_n}\frac{a}{R_g}(\frac{R}{R_g})^{-2}$$

$$\approx 2.4\times10^{22}\frac{\zeta}{\zeta_n}\frac{a}{R_g}(\frac{R}{R_g})^{-2} eV \qquad (12)$$

For the relativistic SMOs, $\zeta \approx \zeta_s \approx (100\sim200)$, $a/R_g \geq 1/2$, $R/R_g \approx (1\sim10)$, then $E_p^{(\max)} > 10^{22}$ eV. Based on these convincing order of magnitude estimated, it seems very natural to suggest that the SMOs (i.e. AGNs) with magnetic monopoles may be the sources of the UHECRs.

3.2. *Production and Acceleration of the UHECR: Physical Process*

We note that the energies of the particles produced by the RC effect are already very high (~1GeV) to begin with. Following the RC effect, other charged particles may also be generated during the subsequent cascades and multiplications such as $\pi^0 \to \gamma+\gamma$, $\pi^0 \to \gamma+\gamma+\gamma$, $\gamma+\gamma \to e^+ + e^-$ and so on. These particles will be further accelerated in the strong electromagnetic fields of the SMOs for a rather long period of time when they escape from the object. Since the effect of gravity is rather insignificant, we can estimate the final energy by considering only the electromagnetic fields.

The main physical process for the production and acceleration of the UHECR is as follows:

RC effect: $p \to \pi^0, (e^{\pm}, \mu^{\pm}) \to (e^{\pm}, \mu^{\pm})$; then $(e^{\pm}, \mu^{\pm})$ are accelerated by the induced electric field. The released high-energy electrons and positrons from the RC effect are further accelerated to much higher energy (for example, over (10^9~10^{10}) GeV) and will radiate synchrotron photons in the magnetic field.

$$e^+ \xrightarrow[Mag.field]{} e^+ + \gamma$$

The power of synchrotron radiation is

$$P = 1.6\times10^{-15}\gamma^2\beta^2H^2\sin^2\alpha \qquad ergs/s \qquad (13)$$

where $\beta = v/c$, v, γ are the velocity and the Lorentz factor of the charged particle, α is the angle of the particle velocity relative to the magnetic field. The peak frequency of the synchrotron radiation is

$$\omega_{peak} \approx \omega_L\gamma^2\sin\alpha \approx 1\times10^{13}HE_{GeV}^2(e^+)\sin\alpha \ (s^{-1}) \qquad (14)$$

where ω_L is the classical cyclotron frequency for the positron in the magnetic field, $\omega_L = eH/m_e c$.

The energy of most synchrotron photons is about

$$E_\gamma^{(syn)} = \hbar\omega_{peak} \approx 4(\frac{H}{10gauss})[\frac{E(e^+)}{10^5GeV}]^2 \ GeV \qquad (15)$$

A proton and an anti-proton pair with rather high energy can be produced by the interaction of the energetic synchrotron photons with thermal photons.

$$\gamma^{(syn)} + \gamma^{(ther)} \rightarrow p + \overline{p} \tag{16}$$

as long as

$$E_{\gamma}^{(syn)} \cdot E_{\gamma}^{(therm)} > (m_p c^2)^2 . \tag{17}$$

The temperature near the surface of the SMOs is about 100 K, we may take the energy for the thermal photons as $10^{-2} eV$. The reaction (16) can happen when the energy of the synchrotron photon $E_{\gamma}^{(syn)} > 10^{11} GeV$. In turn, the energetic synchrotron photons with energy of 10^{11}GeV may be supplied by the energetic positrons (or electrons) with energy $E_{e^{\pm}} \geq (10^9 \sim 10^{10}) GeV$, and magnetic field ($10^3 \sim 10^4$) gauss (see the Eq. (6) and Eq. (15)).

Finally, the energetic proton will be further accelerated by the induced electric field outside the SMO. Because the power of synchrotron radiation is inversely proportional to the square of the mass, the energy loss rate for the protons is much slower than that for the electrons, so that they can travel much farther and still with high energy. Consenquently, the major traveling particles are the protons and will be further accelerated by the strong electromagnetic fields of the SMO for a rather long period of time when they escape from the SMO.

3.3. *Estimated Flux near the Earth*

According to our model, we can estimate the flux of UHECRs near the Earth. The production rate of the positrons by the RC effect is:[27-29]

$$S_{e^+} = \frac{4\pi}{3} R^3 r_{RC} \approx 7.5\times10^{47} \left(\frac{M}{10^8 M_{Sun}}\right)^{-1} \left(\frac{R}{R_g}\right)^{-3} \frac{\zeta}{\zeta_n} \xi \quad sec^{-1} \tag{18}$$

The released positrons will be rapidly accelerated to higher energy over 10^{10}GeV in the induced electric field. The accelerated positron can radiate synchrotron photons with energy higher than 10^{11}GeV. The production rate for the energetic synchrotron photons per positron is given by

$$\frac{dN_{1\gamma}}{dt} \sim \frac{\overline{P}}{\hbar\omega_{peak}} = 5.5\times10^4 H(R) \frac{R^2}{r^2} \tag{19}$$

The total number of the energetic synchrotron photons per positron may be obtained by integrating Eq. (19)

$$N_{1\gamma} \approx \int_R^{\infty} \frac{dN_{1\gamma}}{dt} \frac{dr}{c} \approx 1.4\times10^{14} \frac{\zeta}{\zeta_n} \frac{R_g}{R} . \tag{20}$$

The total number of such energetic synchrotron photons created by all the positrons produced in one second is then

$$N_\gamma = S_{e^+} N_{1\gamma} = 1.0\times10^{62}\left(\frac{M}{10^8 M_{Sun}}\right)^{-1}\left(\frac{R}{R_g}\right)^{-4}\left(\frac{\zeta}{\zeta_n}\right)^2 \xi \quad sec^{-1}. \qquad (21)$$

These very high-energy photons (see the condition (17)) will bump thermal photons to produce protons and antiprotons. The cross section is

$$\sigma(E_1, E_2) = \frac{\pi r_{0p}^2}{2}(1-\beta^2)[2\beta(\beta^2-2)+(3-\beta^4)\ln\frac{1+\beta}{1-\beta}] \qquad (22)$$

$$\approx \sigma_T\left(\frac{m_e}{m_p}\right)^2 \frac{m_p^2 c^4}{E_\gamma^{(syn)} E_\gamma^{(Therm)}} \approx 1\times10^{-31}\ cm^2 \qquad (23)$$

where $\beta = [1 - \frac{m_p^2 c^4}{E_\gamma^{(syn)} E_\gamma^{(therm)}}]^{1/2}$, $r_{0p} = e^2/(m_p c^2) = \frac{m_e}{m_p} r_0$,

The production rate of ultra high energy protons by the process (16) is,

$$\frac{dN_p}{dt} = n_\gamma n_{th} c <\sigma> V = N_\gamma n_{th} c <\sigma>$$

$$\approx 1.0\times10^{32} cT^3\left(\frac{M}{10^8 M_{Sun}}\right)^{-1}\left(\frac{R}{R_g}\right)^{-4}\left(\frac{\zeta}{\zeta_n}\right)^2 \xi, \qquad (24)$$

where $n_\gamma, n_{\gamma,th}$ are the number density for the energetic synchrotron photons and the background thermal photons respectively ($n_{\gamma,th} = aT^4/5k_BT \approx 10T^3 cm^{-3}$, the temperature near the central object is about 100K). The UHECR (proton) flux received at the Earth from one such SMO may be estimated as

$$S_1 = \frac{dN_p/dt}{4\pi D^2} A \approx 0.2\left(\frac{D}{20Mpc}\right)^{-2}\frac{A}{100km^2}\left(\frac{M}{10^8 M_{Sun}}\right)^{-1}\left(\frac{R}{10R_g}\right)^{-4}\left(\frac{\zeta}{\zeta_n}\right)^2\frac{\xi}{10^{-12}}\ (Yr)^{-1} \qquad (25)$$

where D is the distance of the SMO. A is the received area of the detector, for example, A =100 km^2 for AGASA.

Since the GZK cut off limit the possible sources of the UHECRs within the range of 50 Mpc, so we only consider the AGN in these region. The observation reveals that the number of the AGN in this range is about (10^3~10^4). Then the total estimated flux of the UHECR is about $S_{tot} \geq 2\times10^3/10Year$.

The accumulated events observed by AGASA is

$$S_{AGASA} \sim \frac{600}{10 years} \sim 2\times10^{-6} s^{-1}. \qquad (26)$$

As we mentioned before, the insufficient flux problem is a challenge to the Top-Down Model whereas the ultra high-energy is a big problem to the Bottom-Up Model. Although the estimated flux from our model is a little beyond the flux of the observed events, nevertheless, considering the possible loss from the observed UHECRs during their passage to the Earth, this result is reasonable.

4. Conclusion

Introducing the RC effect as the primary energy source for the acceleration of high-energy charged particles, a mechanism for the origin of the UHECRs was suggested. We would like to emphasize that an accretion disk around the central compact stellar object may still exist. The primary and secondary charged particles released from the RC effect, such as protons and nuclei from the SMOs or accretion disks will be further accelerated to ultra high-energy by the strong electromagnetic fields of the SMOs with magnetic monopoles. These charged particles travel along the spiral orbits whose gyro-radii become larger since the magnetic field decrease with distance during the passage. Our quick calculation as given above is of course just estimation. For more exact computation, we should establish the appropriate equation of motion for the particles. Our proposal is of course just one of the probable candidates in exploring the mystery of the exotic UHECRs by properly choosing some novel ingredients such as the SMOs with monopoles. This model will be tested and compared with the accumulation of the observed data such that further improvement of our model on the basis of a more detailed study of the SMOs and the UHECRs can be accomplished.

Acknowledgments

We are grateful to Dr. Xin-Lian Luo for helpful discussions. This research was supported by the National Natural Science Foundation of China (grants 10173005 and 19935030) and the Director Foundation of National Education Ministry (2000028417).

References

1. Linsley J., *Phys. Rev. Lett.* **10**, 146 (1963). In Proc.8th *International Cosmic Ray Conference* **4**, 295 (1963) (the Volcano Ranch experiment).
2. Lawrence M. A., Reid R. J. O., and Watson A. A., *J. Phys. G Nucl. Part. Phys.* **17**, 733 (1991), and references therein; see also http://ast.leeds.ac.uk/haverah/hav-home.html. (the Haverah Park[2] experiment)
3. Efimov N. N. et al., in Proc. *International Symposium on Astrophysical Aspects of the Most Energetic Cosmic Rays,* eds. Nagano M. and Takahara F. (World Scientific, Singapore, 1991), p.~20; Afnasiev B.N., in Proc. *International Symposium on Extremely High Energy Cosmic Rays:*

Astrophysics and Future Observatories, ed. Nagano M. (Institute for Cosmic Ray Research, Tokyo, 1996), p.32. (the Yakutsk experiment)
4. Takeda M. et al., *Phys. Rev. Lett.* **81**, 1163 (1998). (AGASA 1)
5. Takeda M. et al., *Astrophys. J.* **522**, 225 (1999). (AGASA 2)
6. Hayshida N. et al., preprint (astro-ph/0008102). (AGASA 3)
7. Brownlee R. G. et al., *Can. J. Phys.* **46**, S259 (1968); Winn M. M. et al., *J. Phys.* **G12**, 653 (1986); see also http://www.physics.usyd.edu.au/hienergy/sugar.html (SUGAR)
8. Bird D.J. et al., *Phys. Rev. Lett.* **71**, 3401 (1993); *Astrophys. J.* **424**, 491 (1994); *Astrophys. J.* **441**, 144 (1995). (Fly's Eye)
9. Corbat S.C. et al., *Nucl. Phys. B* (Proc. Suppl.) **28B**, 36 (1992); Bird D.J. et al., in Proc. *24th International Cosmic Ray Conference* (IstitutoNazionale Fisica Nucleare, Rome, Italy, 1995), Vol.**2**, 504; Vol.**1**, 750; M.~Al-Seady et al., in Proc. *26th* International *Cosmic Ray Conference*, (Utah, 1999), p.~191; see also http://bragg.physics.adelaide.edu.au/astrophysics/HiRes.html.
10. Teshima M. et al., *Nucl. Phys. B* (Proc. Suppl.) **28B**, 169 (1992); Hayashida M. et al., in Proc. *26th International Cosmic Ray Conference*, (Utah, 1999), p.~205; see also http://www-ta.icrr.u-tokyo.ac.jp. Cronin J.W., *Nucl. Phys. B* (Proc. Suppl.) **28B**, 213 (1992); *The Pierre Auger Observatory Design Report (2nd edition)*, March 1997; see also http://http://www.auger.org and http://www-lpnhep.in2p3.fr/auger/welcome.html. Yoshida S. and Dai H., *J. Phys.* **G24**, 905 (1998). M.~Aglietta et al. (EAS-TOP collaboration), *Phys. Lett.* **B337**, 376 (1994). E.~Andres et al., e-print astro-ph/9906203 T.K. Gaisser. 1990, *Cosmic Rays and Particle Physics* (Cambridge: Cambridge Univ. Press) Hillas A. M., *Nature* **395**, 15 (1998). Hillas A. M., *Ann. Rev. Astron. Astrophys.* **22**, 425 (1984).
11. Kalashev O.E., Kuzmin V. A., and Semikoz D.V., astro-ph/9911035. (fermi)
12. Norman C.T., Melrose D.B., and Achterberg A., *Astrophys. J.* **454**, 60 (1995). (shock)
13. Greisen K., *Phys. Rev. Lett.* **16**, 748 (1966).
14. Zatsepin G. T. and Kuz'min V. A., *Soviet Phys.-Jetp Lett.* **4**, 78 (1966).
15. Gandhi R., Quigg C., Reno M. H., and Sarcevic I., *Astropart. Phys.* **5**, 81 (1996).
16. Burdman G., Halzen F., and Gandhi R., *Phys. Lett.* **B417**, 107 (1998).
17. Stanev T. and Vankov H. P., *Phys. Rev.* **D55**, 1365 (1997).
18. Blandford R. D., *Particle Physics and the Universe*, eds. L. Bergstrom, P. Carlson, and C. Fransson (Physica Scripta, World Scientific, 1999).
19. J. W. Cronin, *Nucl. Phys.* **B28** (Proc. Suppl.), 213 (1992); *Rev. Mod. Phys.* **71**, 175 (1999).
20. Berezinskii V. S. et al., *Astrophysics of Cosmic Rays,* (Amsterdam: North Holland) (1990).
21. Olinto A.V., *Phys. Rept.* **333-334**, 329 (2000).

22. Bahcall J.N., *Neutrino Astrophysics*, Cambridge Univ. Press (New York, 1989).
23. Bonazzola S. and Peter P., *Astropart. Phys.* **7**, 161 (1997).
24. Bhattacharjee P. and Sigl G., *Phys. Rep.* **327**, 109 (2000).
25. Peng Q-H., Li Z.-Y., and Wang D.-Y., *Scientia Sinica* **28**, 970 (1985); Peng Q.-H., Wang D.-Y., and Li Z.-Y., *Acta Astrophys. Sinaca* (in Chinese) **6**, 249 (1986).
26. Peng Q.-H., *Astrophys. Sp. Sci.* **154**, 271 (1989).
27. Wang D.-Y. and Peng Q.-H., *Adv. Space Res.* **6w**, 177 (1986); D.-Y. Wang, Q.-H. Peng, and Chen T.-Y., *Astrophys. & Sp. Sci.* **118**, 379 (1986).
28. Peng Q.-H. and Chou C.-K., *Astrophys. & Sp. Sci.* **257**, 149 (1998).
29. Peng Q.-H. and Chou C.-K., *Astrophys. J.* **551**, L23 (2001).
30. A. H. Guth, "10 Seconds After the Big Bang", a talk presented at The 2nd Moriond Astrophysics Meeting, (XVIIth, Rencontre de Moriond, Les Arcs-Savoie, France, Narch), 14 (1982).
31. G. t'Hooft, *Nucl. Phys.* **B79**, 276 (1974).
32. A. M. Polyakov, *ZhETF Pis'ma* **20**, 430 (1974).
33. Katherine Freese and Eleonora Krasteva, *Phys. Rev.* **D59**, 063007 (1999).
34. V. Rubakov, *Nucl. Phys.* **218**, 240 (1983).
35. E. N. Parker, *Astrophys. J.* **160**, 383 (1970).
36. G. Lazarides et al., *Phys. Lett.* **B100**, 21 (981).
37. X.-Q. Li, *Astrophys. Sp. Sci.* **123**, 125 (1986).

Neutrinos, Oscillations and Nucleosynthesis

Bruce McKellar

NEUTRINOS, OSCILLATIONS AND NUCLEOSYNTHESIS

B. H. J MCKELLAR
Research Center for High Energy Physics,
School of Physics,
University of Melbourne,
Victoria, Australia, 3010
E-mail: b.mckellar@physics.unimelb.edu.au

After reviewing the conventional results of Big Bang Nucleosynthesis, I describe how neutrino oscillations can change the conventional results. Now that we know that neutrinos do oscillate, we need to revisit these classical results.

1. Introduction

The role of weak interactions and, in particular, interactions involving neutrinos, in determining the abundance of the light elements is now well known. Before the advent of LEP, Big-Bang Nucleosynthesis (BBNS) was used as a neutrino (species) counter, to determine the number of light neutrino species. Even after LEP this is still a useful neutrino counter, as sterile as well as active neutrino species are counted. Since the suggestions that neutrinos may oscillate, and especially since the concentration on neutrino oscillations in the early 1980s, there have been many suggestions that neutrino oscillations, or other exotic neutrino neutrino properties, could influence the abundance of the light elements. This influence could take a form which would allow the determination of these exotic properties, but more likely it would weaken the conclusions drawn from the standard cosmological model.

Our group at Melbourne has contributed to these aspects of exotic neutrino physics since the early 1980s. Henry Granek and I made a number of suggestions of ways in which neutrino oscillations could mask the usual deduction of the number of neutrino species[1,2], and studied the relevant kinetic equations[3]. Mark Thomson and I were one of the groups that realized that, in the high neutrino density in the early universe, one would expect MSW oscillations of neutrinos induced by the neutrino background.[4] We then showed how to generalize the "polarization vector" picture of neutrino oscillations to develop the kinetic equations describing the neutrino system. These are highly non-linear equations, and a number of subtle effects emerge in the solutions. It was pointed out by Volkas, Foot and their collaborators [6,7,8,9] that this allowed a large neutrino asymmetry to be developed in certain circumstances, and that this in turn had major effects on big-bang nucleosynthesis, shifting the number of neutrino species deduced from the analysis by ± 1.

In this paper, I will first review the standard physics of Big-Bang Nucleosynthesis, so that we can better understand how this can be modified by changes of the neutrino properties. Then I will describe how this standard cosmological model is modified by neutrino oscillations, by a non-zero chemical potential for the neutrinos, and by interactions between neutrinos.

I will give a somewhat idiosyncratic picture of the field, and only a few references, confident that if I whet your appetite for more, you have available a comprehensive, recent review of the subject of neutrinos in cosmology by Dolgov[10].

2. The standard first 3 minutes

The light element abundance which is most easily determined is that of ^{4}He, although that does not mean that the determination of the primordial helium mass fraction Y from the observations is without controversy.

The He abundance is weakly dependent on the baryon excess. It depends on the number of neutrinos in 2 ways: The expansion rate and the reaction rate. The expansion rate is determined by ALL neutrinos, and reaction rate by only electron neutrinos. Indeed the expansion rate counts both active neutrinos (neutrinos which couple to the W and Z bosons with the usual strength), and sterile neutrinos (neutrinos whose coupling to the W and Z is so small that the neutrinos have not yet been directly observed in weak interactions), simply because it is determined by the energy density.

The expansion time scale is just $t_e = H^{-1}$, where H is the Hubble parameter $\dot{R}/R$, and R is the scale factor in the Robertson-Walker metric

$$ds^2 = dt^2 - R^2(t)\left\{\frac{dr^2}{1-kr^2} + r^2 d\Omega^2\right\}, \tag{1}$$

and the Hubble parameter is determined from the Einstein equations to be

$$H = \sqrt{\frac{8\pi}{3}G_N \rho} = \sqrt{\frac{8\pi}{3}G_N\left(1 + \frac{7}{4} + \frac{7}{8}N_\nu\right)}T^2. \tag{2}$$

We have assumed that the universe is radiation dominated, and that each neutrino species (of which there are N_ν) contributes $(7/16)aT^4$ to the energy density, the photons contribute aT^4, and electrons and positrons each contribute $(7/8)aT^4$. The constant a is $\pi^2/15$, in units where Boltzmann's constant, k_B, the velocity of light, c, and $\hbar$, Planck's constant all unity.

The weak time scale is inversely related to the weak reaction rate for $n + e^+ \to p + \bar{\nu}_e$ (and the other weak reaction rates)

$$t_W = \Gamma^{-1}_{\text{reaction}}$$
$$\Gamma_{\text{reaction}} \approx \frac{7\pi}{30}\left(1 + 3g_A^2\right) G_F^2 T^5 \qquad (3)$$

The weak interaction processes can maintain the protons and neutrons in thermal equilibrium while the weak time scale is shorter than the expansion time scale. As the universe cools, the weak time scale grows until it exceeds the expansion time scale, at a temperature T^* determined from $t_e = t_W$, or

$$\frac{7\pi}{30}\left(1 + 3g_A^2\right) G_F^2 (T^*)^5 = \sqrt{\frac{8\pi}{3} G_N \rho} = \sqrt{[(8\pi)/3] G_N \left(1 + \frac{7}{4} + \frac{7}{8} N_\nu\right)} (T^*)^2 \qquad (4)$$

The value of this "decoupling temperature" in turn determines the helium abundance.

The primordial helium abundance is essentially determined by the ratio of neutrons to protons at decoupling. Nuclear reactions then eventually combine almost all of the neutrons into ^{4}He nuclei. To determine the mass fraction of neutrons, $X_n = (N_n)/(N_p + N_n)$, we consider the reactions

$$p + e \leftrightarrow n + \nu_e$$
$$n + e^+ \leftrightarrow p + \bar{\nu}_e,$$

neglecting for the moment the decay of the neutrons. The rate of conversion of neutrons to protons is

$$\Gamma(n \to p) = \Gamma(n\nu_e \to pe) + \Gamma(ne^+ \to p\bar{\nu}_e) \qquad (5)$$

and that for conversion of protons to neutrons is

$$\Gamma(p \to n) = \Gamma(pe \to n\nu_e) + \Gamma(p\bar{\nu}_e \to ne^+). \qquad (6)$$

The rate equation describing the change in the neutron number is

$$\frac{dN_n}{dt} = -\Gamma(n \to p)N_n + \Gamma(p \to n)N_p, \qquad (7)$$

whence, as $N_p + N_n$, the total number of nucleons, is a constant,

$$\dot{X}_n = -\Gamma(n \to p)X_n + \Gamma(p \to n)(1 - X_n). \qquad (8)$$

The equilibrium mass fraction is thus

$$X_n = \frac{\Gamma(p \to n)}{\Gamma(n \to p) + \Gamma(p \to n)} \qquad (9)$$

A typical reaction rate for a weak interaction is

$$\Gamma(\ell_i N_i \to \ell_f N_f) = (1+g_A^2)G_F^2 \int \frac{p_i^2 dp_i}{2\pi^3} p_f E_f g(E_i)\,[1-g(E_f)] \qquad (10)$$

where $g(E) = \left(e^{E/T}+1\right)^{-1}$ is the Fermi-Dirac distribution at zero chemical potential, and $1-g(E) = \left(1+e^{-E/T}\right)^{-1}$ is the Pauli blocking factor.

For $n \to p$, $E_i = p_i$, $E_f = p_i + \Delta m$, and
For $p \to n$, $E_i = p_f + \Delta m$, $E_f = p_f$, so

$$\Gamma(n \to p) = \int \mathcal{P}_n(q)dq \left(e^{q/T}+1\right)^{-1} \left(1+e^{-(q+\Delta m)/T}\right)^{-1}, \qquad (11)$$

$$\Gamma(p \to n) = \int \mathcal{P}_p(q)dq \left(e^{(q+\Delta m)/T}+1\right)^{-1} \left(1+e^{-q/T}\right)^{-1}, \qquad (12)$$

$$\approx e^{-\Delta m/T}\Gamma(n \to p) \qquad (13)$$

where the approximation is good when the phase space for n conversion $\mathcal{P}_n$ is the same as the phase space for the p conversion, $\mathcal{P}_p$, or in other words when $T \gg \Delta m$. In this approximation

$$X_n = 1/(1+e^{\Delta m/T^*}). \qquad (14)$$

For 3 neutrino flavours, the decoupling temperature is $T^* = 0.8 \times 10^{10}$K, which yields[a] $X_n = 0.14$. From the Hubble relation we see that the temperature of 10^{10}K corresponds to a time of about 1 sec. Why then was Weinberg's book entitled *The First Three Minutes*, and not *The First Second*?

While we expect the nuclear reactions converting neutrons to helium nuclei to now proceed rapidly, the first of these is

$$n + p \leftrightarrow d + \gamma \qquad (15)$$

and there are so many photons compared to nucleons, $\eta = n_{\text{nucleon}}/n_\gamma \approx 10^{-10}$, that the reaction is driven to the left, photodisintegrating any deuterons which are formed, until the temperature T_d for which $\eta e^{Q/T_d} = 1$, where Q is the Q value of the reaction. This is a temperature of $T_d \approx 0.1$MeV. During the time from T^* to T_d, which is about 110 sec, neutrons will decay, and the neutron mass fraction is reduced.

The final Helium mass fraction $Y = 2X_n = 0.25$, depends logarithmically on the value of η — an increase in η by a factor of 2 adds about 0.01 to Y. A similar shift in Y occurs when the number of neutrino flavours is

[a]Strictly, $T^* \gg \Delta m$ is not satisfied, but this can be accommodated by more careful calculations, which do not materially change the result.

increased by 1, $\Delta Y \approx (0.013 \pm 0.001)\Delta N_\nu$. If we are to use Y to count the number of neutrino species we need determine η with reasonable precision.

The data from the Cosmic Microwave Background Radiation give $10^{10}\eta = 5.7 \pm 1$, and the analysis of the primordial abundance of deuterium gives $10^{10}\eta = 6.5 \pm 2$.

The recent data on the primordial helium abundance cluster around two centres, $Y = 0.234 - -0.235$ [11,12,13] and $Y = 0.245$ [14]. The limits on N_ν obtained from this data are then

$$\text{for} 10^{10}\eta = 6.7 \qquad 1.5 < N_\nu < 2.9 \tag{16}$$

$$\text{for} 10^{10}\eta = 5.7 \qquad 1.7 < N_\nu < 3.2 \tag{17}$$

$$\text{for} 10^{10}\eta = 4.7 \qquad 1.8 < N_\nu < 3.5. \tag{18}$$

It is apparent that the Helium abundance is not a very precise counter of the number of neutrino species. We can say that $N_\nu < 3.5$ at the 1σ level. Indeed it is possible that even 3 neutrino species may be excluded if the higher values of η are confirmed.

How robust is the BBNS bound on N_ν in the presence of a neutrino asymmetry, and in the presence of neutrino oscillations? That is the question to which we in Melbourne have been studying for some time, and which I now describe.

3. The Asymmetry Effect

A neutrino asymmetry means that the number density of neutrinos of the flavour f, $n(\nu_f)$ and the number density of antineutrinos of the same flavour, $n(\bar{\nu}_f)$, are not in balance, so that (for example) $n(\nu_e) \gg n(\bar{\nu}_e)$. The reaction rates governing the $n - p$ equilibrium, Eqns. (5) and (6), are such that the excess of ν_e enhances $\Gamma(n\nu_e \to pe)$, and the deficit of $\bar{\nu}_e$ enhances $\Gamma(ne^+ \to p\bar{\nu}_e)$, so both components of $\Gamma(n \to p)$ are increased. Similarly, both components of $\Gamma(p \to n)$ are suppressed. Thus the equilibrium is shifted in the direction of fewer neutrons, and less ^{4}He. A large positive value of the electron neutrino asymmetry $L_e = n(\nu_e) - n(\bar{\nu}_e)$ will allow for a smaller Helium abundance Y than one would have expected on the basis of the number of neutrino species. Alternatively, if one fits N_ν to the data on the assumption that $L_e = 0$, and in fact L_e is large and positive, the number of neutrino species deduced from the analysis is less than the actual number of species.

It is too early to speculate that we may be in an "N_ν crisis" on the basis of the present data, but there is the possibility that such a crisis could develop as the data on Y and η become more precise. Should that happen, it is reassuring to know that a finite value of L_e would save the situation. It is even more reassuring to realize that such a value L_e could be generated as the result of neutrino oscillations.

4. The Simplified Neutrino Kinetic Equations

The detailed kinetic equations for neutrinos which undergo oscillations, and which interact with a background of charged leptons and neutrinos were written down in detail by Thomson and McKellar, and others. For the moment we will consider a simplified form of the equations, restricting our consideration to a two-neutrino system, and to the simplest possible interactions with the background. We use a density matrix representation of the neutrino system, so that, if $|\nu\rangle$ is the neutron state vector, the density matrix is $\rho = |\nu\rangle\langle\nu|$. In the two component system, this is a 2×2 matrix and can be written as

$$\rho = \frac{1}{2}\left(P_0 + \mathbf{P}\cdot\sigma\right). \tag{19}$$

Naively, we expect that the condition that ρ describe a pure state requires $P_0 = \mathbf{P}^2 = 1$, but we need to consider the dependence of the density matrix on the momentum y, and well as its flavour dependence, and these conditions no longer apply. In particular the trace of ρ involves an integral over momenta, as well as a matrix trace.

The simplified equations of motion are

$$\frac{\partial P_0}{\partial t} = R \tag{20}$$

$$\frac{\partial \mathbf{P}}{\partial t} = \mathbf{V}\times\mathbf{P} - D\mathbf{P}_T + R\hat{z} \tag{21}$$

where $\mathbf{P}_T = \mathbf{P} - (\mathbf{P}\cdot\hat{z}\hat{z})$ is the transverse part of $\mathbf{P}$. Both vacuum oscillations and the MSW type oscillations arising from the interactions of the neutrinos with the background leptons and neutrinos enter $\mathbf{V}$, and only interactions are responsible for the factors D and R. The vector $\mathbf{V}$ gives coherent effects in the evolution equation, and as such must contain only terms proportional to G_F. The term D simply takes account of the scattering of the neutrinos by the background, is proportional to the sum of the reaction rates for each background species, and as such is proportional to $G_F^2 T^5 y$. Both $\mathbf{V}$ and D involve interactions with any neutrino background which is present, and thus may involve the components of $\mathbf{P}$, rendering the equations non-linear.

The relaxation term R is a Boltzmann collision term, representing a balance between the scattering into and out of the states of interest. It can be represented in the relaxation time approximation as being proportional to the the deviation of the distribution from equilibrium.

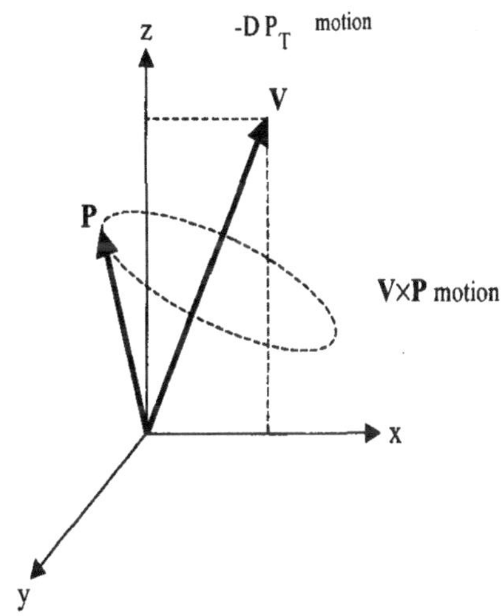

Figure 1. The motion of the polarization vector representation of the neutrino state.

Figure 1 shows the geometry of the polarization vector and its motion in the case $R = 0$. The $\mathbf{V} \times \mathbf{P}$ term gives rise to precession of $\mathbf{P}$ about $\mathbf{V}$. As the probability of the two types of neutrino are $\frac{1}{2}(1 \pm P_z)$, this corresponds to neutrino oscillations. The "damping term" $-D\mathbf{P}_T$ reduces the magnitude of the transverse part of $\mathbf{P}$. If that were the only term in the equation of motion, $\mathbf{P}$ will align with the z axis and the state becomes a mixed state. This term in the equation of motion is a decoherence term, and when it dominates, the oscillations are damped out.

If the collision rate is high, the equilibrium distribution (in momenta) will be reached rapidly, and we can then ignore the relaxation term R, and regard the polarization vector as proportional to the equilibrium Fermi-Dirac distribution with zero chemical potential[b], $f_{\mathrm{eq}}(y)$. In this circumstance it is convenient to rewrite the Eq. (21) as

$$\frac{\partial \mathbf{P}}{\partial t} = \mathbf{K}\mathbf{P} \tag{22}$$

---

[b] y is the neutrino momentum in units of the temperature, and we have assumed that $T \gg m_\nu$, so that $E_\nu = p_\mu$ to a very good approximation.

where the matrix $\mathbf{K}$ is

$$\mathbf{K} = \begin{pmatrix} -D & -V_z & 0 \\ V_z & -D & -V_x \\ 0 & V_x & 0 \end{pmatrix} \tag{23}$$

As the coefficients D and $\mathbf{V}$ may depend on $\mathbf{P}$, they will in general be time dependent. Diagonalize $\mathbf{K}(t)$ at the time t, and call the result $\mathbf{K}_d(t)$. If the instantaneous eigenvectors of $\mathbf{K}(t)$ are $\mathbf{Q}(t)$, write $Q(t) = \mathbf{U}(t)\mathbf{P}(t)$. The equation of motion of $\mathbf{Q}$ is

$$\frac{\partial \mathbf{Q}}{\partial t} = \mathbf{K}_d\mathbf{Q} - \mathbf{U}\frac{\partial U}{\partial t}\mathbf{Q} \tag{24}$$

suggesting the use of an adiabatic approximation

$$\mathbf{P}(t) \approx \mathbf{U}^{-1}(t)\exp\left(\int_0^t \mathbf{K}_d(t')dt'\right)\mathbf{U}(0)\mathbf{P}(0), \tag{25}$$

which is a valid approximation when $(\partial\mathbf{U})/(\partial t)$ can be neglected. The circumstances in which this is valid have been discussed in detail by Volkas and collaborators, but it is a reasonable approximation at high temperatures, when the collision terms dominate (because of the T^5 dependence). Then

$$\mathbf{K}_d \approx \mathrm{diag}\{-\mathrm{D} + \mathrm{i}|\mathbf{V}|, -\mathrm{D} - \mathrm{i}|\mathbf{V}|, -\mathrm{V}_\mathrm{x}^2/\mathrm{D}\} \tag{26}$$

Now we can follow the time development of the density matrix and see it can lead to marked fluctuations in L_e. We develop the kinetic equations for the antineutrinos, leading to a polarization vector $\bar{\mathbf{P}}$ which represents the antineutrino density matrix. some manipulation then gives

$$\frac{\partial L_e}{\partial t} = \frac{T^3}{2n_\gamma}\int_0^\infty V_x\left(P_y - \bar{P}_y\right)\frac{y^2 f_{\mathrm{eq}}(y)}{2\pi^2}dy, \tag{27}$$

and in the high temperature approximation this becomes

$$\frac{\partial L_e}{\partial t} = \frac{T^3}{2n_\gamma}\int_0^\infty \left(\frac{V_x^2}{D}P_z - \frac{\bar{V}_x^2}{D}\bar{P}_z\right)\frac{y^2 f_{\mathrm{eq}}(y)}{2\pi^2}dy \tag{28}$$

Even without doing calculations, it is clear that this can exhibit sign changes, and can lead to rapid changes in the lepton asymmetry in either direction. This is illustrated in the Fig. 2 (from the first of Ref. 9).

Note that the time evolution is right to left in the figure, so that the lepton asymmetry starts at the level of the baryon asymmetry (10^{-10}), dips to very small values and then grows rapidly to order 10^{-6}, then slowly and rapidly to order 3/8 in the nucleosynthesis region. This example shows that it is quite possible for neutrino oscillations to have very strong effects

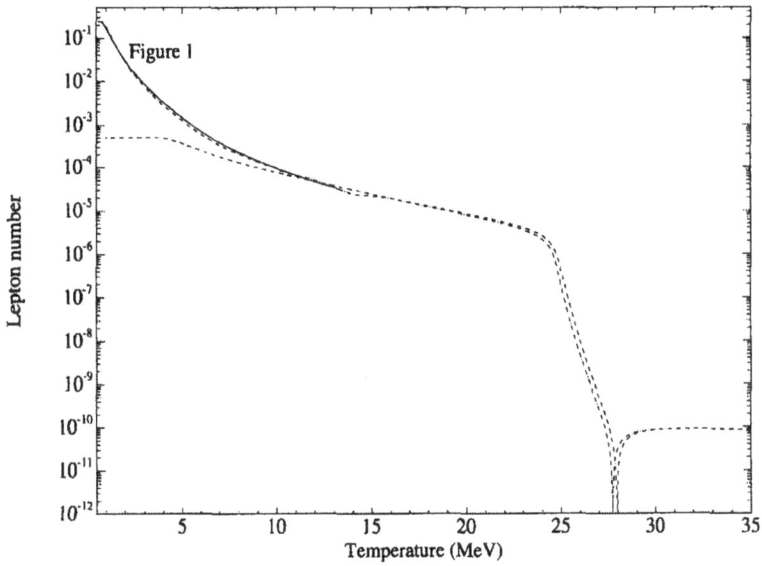

Figure 2. A possible variation of lepton number with temperature. The various lines represent exact and approximate solutions of the quantum kinetic equations. The difference between the curves is unimportant for the present discussion.

on BBNS, and that they need to be taken into account in any contemporary discussion of the abundances of light nuclei. The particular oscillations involved in this example were active-sterile oscillations, which are now somewhat less fashionable than they were in 2000, but they cannot be completely ruled out, even now.

The final sign of L depends on the initial conditions, so it is possible that a domain structure could develop with inhomogeneities in L, and even the sign of L could vary from domain to domain. This could be a justification for the models of nucleosynthesis in an inhomogeneous medium.

5. Neutrino Oscillations and Limits on the Degeneracy Parameter

Even oscillations which do not involve sterile neutrinos can have significant effects on the standard BBNS analysis. The lepton asymmetry is directly related to the degeneracy parameter, which is the ratio of the chemical potential to the temperature, $\xi_\ell = \mu_\ell/T$ — when $xi_\ell \neq 0$, as $xi_\ell = -\xi_{\bar{\ell}}$, $L \neq 0$. In turn $\xi_\ell = 0$ can influence BBNS in two ways. Replacing the $\xi_e = 0$

Fermi-Dirac distribution in Eq. (11) and Eq. (12) with the distribution

$$f_0(y,T) = \frac{1}{e^{(y/T-\xi_e)}+1} \tag{29}$$

we see that $X_n \propto e^{-\xi_e}$. Moreover a non-zero ξ_ℓ for any flavour increases the energy density of that neutrino species, and thus increase the expansion rate, and thus X_n.

An additional radiation density in degenerate neutrinos also gives a delayed transition to a matter dominated universe, boosting the amplitude of the first acoustic peak in the CMBR power spectrum. Moreover there are observable effects in the cosmic large scale structure (LSS).

A recent standard analysis of BBNS, CMBR and LSS [15] gives the limits

$$-0.01 \leq \xi_e \leq 0.22 \qquad |\xi_{\mu,\tau}| \leq 2.6, \tag{30}$$

but the standard analysis omits oscillations. Lunardini and Smirnov [16] argued that flavour oscillations would equilibrate the neutrinos. Dolgov *et al* [17] have studied the effects of oscillations in more detail, and have shown that three flavour oscillations with currently fashionable values of masses and mixing angles, force equilibrium of the flavour states before nucleosynthesis, and thus the more stringent limits on ξ_e apply also to the other flavours.

Figure 3, from Ref. 17 shows the basic physics. Remember that $P_z = 0$

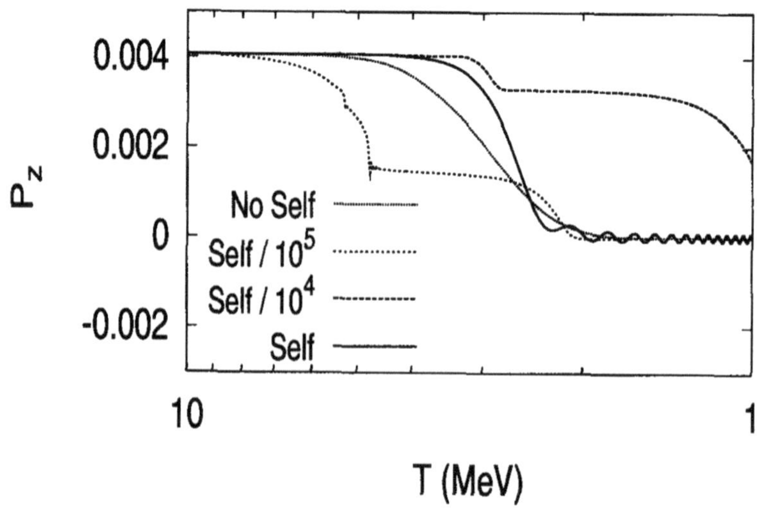

Figure 3. Evolution of P_z with various interactions.

corresponds to the two flavours having the same number density, and thus to flavour equilibrium. Vacuum oscillations, MSW oscillations induced by the $e^+ - -e^-$ background, MSW oscillations induced by the neutrino background ("self") and collisional damping are all included. The detailed discussion shows that all are important. However without collisional damping the flavour equilibrium is still a coherent flavour superposition, rather than a decoherent mixture.

Three flavour oscillations are yet more complex. The "polarization vector" picture can still be applied to the analysis if one expands the density matrix in the Gell-Mann matrices and defines the "vector product" $(\mathbf{V} \times \mathbf{P})_\alpha = f_{\alpha\beta\gamma} V_\beta P_\gamma$, as suggested in Ref. 4. Dolgov *et al* did not choose this route, and solved the necessary matrix equations. They found that it was possible to avoid driving the system to flavour equilibrium by carefully choosing the initial conditions. However this choice required the equality of the magnitudes of the degeneracy parameters.

Thus either because of the oscillations, or of the special choice of initial conditions, the chemical potentials of all of the neutrino flavours are equal in magnitude, and so limits obtained on the magnitude of the electron neutrino degeneracy apply to all degeneracy parameters. Without taking account of the value of N_ν, the limit on the degeneracy parameter is

$$|\xi| \leq 0.22 \tag{31}$$

for all flavours. If one restricts $N_\nu < 4.2$, then $|\xi_e| \leq 0.1$, and can be transferred to all flavours.

This has relevance for the MAP and PLANCK experiments — the limits obtained from BBNS, CMBR and LSS when neutrino oscillations are taken into account restrict the values of $|\xi|$ so severely that neutrino degeneracy can be ignored in the analysis of these experiments.

6. The Neutrino Kinetic Equations

I want to conclude this survey of the effects of neutrino oscillations in cosmology by reminding you that the full quantum kinetic equations of Ref. 5 contain terms which have been neglected in all analysis so far. In the polarization vector picture the full equations are

$$\frac{\partial P_0(y)}{\partial t} = R_\ell(y) + R_{\ell'}(y) \tag{32}$$

$$\frac{\partial \mathbf{P}(y)}{\partial t} = \mathbf{V}(y) \times \mathbf{P}(y) + [R_\ell(y) - R_{\ell'}(y)]\,\hat{z} - D(k)\mathbf{P}_T(k) + \int dy' d(y,y')\mathbf{P}_T(y') - C(y)P_0(y) + \int dy' c(y,y')P_0(y') \tag{33}$$

where the various coefficients are given in detail in Ref. 5. The density matrix $\rho(y)$ is diagonal in the momentum y of the neutrinos, and describes the mixing of the species ℓ and ℓ'. There is of course a similar equation for $\bar{\rho}$, the density matrix of the mixed antineutrinos.

It suffices to comment on the differences between the complete quantum kinetic equation, and the approximate forms which have been used to date. Firstly, the relaxation term which appears in the equation for P_0 differs from the relaxation term in the equation for $\mathbf{P}$. However this term is further approximated in all of the present calculations, and represented in the relaxation time approximation. If the relaxation times of the two species differ significantly, one will dominate and the relaxation term in the two equations will be approximately the same.

Next notice the damping terms which are non-diagonal in the momenta. The evolution of the scalar and vector components of the density matrix are linked, firstly by the terms in C and c which appear in the equation for $\dot{\mathbf{P}}$, and they are also coupled through the appearance of $n_\ell = (P_0 + P_z)/2, n_{\ell'} = (P_0 - P_z)/2$, and $\mathbf{P}_T$ in the various coefficients. Indeed $n_{\bar{\ell}}, n_{\bar{\ell}'}$, and $\bar{\mathbf{P}}_T^*$ also appear in the coefficients, linking the equations for neutrinos and antineutrinos. ($\mathbf{P}_T^* = P_x\hat{x} - P_y\hat{y}$ is the reflection of $\mathbf{P}_T$ in the $x - z$ plane.)

The antineutrino term features prominently in $C(y)$, which in an approximation similar to the relaxation time approximation can be written as $C(y) \approx \mathcal{C}\mathbf{P}_T^*$, creating a damping-like term which however is not necessarily "damping" in the sense of giving a negative time derivative. Moreover this term gives an explicit coupling of the equations for $\mathbf{P}$ and $\bar{\mathbf{P}}$.

This additional coupling term should be examined in more detail in future work, now that we know that neutrino oscillations are important in cosmology. After all, we have found so much interesting physics already that it would be a good idea to study the equations in their full detail!

References

1. B. H. J. McKellar and H. Granek, in *Proceedings of the Workshop on Neutrino Mass* (V. Barger and D. Cline, eds), *University of Wisconsin Report No 186*, 1980, pp. 165-167.
2. B. H. J. McKellar and H. Granek, in *Neutrino Mass and Gauge Theories of Weak Interactions* (ed. V. Barger and D. Cline, American Institute of Physics 1983) pp 91 - 99.
3. H. Granek and B.H.J. McKellar, *Aust. J. Phys.* **44**, 591 (1991).
4. M.J. Thomson and B.H.J. McKellar, *Phys. Lett.* **B259**, 113 (1991).
5. M.J. Thomson and B.H.J. McKellar, University of Melbourne Report No UM-P 89-108 (1989); B. H. J. McKellar and M. J. Thomson, in *In celebration of*

of the discovery of the neutrino (C. E. Lane and R. I. Steinberg, eds) *World Scientific, Singapore* 1993, p 169; M.J. Thomson and B.H.J. McKellar, *Phys. Rev.* **D49**, 2710 (1994).
6. R. Foot, M. J. Thomson and R. R. Volkas, *Phys. Rev.* **D53**, 5349 (1996).
7. R. Foot and R. R. Volkas, *Phys. Rev* **D55**, 5147 (1997)
8. N. F. Bell, R. R. Volkas and Y. Y. Wong, *Phys. Rev.* **D59**, 113001 (1999).
9. Recent reviews of this work, with further references, have been given in R. R. Volkas, *Prog. Part. Nucl. Phys.* **48**, 161 (2002) and R. R. Volkas *Nucl. Phys. Proc. Suppl.* **91**, 487 (2000).
10. A. D. Dolgov, arXiv:hep-ph/0202122 (2002).
11. K. A. Olive, E. Skillman and G. Steigman, arXiv:astro-ph/9611166.
12. A. Peimbert, M. Peimbert and V. Luridiana, arXiv:astro-ph/0107189.
13. R. Gruenwald, G. Steigman and S. M. Viegas, arXiv:astro-ph/0109071.
14. T. X. Thuan and Yu. I. Izotov, arXiv:astro-ph/0003234.
15. S. H. Hansen, G. Mangano, A. Melchiorri, G. Miele and O. Pisanti, *Phys. Rev.* **D65**, 023511 (2002).
16. C. Lunardini and A. Yu. Smirnov, *Phys. Rev.* **D64**, 073006 (2001).
17. A. D. Dolgov, S. H. Hansen, S. Pastor, S. T. Petcov, G. G. Raffelt, and D. V. Semikoz, arXiv:hep-ph/0201287 (2002).

Supernova Neutrinos and their Implications for Neutrino Parameters

Katsuhko Sato

SUPERNOVA NEUTRINOS AND THEIR IMPLICATIONS FOR NEUTRINO PARAMETERS

K. SATO

Department of Physics, School of Science, the University of Tokyo, 113-0033 Hongo, Bunkyo-ku, Tokyo, Japan

Research Center for the Early Universe, School of Science, the University of Tokyo, 113-0033 Hongo, Bunkyo-ku, Tokyo, Japan

K. TAKAHASHI AND S. ANDO

Department of Physics, School of Science, the University of Tokyo, 113-0033 Hongo, Bunkyo-ku, Tokyo, Japan

Core-collapse driven supernovae are the most luminous neutrino source in the universe. If neutrinos have finite masses and convert each other, the time profile and energy spectrum of the neutrino burst from supernovae are greatly modified. We review the neutrino conversion in a supernova mantle, and how the burst will be detected by SK (Super-Kamiokande) and SNO (Sudbury Neutrino Observatory) if a supernova appear at Galactic Center. We show that various neutrino oscillation models can be discriminated by the combination of the SK and SNO detections of a future galactic Supernova. We also discuss effects of neutrino oscillation on the supernova relic neutrino observations.

1. Introduction

Recently SK (Super-Kamiokande) collaboration showed that neutrinos have finite masses and oscillate each other from the observations of solar neutrinos[1] and atmospheric neutrinos[2]. More recently SNO (Sudbury Neutrino Observatory) collaboration confirmed the neutrino oscillation by combining the SK data with SNO data,[3] and clearly showed that electron type neutrinos are converted into muon type or tau type neutrinos from the solar neutrino observation by detecting neutral current events.[4] Although the neutrino oscillation is confirmed, still the values of the mass squared differences and mixing angles are not firmly established. For the observed ν_e suppression of solar neutrinos, four solutions were proposed : large mixing angle (LMA), small mixing angle (SMA), low Δm^2 (LOW), and vacuum oscillation (VO). Although SMA solution has been almost ruled out and

LMA solution is the most probable model, the other models have not yet been ruled out. For θ_{13}, the mixing angle between mass eigenstate ν_1, ν_3, only upper bound is known from reactor experiment[5] and combined three generation analysis[6]. Also the nature of neutrino mass hierarchy (normal or inverted) is still a matter of controversy.

In 1987, Supernova 1987A appeared in Large Magellanic cloud. Huge water Čherenkov counters embedded deeply in the Earth, Kamiokande[7] and IMB[8], detected the neutrino burst from the SN1987A. Core-collapse driven supernovae are the most powerful neutrino source in the universe. It is very natural to consider to use the neutrino burst from supernova as a tool to probe neutrino properties. Fortunately huge water Čherenkov counters, SK and SNO are running at present. If a supernova appears at Galactic Center, more than 10 thousand events will be detected by SK, and a few hundreds events by SNO, as discussed in the following sections.

There have been some studies on future supernova neutrino detection taking neutrino oscillation into account. Dighe and Smirnov[9] estimated qualitatively the effects of neutrino oscillation in a collapse-driven supernova on the neutronization peak, the distortion of energy spectra, and the Earth matter effects. They concluded that it is possible to identify the solar neutrino solution and to probe the mixing angle θ_{13}. Dutta et al.[10] showed numerically that the events involving oxygen targets increase dramatically when there is neutrino mixing. Takahashi et al.[11,12,13,14] investigated the effects of three flavor oscillation on the supernova neutrino spectra, taking into account the constraints on the neutrino mixing and masses imposed by solutions consistent with the solar and atmospheric neutrino problems in detail. They proposed a method to discriminate quantitatively the solutions of the solar neutrino problem.

In Section 2, we briefly describe the features of supernova neutrinos and the neutrino conversion on their way out to the surface of the star. Then, we show neutrino energy spectra at detectors and time evolution of neutrino number luminosity in Section 3, and discuss the discrimination of the oscillation models using these result in Section 4. The earth effect is discussed in Section 5.

Detection of the cosmological neutrino background originated by past supernovae is one of the interesting target for neutrino astronomy. Detectability of this supernova relic neutrinos have been investigated by many people. Recently Ando et al.[15] investigated this problem in detail taking into account the neutrino oscillation, which is shown in Section, 6.

2. Supernova Neutrinos and Conversion Probabilities

Supernova neutrino emission process can be divided into two distinct phases;[16] the neutronization burst and nearly thermal neutrino emission. Almost all of the binding energy,

$$E_b \simeq \frac{GM_{\rm NS}^2}{R_{\rm NS}} \simeq 1.5 \sim 4.5 \times 10^{53} \rm erg, \tag{1}$$

is radiated away as neutrinos, while a small fraction of which ($\sim 2 \times 10^{51}$ erg) are emitted during the first phase. While only ν_e is emitted during the first phase, neutrinos and anti-neutrinos of all types are emitted during the second phase with roughly the same luminosity. The average energies are different between flavors:

$$\langle E_{\nu_e} \rangle \simeq 13 \rm MeV \tag{2}$$

$$\langle E_{\bar{\nu}_e} \rangle \simeq 16 \rm MeV \tag{3}$$

$$\langle E_{\nu_x} \rangle \simeq 23 \rm MeV, \tag{4}$$

where ν_x means either of ν_μ, ν_τ, and their anti-neutrinos. Time-integrated energy spectra and time evolution of neutrino flux based on a realistic model of a collapse-driven supernova by the Lawrence Livermore group[17] are shown in Fig. 1 and Fig. 2. (See Totani et al.[18] for detail.)

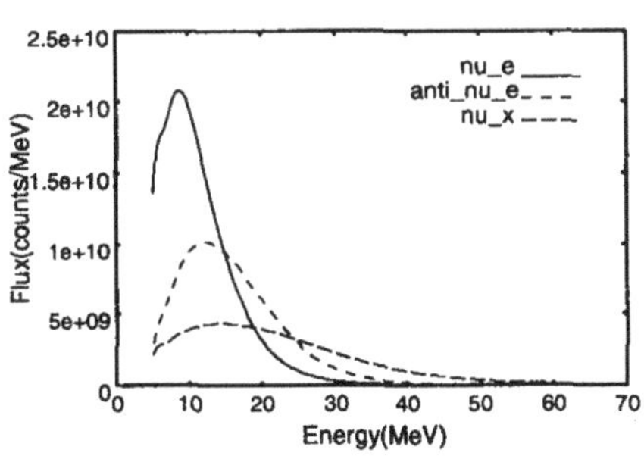

Figure 1. Original energy spectra of neutrinos. Solid, dashed, and long-dashed lines correspond to ν_e, $\bar{\nu}_e$, and ν_x, respectively.

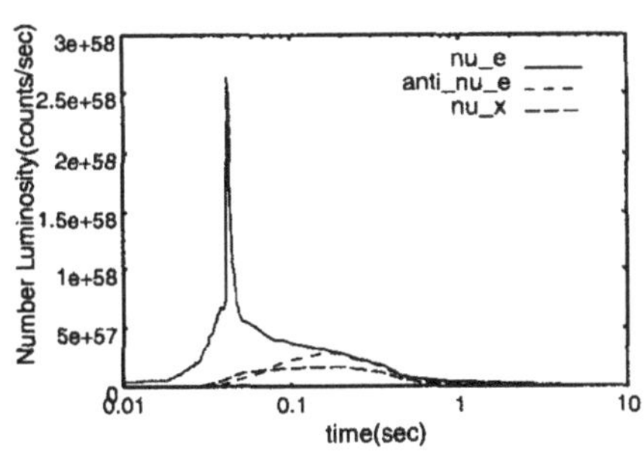

Figure 2. Time evolution of the original neutrino number luminosity. Solid, dashed, and long-dashed lines correspond to ν_e, $\bar{\nu}_e$, and ν_x, respectively.

These neutrinos, which are produced in the high dense region of the iron core, interact with matter before emerging from the supernova. The presence of non-zero masses and mixing in vacuum among various neutrino flavors results in strong matter dependent effects, including conversion from one flavor to another. In supernova, the conversions occur mainly in the resonance layers. The resonance matter density can be written as

$$\rho_{res} \sim 1.4 \times 10^6 (\frac{\Delta m^2}{1\mathrm{eV}^2})(\frac{10\mathrm{MeV}}{E})(\frac{0.5}{Y_e}) \cos\theta \ \mathrm{g/cc}, \tag{5}$$

whereΔm^2 is the mass squared difference, E is the neutrino energy, and Y_e is the mean number of electrons per baryon. In normal mass hierarchy scheme ($m_3 > m_2 > m_1$), the system has two resonances in neutrino sector: one at higher density(H-resonance) and the other at lower density(L-resonance). On the other hand, anti-neutrino sector has no resonance.

The dynamics of conversions in each resonance is determined by the adiabaticity parameter γ,

$$\gamma \equiv \frac{\Delta m^2}{2E} \frac{\sin^2 2\theta}{\cos 2\theta} \frac{n_e}{dn_e/dr}, \tag{6}$$

where θ is mixing angle, and n_e is the electron number density. The flip probability P_f, the probability that a neutrino in one matter eigenstate jumps to the other matter eigenstate, is,

$$P_f = \exp(-\frac{\pi}{2}\gamma), \tag{7}$$

as given by the Landau-Zener formula.[19] Adiabatic resonance corresponds to $\gamma \gg 1$. Note that adiabaticity of resonance depends on the mixing angle and the squared mass difference, that is,[9]

$$\text{H resonance} \longrightarrow \theta_{13}, \Delta m^2_{13}, \tag{8}$$

$$\text{L resonance} \longrightarrow \theta_{12}, \Delta m^2_{12}. \tag{9}$$

The neutrino spectra observed at the detectors can be dramatically different from the original spectra according to the adiabaticity of these two resonances. If neutrino oscillation occurs, for example, between ν_e and ν_x, observed energy spectrum will be a mixture of original ν_e and ν_x spectra and the average energy of ν_e will be higher than the original ν_e average energy.

Takahashi et al.[11,12,13,14] used the massive star density profile calculated by Woosley and Weaver[20] to calculate the time evolution of neutrino wave functions. The progenitor mass was set to be $15M_\odot$, and the metallicity was set to be the same as that of Sun.

Further, multiplying these probabilities by original neutrino luminosity, neutrino luminosity at the Earth can be obtained. Takahashi et al.[11,12,13,14] used a realistic model of a collapse-driven supernova by the Lawrence Livermore group[17] shown in Fig. 1 and Fig. 2 to calculate the neutrino luminosity and energy spectrum including neutrino oscillation effects.

Table 1. Sets of mixing parameter for calculation

model	$\sin^2 2\theta_{12}$	$\sin^2 2\theta_{23}$	$\sin^2 2\theta_{13}$	$\Delta m^2_{12}(\mathrm{eV}^2)$	$\Delta m^2_{13}(\mathrm{eV}^2)$	$\nu_\odot$ problem
LMA-L	0.87	1.0	0.043	7.0×10^{-5}	3.2×10^{-3}	LMA
LMA-S	0.87	1.0	1.0×10^{-6}	7.0×10^{-5}	3.2×10^{-3}	LMA
SMA-L	5.0×10^{-3}	1.0	0.043	6.0×10^{-6}	3.2×10^{-3}	SMA
SMA-S	5.0×10^{-3}	1.0	1.0×10^{-6}	6.0×10^{-6}	3.2×10^{-3}	SMA

In table 1, the sets of mixing parameters are shown . Here θ_{12} and Δm^2_{12} correspond to the solutions of solar neutrino problem and θ_{23} and Δm^2_{13} correspond to the solution of atmospheric neutrino problem. The value of θ_{13} is taken to be consistent with current upper bound from reactor experiment[5]. These models are named after their values of mixing angle: LMA-L means that θ_{12} is set to be LMA of solar neutrino problem and θ_{13} is large.

3. Expected Event Rates in SuperKamiokande and SNO

3.1. *Event Rates at SuperKamiokande*

SuperKamiokande is a water Čherenkov detector with 32,000 ton pure water based at Kamioka in Japan. The relevant interactions of neutrinos with water[21] are as follows:

$$\bar{\nu_e} + p \rightarrow n + e^+ \quad \text{(CC)} \tag{10}$$

$$\nu_e + e^- \rightarrow \nu_e + e^- \quad \text{(CC and NC)} \tag{11}$$

$$\bar{\nu_e} + e^- \rightarrow \bar{\nu_e} + e^- \quad \text{(CC and NC)} \tag{12}$$

$$\nu_x + e^- \rightarrow \nu_x + e^- \quad \text{(NC)} \tag{13}$$

$$\nu_e + O \rightarrow F + e^- \quad \text{(CC)} \tag{14}$$

$$\bar{\nu_e} + O \rightarrow N + e^+ \quad \text{(CC)} \tag{15}$$

where CC and NC stand for charged current and neutral current interactions, respectively. The lower limit of detection is $\sim$ 5MeV, and the energy resolution is $\sim$ 15% for an electron with energy 10MeV. In these interactions, the $\bar{\nu_e}p$ CC interaction [Eq.(10)] has the largest contribution to the detected events at SK. Hence the energy spectrum detected at SK (including all the reactions) is almost the same as the spectrum derived from the interaction Eq.(10) only.

Fig. 3 and Fig. 4 show energy spectrum and time evolution of number luminosity of positrons and electrons expected to be detected at SuperKamiokande, respectively. The distance is assumed as d= 10kpc.

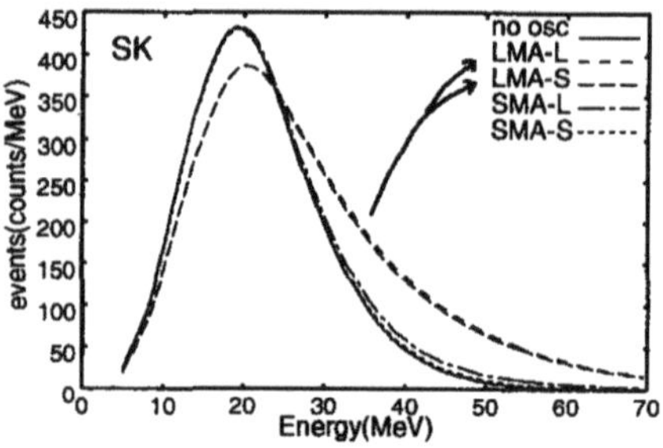

Figure 3. Energy spectrum of positrons and electrons expected to be detected at SuperKamiokande. Solid, dashed, long-dashed, dash-dot-dash, and dotted lines correspond to no oscillation, model LMA-L, LMA-S, SMA-L, and SMA-S, respectively.

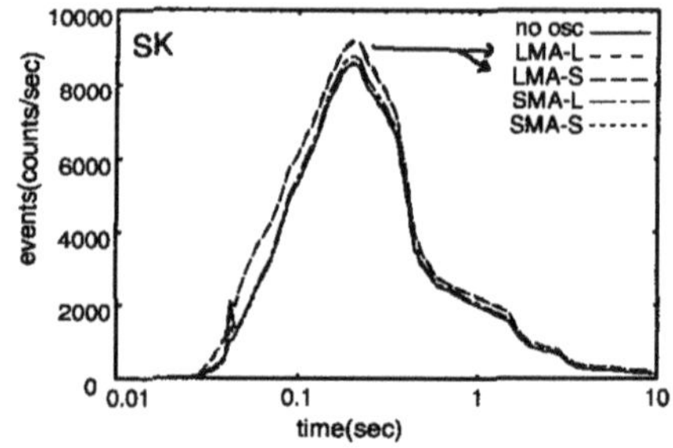

Figure 4. Time evolution of number luminosity of positrons and electrons expected to be detected at SuperKamiokande. Solid, dashed, long-dashed, dash-dot-dash, and dotted lines correspond to no oscillation, model LMA-L, LMA-S, SMA-L, and SMA-S, respectively.

3.2. *Event Rates in SNO*

Sudbury Neutrino Observatory(SNO) is a water Čherenkov detector based at Sudbury, Ontario. SNO is unique in its use of 1000 tons of heavy water, by which both the charged-current and neutral-current interactions can be detected. The interactions of neutrinos with heavy water[22] are as follows,

$$\nu_e + d \rightarrow p + p + e^- \quad \text{(CC)} \tag{16}$$

$$\bar{\nu_e} + d \rightarrow n + n + e^+ \quad \text{(CC)} \tag{17}$$

$$\nu_x + d \rightarrow n + p + \nu_x \quad \text{(NC)} \tag{18}$$

$$\bar{\nu_x} + d \rightarrow n + p + \bar{\nu_x} \quad \text{(NC)} \tag{19}$$

The two interactions written in Eqs.(16) and (17) are detected when electrons emit Čherenkov light. These reactions produce electrons and positrons whose energies sensitive to the neutrino energy, and hence the energy spectra of electrons and positrons give us the information on the original neutrino flux.

Two neutral current interactions, which produce neutrons, are detected by observing the photons emitted at the neutron absorption. Photons give energy to electrons, then the Čherenkov light from the electrons is detected. Moreover, there is a possibility to distinguish the two CC interactions by detecting neutrons because the detection of the neutron and the positron at the same time indicates the interaction in Eq.(17).

Fig. 5 and Fig. 6 show energy spectrum and time evolution of number luminosity of positrons and electrons, produced by the two CC interactions expected at SNO, respectively.

The SNO detector has also 7,000 tons of light water which can be used to detect neutrinos. This can be considered to be a miniature of SuperKamiokande (32,000 tons of light water). Then the number of events detected by light water at SNO is 7/32 of that at SuperKamiokande.

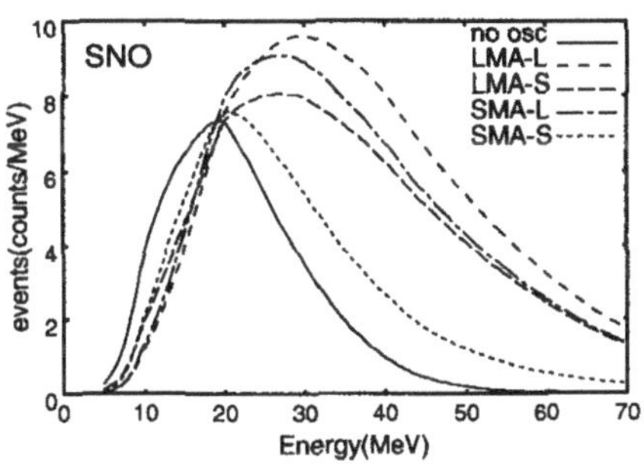

Figure 5. Energy spectrum of positrons and electrons expected to be detected at SNO taking only CC events into account. Solid, dashed, long-dashed, dash-dot-dash, and dotted lines correspond to no oscillation, model LMA-L, LMA-S, SMA-L, and SMA-S, respectively.

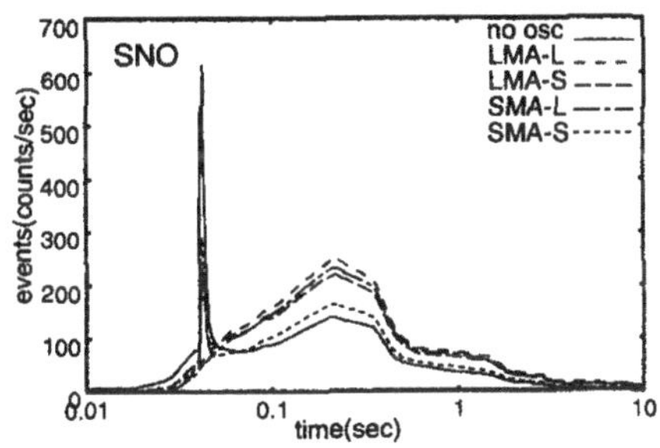

Figure 6. Time evolution of number luminosity of positrons and electrons expected to be detected at SNO taking only CC events into account. Solid, dashed, long-dashed, dash-dot-dash, and dotted lines correspond to no oscillation, model LMA-L, LMA-S, SMA-L, and SMA-S, respectively.

4. Discrimination of the Oscillation Models

As can be seen in Fig. 3 and 5, when there is neutrino oscillation, neutrino spectra are harder than those in absence of neutrino oscillation. This is because average energies of ν_e and $\bar{\nu}_e$ are smaller than those of ν_x and neutrino oscillation produces high energy ν_e and $\bar{\nu}_e$ which was originally ν_x.

It is worth noting that number of events during neutronization burst phase is highly suppressed in model LMA-L and SMA-L. This is because, due to large value of θ_{13} in these two models, H resonance occurs adiabatically and ν_e produced at the center of supernova is detected as ν_x which has small cross section. But the number of events during neutronization burst will be too small to extract statistically significant information.

It is possible that He and H layers of progenitor star are missing when supernova burst occurs, and density decreases abruptly outside the O+C layer. Then L resonance would occur nonadiabatically to some extent even in case of LMA, and differences between LMA and SMA would become smaller.

As mentioned in the previous section, neutrino oscillation makes ν_e and $\bar{\nu}_e$ spectra harder. Therefore, the ratio of high-energy events to low-energy events will be a good measure of neutrino oscillation effects. We calculated the following ratio of events at both detectors:

$$R_{SK} \equiv \frac{\text{number of events at } 30 < E < 70\text{MeV}}{\text{number of events at } 5 < E < 20\text{MeV}} \tag{20}$$

$$R_{SNO} \equiv \frac{\text{number of events at } 25 < E < 70\text{MeV}}{\text{number of events at } 5 < E < 20\text{MeV}} \tag{21}$$

The plots of R_{SK} vs. R_{SNO} are shown in Fig. 7. The errorbars include only statistical errors. At first glance, it seems to be possible to distinguish all the models including the no oscillation case. But there are other ambiguities besides statistical errors.

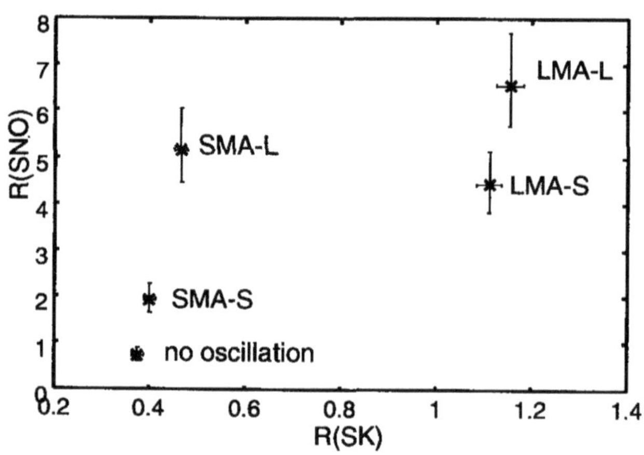

Figure 7. The plot of R_{SK} vs. R_{SNO} for all the models. The error-bars represent the statistical errors.

One is the mass of the progenitor star. Supernovae with different progenitor masses may result in different original neutrino spectra and neutrino oscillation effects. Studies on this point are now in progress. But dependence of shape of neutrino spectra on progenitor mass is not so large [23] and we would be able to distinguish the models. The difference among the following three groups will still be clear: (1)LMA-L and LMA-S, (2)SMA-L, and (3)SMA-S and no oscillation.

Another ambiguity is the effect of stellar lotation. Massive stars, which would be the progenitor of collapse-driven supernovae, are rapid rotators. It is natural to consider that proto neutron stars are oblate. Recently Shimizu et al.[24] investigated jet like explosion induced by asymmetric neutrino radiation from oblate proto neutron stars. In order to obtain realistic neutrino burst model, further detailed investigations are necessary.

5. Earth Effects on Supernova Neutrinos and their Implications for Neutrino Parameters

The Earth effects have been studied by several authors. Takahashi and Sato[13] analyzed numerically the time-integrated energy spectra in a simple case and found the possibility to probe the mixing angle θ_{13}. Lunardini and Smirnov[25] performed a study of the Earth effects taking the arrival directions of neutrinos into account and showed that studies of the Earth effects will select the solution of the solar neutrino problem, probe the mixing U_{e3} and identify the hierarchy of the neutrino mass spectrum.

Takahashi and Sato[14] performed a detailed study of the Earth matter effects on supernova neutrinos and show that the detection of Earth matter effects allows us to probe Δm_{12}^2 more accurately than by solar neutrino observations.

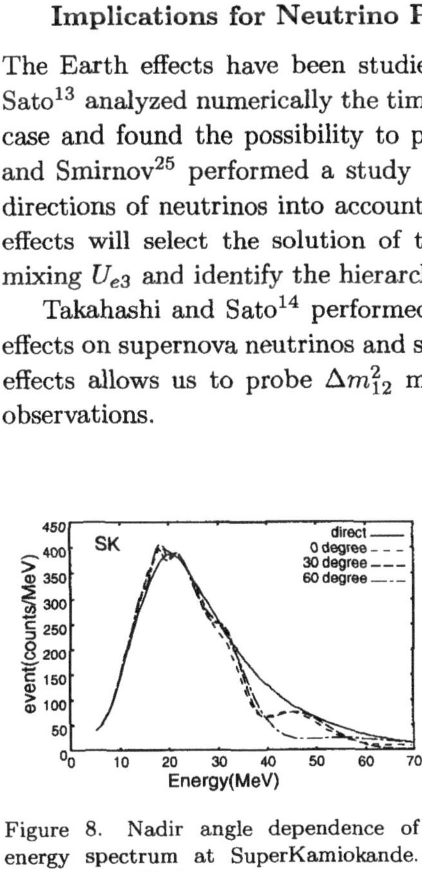

Figure 8. Nadir angle dependence of energy spectrum at SuperKamiokande. Neutrino oscillation parameters are set to the values in model LMA-S except $\Delta m_{12}^2 = 2 \times 10^{-5}\mathrm{eV}^2$.

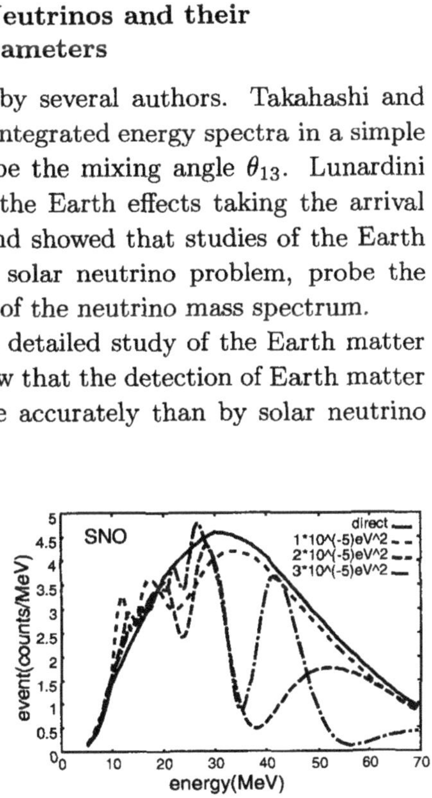

Figure 9. Δm_{12}^2 dependence of energy spectrum at SNO taking only CC events into account. Neutrino oscillation parameters are set to the values in model LMA-S except Δm_{12}^2. Nadir angle is set to 0 degree.

Fig. 8 shows the nadir angle dependence of neutrino spectrum at SK, SNO and LVD, respectively. Neutrino oscillation parameters are set to the values in model LMA-S except $\Delta m_{12}^2 = 2 \times 10^{-5}\mathrm{eV}^2$. Since the oscillation lengths are comparable to the radius of the Earth, the Earth effects are strongly dependent on the nadir angle.

Fig. 9 shows the Δm_{12}^2 dependence of energy spectrum at SNO. Neutrino oscillation parameters are set to the values in model LMA-S except Δm_{12}^2. Nadir angle is set to 0 degree. The larger Δm_{12}^2 results in higher-frequency oscillation in the energy spectra with respect to the energy. This is because the neutrino oscillation length is proportional to the inverse of Δm^2.

As can be seen, there is significant dependence of the Earth effect on Δm_{12}^2. Making use of this dependence, we can probe Δm_{12}^2 by the observations of the supernova neutrino spectra more accurately than by the observations of solar neutrino, if neutrino oscillation parameters are as in model LMA-S.

The direction of the supernova is given by direct optical observations or by the experimental study of the neutrino scattering on electrons,[26,27] which is discussed in the following section. Comparing the observed neutrino spectra to the predicted spectra of various values of Δm_{12}^2, we can determine the value of Δm_{12}^2.

6. Supernova Relic Neutrinos and Neutrino Oscillation

It is generally believed that the core-collapse supernova explosions have traced the star formation history in the universe and have emitted a great number of neutrinos, which should make a diffuse background. This supernova relic neutrino (SRN) background is one of the targets of the currently working large neutrino detectors, SK and SNO. Comparing the predicted SRN spectrum with the observations by these detectors provides us potentially valuable information on the nature of neutrinos as well as the star formation history in the universe. This SRN background has been discussed in a number of previous papers.[28–37] The work after Totani et al.[34] takes into account the realistic star formation history inferred from various observations and theoretical modeling of galaxy formation, to calculate the SRN flux. Totani et al.[34] calculated the energy flux of SRN at SK detector and compared it with neutrinos emitted by other sources (solar, atmospheric, and reactor neutrinos and so on). Then they concluded that the visible event rate of SRN at SK is 1.2 yr^{-1} in the energy range from 15 to 40 MeV. On the other hand, Kaplinghat et al.[37] calculated the upper limit of SRN and discussed the possibility of the detection at SK detector comparing to the other backgrounds. Their result is that at the energy range from 15 to 40 MeV, where no significant background had been considered to exist in previous studies, there is a huge background of the invisible muon decay and the detection of SRN is difficult. They also discussed the effects of neutrino oscillation and briefly mentioned that *if* the SRN flux is in the vicinity of their upper bound and all three flavors are maximally mixed, it may be detectable as a distortion of the expected muon background.

Ando et al.[15] investigated the SRN flux and the event rate at SK for various neutrino oscillation models taking into account of a realistic neutrino spectrum[18] (Fig. 1). If neutrino oscillation occurs, $\bar{\nu}_{\mu,\tau}$'s are converted into $\bar{\nu}_e$'s which are mainly detected at SK detector. Because $\bar{\nu}_{\mu,\tau}$'s interact with matter only through the neutral-current reactions in supernovae, they are weakly coupled with matter compared to $\bar{\nu}_e$. Thus the neutrino sphere of $\bar{\nu}_{\mu,\tau}$'s is deeper in the core than $\bar{\nu}_e$'s and their temperatures are higher than $\bar{\nu}_e$'s. Therefore neutrino oscillation enhances the mean $\bar{\nu}_e$'s energy and enhances event rate at SK detector.

In Fig. 10, the event rate of various neutrino oscillation models is shown assuming the supernova rate model estimated by observations of star formation rate.[38] Clear difference of the LMA model from SMA or no oscillation models can be seen, especially at the high energy tail. This property results from the flux dependence on oscillation models. As a result, if we can detect SRN events above ~ 10 MeV, we can discriminate the LMA model from the SMA or no oscillation models in any supernova rate models and any sets of cosmological parameters.

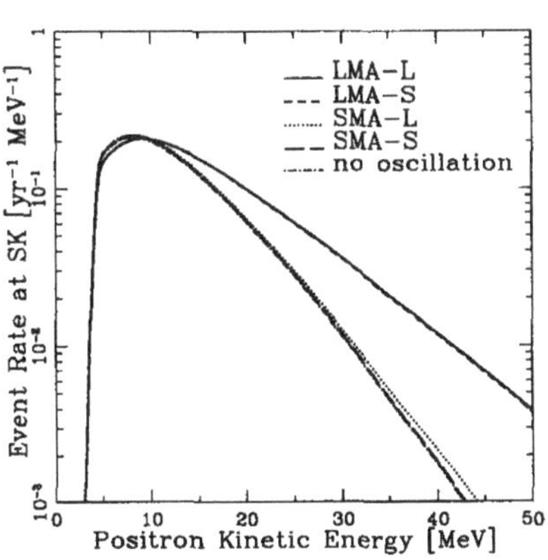

Figure 10. Event rate of $\bar{\nu}_e$'s at SK for the neutrino oscillation models.

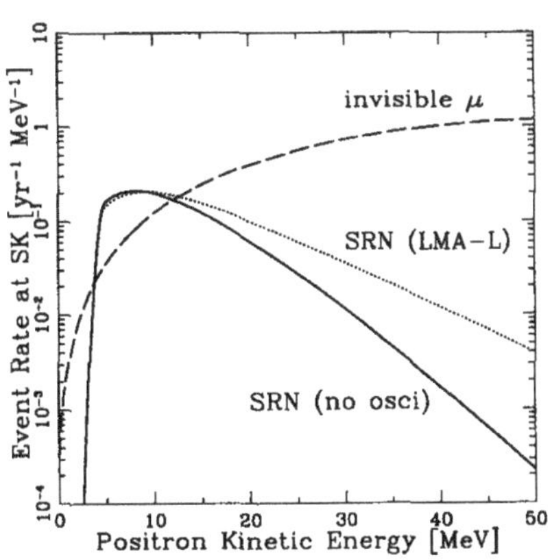

Figure 11. Event rate at SK detector of SRN and invisible μ decay products. Two oscillation models are shown (no oscillation and LMA-L)

There are several background events which hinder the detection of SRN. These includes atmospheric and solar neutrinos, anti-neutrinos from nuclear reactors, spallation products, and decay electrons from invisible muons. We should find the energy region which is not contaminated by these background events and then calculate the detectable event rate of SRN.

By careful examination of these events, we found that there is a narrow energy window from 10 MeV to 27 MeV, which is free from solar, atmospheric, and reactor neutrinos. (Since the solar neutrino events are highly correlated to the solar direction, it can be avoided by directional analysis.)

However, according to Kaplinghat et al.[37] electrons or positrons from invisible muons are the largest background in the energy window from 19 to 35 MeV. This invisible muon event is illustrated as follows. The atmospheric neutrinos produce muons by interaction with the nucleons (both free and bound) in the fiducial volume. If these muons are produced with energies below Čherenkov radiation threshold (kinetic energy less than 53 MeV), then they will not be detected ("invisible muons"), but their decay-produced electrons and positrons will be. Since the muon decay signal will mimic the $\bar{\nu}_e p \to n e^+$ process in SK, it is difficult to distinguish SRN from these events. In Fig. 11, SRN event rate is compared with the invisible muon events. From this figure it is shown that SRN events can be seen only below about 12 MeV.

In practice, there is another serious background, i.e., spallation products induced by cosmic ray muons. Ultra high energy cosmic ray muons spall oxygens in the detector, and radioactive decay processes of these spalled nuclei occur. The event rate of the spallation background is several hundred per day per 22.5 kton. Although most of them can be rejected by the information of preceding muons, even a small fraction can not be. (Roughly, this spallation products produces about 200,000 events per a year. Because expected SRN event rate is less than 1 per a year, we should reject all these 200,000 events. For future detectors this problem is also quite difficult to solve.) Thus, this makes a serious background at the energy range below the maximum energy of beta spectrum of spallation products, 16 MeV.[39,40]

It looks there is no energy window of SRN, but we can detect the SRN events by subtracting the other background events from total detected events. Consider the energy range $17 < (T_e/\text{MeV}) < 25$, where T_e is positron kinetic energy. This range corresponds to $19 < (E_{\bar{\nu}_e}/\text{MeV}) < 27$ by the simple relation, $E_{\bar{\nu}_e} = T_e + 1.8\text{MeV}$. There are two advantages in using this energy region. First, SRN event rate is rather large, and second, the background (invisible muon) event rate is fairly well known by SK observation. The SRN event rate at SK in this energy range, i.e., $0.4 - 0.8\ \text{yr}^{-1}$. In contrast, the event rate of the invisible muon over the same energy range is $3.4\ \text{yr}^{-1}$. When SRN event rate is larger than the statistical error of background event rate, we can conclude that the SRN is detectable as a distortion of the expected invisible muon background event. Unfortunately, only one year observation does not provide any useful information about SRN. However, we can expect that ten-year observation provides several statistically meaningful results. The statistical error of invisible muon events in ten years is $\sqrt{34} = 5.8$, which is smaller than the event rate of LMA models and is larger than that of SMA and no oscillation models. Then these neutrino oscillation models can be distinguished by the observation of the event rate of invisible muon events. (If there is a discrepancy from expected event rate, this is due to SRN events and LMA models are favored.)

In future, it is expected that next generation of water Čherenkov detectors have much larger volume than that of SK. For example Hyper-Kamiokande project is now under consideration. HyperKamiokande detector is planed to be a water Čherenkov detector whose mass is about 1,000,000 tons (about 20 times larger than SK), and its location is near SK detector. We expect that the SRN event becomes about 10 per one year for this detector, and statistically sufficient discussion of SRN is possible even using only one year data.

Acknowledgments

We would like to thank Super Kamiokande people including Y. Totsuka, Y. Suzuki,A. Suzuki, M. Nakahata, and Y. Fukuda for useful discussions and comments. This work was supported in part by grants-in-aid for scientific research provided by the Ministry of Education, Science and Culture of Japan through Research grant no. S14102004 and 14079202.

References

1. S. Fukuda et al., hep-ex/0103033. *Phys. Rev. Lett.* **86**, 5656 (2001).
2. Y. Fukuda et al., *Phys. Rev. Lett.* **82**, 2644 (1999); K. Scholberg, hep-ex/9905016.
3. Q.R. Ahmad et al., SNO Collaboration, *Phys. Rev. Lett.* **87**, 071301 (2001).
4. Q.R. Ahmad et al., SNO Collaboration, *Phys. Rev. Lett.* **89**, 011301 (2002).
5. M. Apollonio et al., *Phys. Lett.* **B466**, 415 (1999).
6. G. L. Fogli et al., hep-ph/0104221; O. Yasuda and H. Minakata, hep-ph/9602386; H. Minakata and O. Yasuda, *Phys. Rev.* **D56**, 1692 (1997).
7. K. Hirata et al., *Phys. Rev. Lett.* **58**, 1490 (1987).
8. R. M. Bionta et al., *Phys. Rev. Lett.* **58**, 1494 (1987).
9. A. S. Dighe, A. Yu. Smirnov, *Phys. Rev.* **D62**, 033007 (2000).
10. G. Dutta, D. Indumathi, M. V. N. Murthy, and G. Rajasekaran, *Phys. Rev.* **D61**, 013009 (1999).

11. K. Takahashi, M. Watanabe, K. Sato and T. Totani, *Phys. Rev.* **D64**, 093004 (2001).
12. K. Takahashi, M. Watanabe and K. Sato, *Phys. Lett.* **B510**, 189 (2001).
13. K. Takahashi and K. Sato, *Phys. Rev.* **D66**, 033006 (2002). (hep-ph/0110105)
14. K. Takahashi and K. Sato, (hep-ph/0205070)
15. S. Ando, K. Sato, and T. Totani, *Astropart. Phys.* (2002), in printing. (astro-ph/0202450)
16. H. Suzuki: Supernova Neutrino in *Physics and Astrophysics of Neutrino*, edited by M. Fukugita and A. Suzuki (Springer-Verlag, Tokyo, 1994).
17. J. R. Wilson, R. Mayle, S. Woosley, T. Weaver, *Ann. NY Acad. Sci.* **470**, 267 (1986).
18. T. Totani, K. Sato, H.E. Dalhed, and J.R. Wilson, *Astrophys. J.* **496**, 216 (1998).
19. L. Landau, *Phys. Z. Sowjetunion* **2**, 46 (1932);
C. Zener, *Proc. R. Soc. London*, **Ser.A137**, 696 (1932).
20. S. E. Woosley and T. A. Weaver, *ApJ. Suppl.* **101**, 181 (1995).
21. Y. Totsuka, *Rep. Prog. Phys.* **55**, 377 (1992); K. Nakamura, T. Kajita and A. Suzuki, *Kamiokande*, in *Physics and Astrophysics of Neutrino*, edited by M. Fukugita and A. Suzuki (Springer-Verlag, Tokyo, 1994).
22. S. Ying, W.C. Haxton and E.M. Henley, *Phys. Rev.* **D40**, 3211 (1989).
23. R. Mayle, Ph.D. Thesis, University of California (1987); R. Mayle, J.R. Wilson, and D.N. Schramm, *Astrophys. J.* **318**, 288 (1987).
24. T.M. Shimizu, T. Ebisuzaki, K. Sato, and S. Yamada, *Astrophys. J.* **552**, 756 (2001).
25. C. Lunardini and A.Y. Smirnov, hep-ph/0106149.
26. J.F. Beacom and P. Vogel, *Phys. Rev.* **D60**, 033007 (1999).
27. *Prog. Theor. Phys.* **107**, 957 (2002).
28. G.S. Bisnovatyi-Kogan, S.F. Seidov, *Ann. N.Y. Acad. Sci.* **422**, 319 (1984).
29. L.M. Krauss, S.L. Glashow, D.N. Schramm, *Nature* **310**, 191 (1984).
30. S.E. Woosley, J.R. Wilson, and R. Mayle, *Astrophys. J.* **302**, 19 (1986).
31. K.S. Hirata (1991) PhD theses, University of Tokyo.
32. Y. Totsuka, *Rep. Prog. Phys.* **55**, 377 (1992).
33. T. Totani and K. Sato, *Astropart. Phys.* **3**, 367 (1995).
34. T. Totani, K. Sato, and Y. Yoshii, *Astrophys. J.* **460**, 303 (1996).
35. R.A. Malavey, *Astropart. Phys.* **7**, 125 (1997).
36. D.H. Hartmann and S.E. Woosley, *Astropart. Phys.* **7**, 137 (1997).
37. M. Kaplinghat, G. Steigman and T.P. Walker, *Phys. Rev.* **D62**, 043001 (2000).
38. C. Porciani and P. Madau, *Astrophys. J.* **548**, 522 (2001).
39. M. Nakahata, private communication.
40. Y. Fukuda, private communication.

Brane World Cosmology: From Superstring to Cosmic Strings

Henry Tye

BRANE WORLD COSMOLOGY: FROM SUPERSTRING TO COSMIC STRINGS

S.-H. HENRY TYE

Laboratory of Elementary Particle Physics, Cornell University, Ithaca, NY 14853

Brane inflation, where branes move towards each other, has been shown to be very robust. Inflation ends when branes collide and heat the universe. Towards the end of the brane inflationary epoch in the brane world, a complex tachyon field appears. As the tachyon rolls down its potential, cosmic strings (but not domain walls or monopoles) are copiously produced during brane collision. These cosmic strings are Dp-branes with $(p-1)$ dimensions compactified. Using the COBE data on the temperature anisotropy in the cosmic microwave background, the cosmic string tension μ is estimated to be around $G\mu \simeq 10^{-7}$. This in turn implies that the anisotropy in the cosmic microwave background comes mostly from inflation, but with a component (of order 10%) from cosmic strings. This can be tested by MAP and PLANCK. This cosmic string effect should also be observable in gravitational wave detectors like LIGO II/VIRGO and LISA.

1. Introduction

The cosmic microwave background (CMB) data[1,2] strongly supports the inflationary universe scenario[3] to be the explanation of the origin of the big bang. However, the origin of the inflaton and its potential is not well understood. Recently, the brane world scenario suggested by superstring theory was proposed, where the standard model of the strong and electroweak interactions are open string (brane) modes while the graviton and the radions are closed string (bulk) modes. In a generic brane world scenario, there are three types of light scalar modes : (1) bulk modes like radions (i.e., the sizes/shape of the compactified dimensions) and the dilaton (i.e., the coupling), (2) brane positions (or relative positions) and (3) tachyonic modes, the last are present in the presence of non-BPS branes or when branes are not BPS relative to each other. In general, the bulk modes have gravitational strength couplings (so too weak to reheat the universe) and so are not good inflaton candidates. Neither are the tachyonic modes, which roll down the potential too fast for inflation. This leaves the relative brane positions (i.e., brane separation) as candidates for inflation. So, natural in the brane world is the brane inflation scenario,[4] in which the inflaton is an open string mode identified with an inter-brane separation, while the inflaton potential emerges from the exchange of closed string modes between branes; the latter is the dual of the one-loop partition function of the open string spectrum, a property well-studied in string theory.[5] The potential is essentially dictated by their gravitational attractive (and the Ramond-Ramond) interaction. As the branes move towards each other, slow-roll exponential inflation takes place. This yields an almost scale-invariant power spectrum for the density perturbation, except there is a slight red-tilt (at a few percent level). As they reach a distance around the string scale, the inflaton potential becomes quite steep so that the slow-roll condition breaks down. At around the same time, a complex tachyon appears, so inflation ends rapidly as the tachyon rolls down its potential; that is, inflation ends when the branes collide, heating the universe that starts the big bang. This brane inflationary scenario may be realized in a number of ways.[6,7] The scenario is simplest when the radion and the dilaton (bulk) modes are assumed to be stabilized by some unknown non-perturbative bulk dynamics at the onset of inflation. Since the inflaton is a brane mode, and the inflaton potential is dictated by the brane mode spectrum, it is reasonable to assume that the inflaton potential is insensitive to the details of the bulk dynamics. In the presence of the branes that contain the standard model, coupling of the tachyon to inflaton and standard model fields allow efficient heating of the universe.[8] As a consequence, the overall brane inflationary picture is very robust. Because the inflaton and the ground state open string modes responsible for defect formation are different, and that the ground state open string modes become tachyonic and develop vacuum expectation values only towards the end of the inflationary epoch, various types of defects (lower-dimensional branes) may be formed. Apriori, defect production after inflation may be a serious problem. Fortunately, it is argued in Refs. 7 and 9 that, following the properties of superstring/brane theory and the cosmological evolution of the universe, the only defects copiously produced are cosmic strings. (Cosmic string properties in the early universe is a well-studied topic.[10]) Here, we elaborate on this point and consider some of its observational consequences:

- Sec. II : In superstring theory, Dp-branes come with either odd p (in IIB theory) or even p (in IIA theory). The collision of a Dp-brane with another Dp-brane at an angle (or with an anti- Dp-brane) yields D$(p-2)$-solitons (i.e., codimension 2). Topologically, a variety of defects may be produced. Because they have even

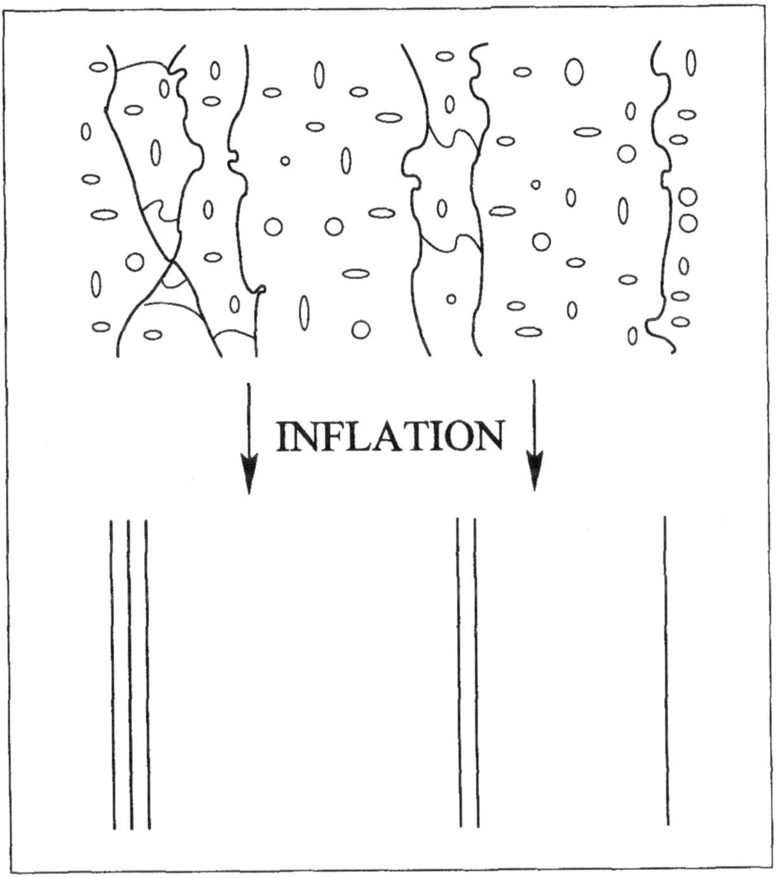

Figure 1. The branes may start out wrinkled and curved, with matter density and defects on them and in the bulk. They may even intersect each other in the uncompactified directions (only one of the 3 uncompactified dimensions is displayed). Generically, there is a density of strings stretched between branes as well. They look like massive particles to the 4-D observer. Branes with zero, one or two dimensions along the uncompactified directions are defects that look like point-like objects, cosmic strings and domain walls, respectively. Fortunately, inflation will smooth out the wrinkles, inflate away the curvature and the matter/defect/stretched-string densities, and red shift the intersections exponentially far away from any generic point on the brane, so one may consider the branes to be essentially parallel in the uncompactified directions, with empty branes and empty bulk.

codimensions with respect to the branes that collide, they have specific properties.[11]

- Sec. III : Cosmologically, since the compactified dimensions tangent to the brane is smaller than the Hubble size, the Kibble mech-

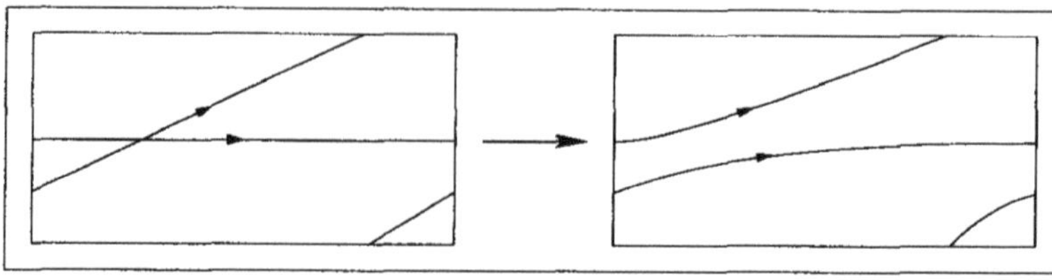

Figure 2. Branes wrapping a torus with sides $\ell_{\parallel}$ and $u\ell_{\parallel}$, where $u < 1$. As an illustration, the two branes (with winding numbers (1,1) and (1,0)) are at an angle θ, but they are separated in directions orthogonal to the torus (the separation being the inflaton). When that separation approaches zero, the two branes collide to become a single (2,1) brane.

anism works only if all the codimensions are tangent to the uncompactified dimensions. As a consequence, only cosmic strings may be copiously produced.[7,9]

- Sec. IV : The cosmic strings may be D1-branes, but most likely, they are Dp-branes wrapping around $(p-1)$-cycles in the compactified dimensions. In this case, the cosmic string tension $\mu = M_s^2/(4\alpha\pi) \simeq 2M_s^2$, where α is the gauge coupling, with $\alpha \simeq 1/25$. Using the density perturbation in the CMB data from COBE[1], the superstring scale is estimated to be $M_s \simeq 2 \times 10^{15}$ GeV.[7] With G being the Newton's constant, this yields $G\mu \simeq 10^{-7}$. This is the canonoical value that we shall use. Considering different brane inflationary scenarios, we have $10^{-6} \geq G\mu \geq 10^{-10}$.
- Sec. V : We briefly review the reason why cosmic strings, as opposed to domain walls and monopoles, are safe from the the overabundance problem. The resulting cosmic string network is expected to evolve to the scaling solution.[12] It is amazing that the superstring property avoids the domain wall and monopole problems, while brane inflation gives a cosmic string tension value just at the range of detectabilty.
- Sec. VI : This implies that the anisotropy in the CMB comes mostly from the inflationary epoch, while a component of order $10^6 G\mu \simeq 10\%$ comes from cosmic strings. This is perfectly consistent with all present CMB data[2], but will be tested by MAP and PLANCK. Gravitational waves from this cosmic string network is likely to be measured by gravitational wave interferometers like LIGO II/VIRGO and LISA.

Although hybrid inflationary models with cosmic string production after inflation can be constructed, they are just possible scenarios among many others. The difference here is that the production of cosmic strings towards the end of inflation seems inevitable in brane inflation. The detection of cosmic string signatures will fix the string scale. Since the string scale is determined by the amplitude of the anisotropy in CMB, we can learn a lot about the superstring/braneworld realization of our universe from cosmological observations.

2. Tachyon Condensation and Defect Formation

The topological properties of defect formation in tachyon condensation is understood in superstring theory.[11] Suppose we have $U(N)$ Chan-Paton bundles on one stack of Dp-branes and $U(N+M)$ Chan-Paton bundles on a second stack of Dp-branes, which is at an angle θ with respect to the first stack. They are also separated by a distance y in some orthogonal directions. In brane inflation, y is essentially the inflaton. Figure 3 shows a schematic view of the collision of branes at a fixed angle. As these two stacks of Dp-branes approach each other towards the end of the inflationary epoch, the ground state open string mode T becomes tachyonic. Open strings ending on all possible pairs of branes will give rise to a $U(M+N) \times U(N)$ gauge fields, and the tachyon field T is in the $(M+N, \bar{N})$ (bi-fundamental) representation of the gauge group. Together with $T^{\dagger}$, they form the "superconnection", and the defect properties may be elegantly described by K-theory.[11] Upon collision, the Higgs mechanism takes place as T develops a vacuum expectation value T_0 :

$$U(N) \times U(N+M) \rightarrow U(N) \times U(M)$$

To heat the universe after inflation, we like to have $M \neq 0$, so that M branes are left after inflation. If we identify them as the standard model branes, the tachyon energy can be transferred to the standard model fields providing an efficient way to heat the universe.[8] To consider the formation of defects, we may take $M = 0$ to simplify the problem, The minimum of the tachyon potential corresponds to the vacuum manifold

$$\mho(N) = \frac{U(N) \times U(N)}{U(N)} \tag{1}$$

which is topologically equivalent to $U(N)$. In the simplest situation where the angle between the two stacks of branes is $\theta = \pi$, we have a stack of N Dp-brane and a stack of anti-Dp-branes (which have opposite RR charges); so their collision results in annihilation.

The spontaneous symmetry breaking will support defects in codimension $2k$, classified by the homotopy groups $\pi_{2k-1}(\mho(N)) = \mathbf{Z}$ (for stable values of N). These defects are simply D$(p-2k)$-branes (and anti-D$(p-2k)$-branes) inside the Dp-branes. In the compactified case, the net RR-charge must be conserved. In particular, D3-brane-anti-D3-brane annihilation yields D1-branes, which appear as cosmic strings in our universe. Actually, following from the fact that vortices (instead of domain walls or monopoles) are formed, the appearance of only cosmic strings is much more general.[7,9] As an illustration, consider D5-branes with two of its dimensions compactified on a two-cycle, while the remaining 3 dimensions span our universe. D5-brane-anti-D5-brane annihilation will produce D3-branes, which will also wrap around the two-cycle, so they also appear as cosmic strings. If there is no one-cycle inside the two-cycle, (e.g., topology of a sphere, as in orbifolding a torus), then no domain wall or monopole-like solitons will appear in our universe. For $N > 1$, D1-branes (for $k = 2$)

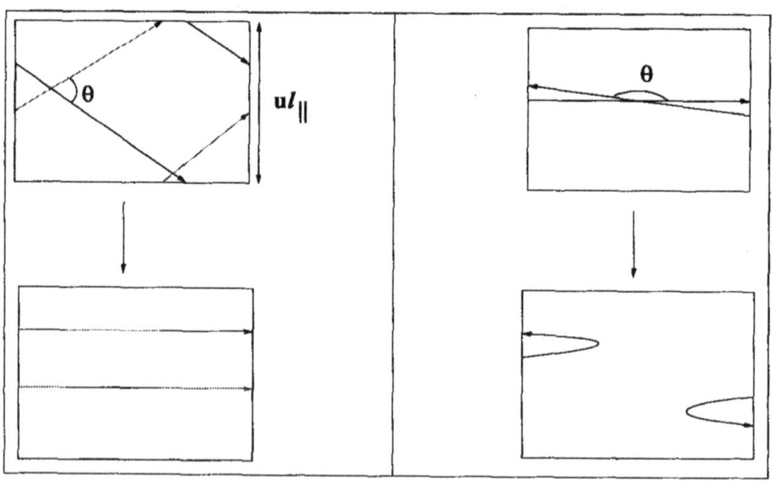

Figure 3. Collision of Branes at angle θ. In the left figures, the two branes wrap around different 1-cycles of a torus with sides $\ell_{\|}$ and $u\ell_{\|}$. For small u, $\theta \simeq 2u$. They are separated in compactified directions orthogonal to the torus. When that separation approaches zero, they collide, resulting in two branes. The resulting brane tensions are cancelled by orientifold planes in the model. In the right figures, when the angle θ is close to π, annihilation takes place.

will also be produced and they may appear as cosmic strings. If there are non-trivial 1-cycles inside the 2-cycles, then they may wrap around the 1-cycles and will appear as monopoles in our universe. However, even though they are allowed by topology, their actual production may be suppressed by cosmology.

3. Cosmological Production of Cosmic Strings

Imagine a string model that describes our universe today. In the early universe, our universe contains branes of all types. The higher-dimensional branes collide to produce lower-dimensional branes and branes that are present today. Consider the last two branes with 3 uncompactified spatial dimensions that are not in today's string model ground state. As they approach each other, the universe is in the inflationary epoch. After inflation ends, the collision of these two branes may produce lower-dimensional branes, which appear as defects.

A large density of such defects produced after inflation may destroy the nucleosynthesis or even overclose the universe (like the old monopole problem). Fortunately, the fact that they can be produced topologically does not necessarily imply that they will be produced cosmologically. To see what types of defects are produced towards the end of inflation, let us assume second order phase transition (a point we shall come back) and estimate the production rate using the Kibble mechanism. During the brane collision, the tachyon acquires the value T_0. Since the vacuum manifold is non-trivial, T can take different values at different spatial point. The existence of the particle horizon implies that T cannot be correlated on scales larger than the horizon length H^{-1}, where H is the Hubble constant. Therefore, non-trivial vacuum configurations, i.e., defects, will be produced, with a density of order one per Hubble volume.

At this time, the particle horizon size H^{-1} is given by

$$H^2 \simeq \frac{8\pi GV}{3} \simeq \frac{V}{3M_P^2} \tag{2}$$

where $M_P = 2.4 \times 10^{18}$ GeV. Here $1/H$ is typically much bigger than the compactification sizes that the branes wrap around,

$$\frac{1}{H} \simeq \frac{M_P}{M_s^2}\frac{(2\pi)^{3/2}}{\theta} \simeq \frac{10^5}{M_s} \gg \ell_{\|} \tag{3}$$

where $V_{\|} \simeq \ell_{\|}^{p-3}$, $\theta \simeq 1/10$ and $M_s\ell_{\|} \simeq 10$. As a consequence of the smallness of $\ell_{\|}$, the Kibble mechanism does not happen in the compactified directions. That is, the codimensions of the defects must lie in the uncompactified dimensions. Since the codimension is always even, and there are only 3 uncompactified dimensions, only the defects with codimension 2 ($k = 1$), i.e., cosmic string-like defects, can be formed via the Kibble mechanism (see Figure 2). They are D$(p-2)$-branes wrapping the same compactified space as the original Dp-branes, with one uncompactified dimension. If the Dp-brane collision can produce D$(p-4)$-branes, their production will also be suppressed since there is less than one Hubble volume in the compactified directions. This implies that the production of domain walls and monopole-like objects by the Kibble mechanism are heavily suppressed, while the production of cosmic strings is not. Generically, there may be closed and stretched cosmic strings, and they form some sort of a cosmic string network that evolves to the scaling solution. In Ref. 7, it is argued that thermal production of any defect is negligible.

4. Cosmic String Tension

If the D1-brane is the cosmic string (i.e., $p=3$), its tension is simply the cosmic string tension:

$$\mu = \tau_1 = M_s^2/(2\pi g_s) \tag{4}$$

However, we expect the string coupling generically to be $g_s \gtrsim 1$. To obtain a theory with a weakly coupled sector in the low energy effective field theory (i.e., the standard model of strong and electroweak interactions with weak gauge coupling constant α), it then seems necessary to have the brane world picture. This argument leads us to consider the Dp-branes for $p > 3$, where the $(p-3)$ dimensions are compactified to volume $V_{\parallel} \sim \ell_{\parallel}^{p-3}$. Now the cosmic strings are D$(p-2)$-branes, with the $(p-3)$ dimensions compactified to the same volume $V_{\parallel}$. Noting that a Dp-brane has tension $\tau_p = M_s^{p+1}/(2\pi)^p g_s$, the tension of such cosmic strings is

$$\mu = \frac{M_s^{p-1} V_{\parallel}}{(2\pi)^{p-2} g_s} = \frac{M_s^2 v_{\parallel}}{2\pi g_s} = \frac{M_s^2}{4\alpha\pi} \simeq 2M_s^2 \tag{5}$$

where

$$g_s \simeq 2 v_{\parallel} \alpha, \qquad \alpha \simeq 1/25 \tag{6}$$

For $v_{\parallel} \sim 10$, $g_s \sim 1$ while α is small. For $N=1$, only this type of cosmic strings are produced topologically. For $N > 1$, the D1-branes may also be allowed topologically, but they are not produced cosmologically. So $\mu \simeq 2M_s^2$ is quite generic.

To get an order of magnitude estimate of M_s, we may use the small θ case, which is argued to be the most likely inflationary scenario. In Ref. 7, the string scale is determined by the anisotropy in CMB[1]

$$\delta_H \simeq 1.9 \times 10^{-5} \quad \Longleftrightarrow \quad M_s \simeq 2 \times 10^{15} GeV \tag{7}$$

This gives

$$G\mu \simeq 10^{-7} \tag{8}$$

Note that the determination of M_s is somewhat sensitive to the details of the brane inflationary scenario and the specific string model realization. It is easy for M_s to vary by a factor of 2 or more. As we shall see, the observability of the cosmic string effect can be very sensitive to this uncertainty. Let us estimate the range of μ within the brane inflationary scenario. For the brane-anti-brane scenario, M_s determined by the COBE data is somewhat smaller. However, the force between the brane-anti-brane system is stronger than that for branes at a small angle. As a consequence, the brane-anti-brane system will need some fine-tuning (so they are separated far enough apart) to give enough inflation. Such fine-tuning is avoided if branes are at a small angle θ, where θ is fixed by the wrapping of the branes. Allowing the various possibilities, we have

$$10^{-6} \geq G\mu \geq 10^{-10} \tag{9}$$

where the lower value is for a brane-anti-brane system, while the upper value comes from decreasing the value of the angle θ by increasing the number of winding along the large dimension. This results in a stack of branes after annihilation.

5. Cosmic String Network Evolution

Here we shall give a very brief review on why the cosmic strings production in the early universe is safe, as opposed to the production of monopoles and domain walls. This is a very well-studied topics,[10] although may not be familiar to some readers.

Let us first consider the production of monopoles. Let us estimate the monopole energy density today as a fraction Ω_M of the total energy density $3H_0^2M_{Pl}^2$ of our universe. The Kibble mechanism would imply an initial density $\rho_{M,initial} \sim H^2M_s^2 \sim M_s^6/M_{Pl}^2$ in the absence of inflation, where the string scale $M_s^2 = 1/\alpha'$. Using $H_0^2/M_{Pl}^2 \simeq 10^{-123}$ and $a_{initial}/a_{today} \simeq 2.7^oK/M_s$, we find

$$\Omega_M = \frac{\rho_{M,initial}}{3H_0^2M_{Pl}^2}\left(\frac{a_{initial}}{a_{today}}\right)^3 \simeq 10^{28}(\frac{M_s}{M_{Pl}})^3 \tag{10}$$

where $2.7^oK/M_{Pl} \simeq 10^{-31}$. Since Ω_M must be less than 1, for $M_s \simeq M_{GUT}$, we see that the *GUT* monopole energy density will be many orders of magnitude too big to be compatible with our universe. This is the well-known monopole problem. In fact, the original motivation of inflation is to solve this over-abundance problem.

Naively, the cosmic string density is also a problem. For cosmic string loops, the naive energy density is similar to that for monopoles, in that the density scales like a^{-3}. However, their shrinking and oscillations radiate gravitational waves, so eventually they vanish. For cosmic strings that stretch across the horizon, the density actually scale like a^{-2}. Fortunately, the probability of such strings interacting is of order unity. The inter-commutation of intersecting cosmic strings and the decay of the resulting cosmic string loops (to gravitational waves) modify the density to decrease like radiation.[12] This adds another power of $2.7^o K/M_s$ to Eq.(10), rendering the cosmic string network energy density to an acceptable level. It is interesting to note that resulting cosmic string network energy density is insensitive to the initial density. The evolution of the cosmic string network after its initial production is well-studied.[12] After the initial production of cosmic strings, they continue to interact among themselves. When two cosmic strings intersect, they reconnect or intercommute (see Figure 3, where the diagrams are reinterpreted for cosmic strings in the uncompactified dimensions). When a cosmic string intersects itself, a closed string loop is broken off. Such a loop will oscillate quasi-periodically and gradually lose energy by gravitational radiation. Its eventual decay transfers the cosmic string energy to gravitational waves. Higher (lower) initial string density brings higher (lower) interaction rate so, not surprisingly, the cosmic string network evolves towards a scaling solution. As a consequence, the physics is essentially dictated by the single parameter $G\mu$.

To see qualitatively the appearance of the scaling solution, consider the evolution of the cosmic string network energy density ρ,[10]

$$\dot{\rho} \simeq -2H\rho - \lambda\rho^{3/2} \tag{11}$$

where $H = \dot{a}/a$ and λ is the cosmic string interaction coupling. In the absence of interaction, $\rho \simeq a^{-2}$, as expected. In the presence of $\lambda \neq 0$, during the radiation dominated era where $a \simeq \sqrt{t}$, we let $\rho \simeq (\alpha(t)t)^{-2}$, so the equation becomes

$$\frac{\dot{\alpha}}{\alpha} = -\frac{1}{2t}(1 - \frac{\lambda}{\alpha})$$

We see that $\alpha = \lambda$ is a stable fixed point and $\rho \simeq t^{-2} \simeq a^{-4}$. The simulations[12] show that the final result yields a scaling solution where, roughly,

$$\Omega_{cosmicstring} = \frac{\rho}{3H_0^2 M_{Pl}^2} \simeq G\mu \tag{12}$$

So the consequence of the cosmic string network is summarized in the value of $G\mu$. Observational bounds require $G\mu < 10^{-6}$, in which case, the cosmic string network is too small to seed the galaxy formation.

Similarly, the naive domain wall energy density scales like $1/a$. Its interaction also renders the density to decrease faster, but not sufficient to resolve the over-abundance problem.[10] So it is quite amazing that the superstring theory property precisely avoids the undesirables.

6. Observational Consequences

In this brane inflationary scenario, we see that the density perturbation (and the CMB anisotropy) comes from two sources: the inflaton fluctuation during inflation and the cosmic string network. Here, let us get a crude estimate of the magnitude of these two components. The COBE data roughly yields $G\mu \simeq 10^{-6}$ if the scaling solution of the cosmic string network is the sole source of the density perturbation.[13] Since $\Delta T/T \propto G\mu$, the density perturbation is given by

$$\delta_H = (1-a)\delta^I + a\delta^{CS}$$

$$a \simeq \frac{G\mu}{10^{-6}} \simeq 10\%$$

where δ^I comes from inflation and δ^{CS} comes from the cosmic string network. In terms of the spectrum of the CMB, namely, C_l (l the partial wave integer) ($C_l \propto \delta_H^2$):

$$C_l \simeq (1-a)^2 C_l^I + 2a(1-a)C_l^{CS} \tag{13}$$

where C_l^I comes from inflation while C_l^{CS} comes from cosmic strings. So, with $G\mu \simeq 10^{-7}$, the cosmic string network contributes around 18 % of the anisotropy spectrum C_l in the CMB data. As shown in Ref. 14, the present CMB data[2] can easily accommodate up to $a = 10\%$, so the cosmic string production towards the end of brane inflation is perfectly compatible with the present CMB data[2], while future data from MAP and PLANCK will be able to test this scenario. It is obviously very important to estimate a more carefully in the cosmic string network in a phenomenologically realistic superstring model. Since a is most sensitive to the string scale M_s, the CMB data may be a very good way to eventually determine the value of M_s. It is also obvious that the density perturbation coming from the cosmic string network should be more carefully determined. At the moment, there is large uncertainties in its spectrum. In any case, the density perturbation coming from the cosmic string network is incoherent, so there is no acoustic peaks that are prominent in the density perturbation coming from inflation. Using COBE data to normalize δ_H at small ls, the presence of δ^{CS} will tend to attenuate the prominence of the acoustic peaks.

As a cosmic string moves with velocity $\mathbf{v}$ across the sky, a shift in the CMB temperature may be observed, $\Delta T/T \simeq 8\pi G\mu v\gamma$.[15] A careful analysis of the MAP data may probe $G\mu \simeq 10^{-10}$. It is important to see what bound on $G\mu$ the data can eventually reach. I was informed [16] that detection may be possible for as small as $G\mu \simeq 10^{-13}$.

Polarization in CMB will also be measured. In particular, the B (i.e., curl) mode due to the tensor mode perturbation will be tested, reaching $\Delta T \simeq 0.5\mu K$ or better. Here the gravitational wave density is much higher than that in a pure inflationary scenario, so passage through space will presumably yield a B mode polarization clearly larger than that comes from a purely inflationary scenario.

The cosmic string network also generates gravitational waves that may be observable. This has been studied extensively in the literature. The stochastic gravitational wave spectrum has an almost flat region that extends from $f \sim 10^{-8}$ Hz to $f \sim 10^{10}$ Hz. Within this frequency range, both LIGO II/VIRGO (sensitive at around $f \sim 10^2$ Hz) and LISA (sensitive at around $f \sim 10^{-3}$ Hz). Following Ref. 17, we obtain

$$\Omega_{gw}h^2 \simeq 0.04G\mu$$

Since LIGO II/VIRGO can reach $\Omega_{gw}h^2 \simeq 10^{-10}$ at $f \simeq 100$ Hz, it can reach $G\mu \geq 2 \times 10^{-9}$.

If the cosmic string network of oscillating loops involve cusps and kinks of the cosmic strings, then strong beams of high-frequency gravitational waves are emitted by these cusps and kinks. This scenario is studied by Damour and Vilenkin.[18] In this scenario, the sharp bursts of gravitational waves should be easily observable since they have very distinctive waveform $t^{1/3}$ (cusps) and $t^{2/3}$ (kinks). LIGO II/VIRGO and LISA may detect them for values down to $G\mu \geq 10^{-13}$. So a lack of the detection of such bursts may actually rule out brane inflation.

Although present pulsar timing measurement is compatible with $G\mu \lesssim 10^{-6}$, a modest improvement on the accuracy may detect a network of cuspy cosmic string loops down to $G\mu \simeq 10^{-11}$. To conclude, if cusps and kinks happen to a small but reasonable fraction of the cosmic strings, gravitational wave detection will be able to critically test the brane inflationary scenario. If detected, it will further shed light on the specific brane inflationary scenario.

7. Summary

Brane inflation is a natural realization of inflation in the brane world scenario. For the string scale close to the GUT scale (the case we are interested in), presumably any phenomenologically realistic string model is dual to another string model that has a brane world interpretation. In this sense, brane inflation is generic. In Refs. 6, 7, and 8, it is shown that the brane inflation scenario is robust, that is, the probability that the universe has an extended inflationary epoch before big bang is of order unity. The power spectrum index of the density perturbation has the usual red-tilt from a scale-invariant spectrum, i.e.,

$$n \simeq 0.97$$

The interesting prediction is that cosmic strings will be copiously produced in any brane inflationary scenario. Other defects, if exist topologically, may also be produced, though with very suppressed rates. In conclusion, we find that brane inflation has interesting predictions beyond that of the slow-roll inflationary scenario, and provides a testing ground for superstring theory to confront experiments. Existing data is perfectly compatible with brane inflation. It is exciting that future experiments will likely provide non-trivial tests of the scenario.

If cosmic string is detected, the cosmic string tension $G\mu$ will be measured. This will shed light on the specific brane inflationary scenario that took place, providing a probe to the brane world picture before inflation. That is, information on the early universe before inflation may not be totally lost.

I thank Gia Dvali, Nick Jones, Sash Sarangi, Gary Shiu, Horace Stoica and Ira Wasserman for collaborations and valuable discussions. Discussions with Robert Caldwell, Jim Cline, Lev Kofman, Andre Linde, Ashoke Sen, Samson Shatashvili and Protty Wu are acknowledged. This material is based upon work supported by the National Science Foundation under Grant No. PHY-0098631.

References

1. G.F. Smoot et. al., *Astrophy. J.* **396**, L1 (1992);
   C.L. Bennett et. al., *Astrophy. J.* **464**, L1 (1996) [astro-ph/9601067].

2. A.T. Lee et. al. (MAXIMA-1), astro-ph/0104459;
C.B. Netterfield et. al. (BOOMERANG), astro-ph/0104460;
C. Pryke, et. al. (DASI), astro-ph/0104490.
3. A.H. Guth, *Phys. Rev.* **D23**, 347 (1981);
A.D. Linde, *Phys. Lett.* **B108**, 389 (1982);
A. Albrecht and P.J. Steinhardt, *Phys. Rev. Lett.* **48**, 1220 (1982).
4. G. Dvali and S.-H.H. Tye, *Phys. Lett.* **B450**, 72 (1999) [hep-ph/9812483].
5. J. Polchinski, *String Theory*, Cambridge University Press, 1998.
6. C. P. Burgess, M. Majumdar, D. Nolte, F. Quevedo, G. Rajesh and R. J. Zhang, *JHEP* **07**, 047 (2001) [hep-th/0105204];
J. Garcia-Bellido, R. Rabadan, F. Zamora, *JHEP* **01**, 036 (2002) [hep-th/0112147];
K. Dasgupta, C. Herdeiro, S. Hirano and R. Kallosh, hep-th/0203019.
7. N. Jones, H. Stoica and S.-H.H. Tye, hep-th/0203163.
8. G. Shiu, S.-H.H. Tye and I. Wasserman, hep-th/0207119;
J.M. Cline, H. Firouzjahi and P. Martineau, hep-th/0207156;
G. Felder, J. Garcia-Bellido, P.B. Greene, L. Kofman, A. Linde and I.Tkachev, *Phy. Rev. Lett.* **87**, 011601 (2001);
G. Felder, L. Kofman and A. Linde, hep-th/0106179.
9. S. Sarangi and S.-H.H. Tye, hep-th/0204074.
10. See e.g., E.W. Kolb and M.S. Turner, *The Early Universe*, Addison-Wesley Publ. Co., 1990;
A. Villenkin and E.P.S. Shellard, *Cosmic strings and other topologiocal defects*, Cambridge University Press, 2000.
11. A. Sen, *JHEP* **9808**, 010 (1998) [hep-th/9805019]; *JHEP* **9809**, 023 (1998) [hep-th/9808141];
E. Witten, *JHEP* **9812**, 019 (1998) [hep-th/9810188];
P. Horava, *Adv. Theor. Math. Phys.* **2**, 1373 (1999) [hep-th/9812135].
12. A. Albrecht and N. Turok, *Phy. Rev. Lett.* **54**, 1868 (1985);
D.P. Bennett and F.R. Bouchet, *Phy. Rev. Lett.* **60**, 257 (1988);
B. Allen and E.P.S. Shellard, *Phy. Rev. Lett.* **64**, 119 (1990).
13. D.P. Bennett, A. Stebbin and F. Bouchet, *Ap. J.* **399**, L5 (1992).
14. C. Contaldi, M. Hindmarsh and J. Magueijo, *Phys. Rev. Lett.* **82**, 2034 (1999) [astro-ph/9809053];
R.A. Battye and J. Weller, *Phys. Rev.* **D61**, 043501 (2000) [astro-ph/9810203];
F.R. Bouchet, P. Peter, A. Riazuelo and M. Sakellariadou, astro-ph/0005022;
A. Gangui, L. Pogosian and S. Winitzki, astro-ph/0112145.
15. N. Kaiser and A. Stebbin, *Nature* **310**, 391 (1984);
J.R. Gott, *Ap. J.* **288**, 422 (1985).
16. P. Wu, private communication.
17. R.R. Caldwell and B. Allen, *Phys. Rev.* **D45**, 3447 (1992).
18. T. Damour and A. Vilenkin, *Phys. Rev.* **D64**, 064008 (2001) [gr-qc/0104026].

Relativistic Braneworld Cosmology

Tetsuya Shiromizu

RELATIVISTIC BRANEWORLD COSMOLOGY

TETSUYA SHIROMIZU

Department of Physics, Tokyo Institute of Technology, Tokyo 152-8551, Japan
and
Advanced Research Institute for Science and Engineering, Waseda University, Tokyo 169-8555, Japan
E-mail: shiromizu@th.phys.titech.ac.jp

I briefly review the basics of the Randall-Sundrum braneworld cosmology and summarise current issues on cosmology, black holes and holography.

1. Introduction

Recent progress in superstring/M-theory provides us a new picture of our universe. We know that the D-brane is key object to look at the whole view of the string theory.[1] From the basic feature of D-brane, it is natural to expect that our universe is a sort of domain wall in higher dimensional spacetimes, that is, the ordinal matter is confined on the brane. This picture for our universe is also attained from the Horave-Witten model[2] which connects between M-theory and heterotic superstring theory. Related to this, so-called large extra dimension scenario has been proposed to solve the gauge hierarchy problem.[3] In this scenario the fundamental scale M of the gravity could be weak scale if the volume V and number n of the extra dimension is large. This is because the four dimensional planck scale $M_{\rm pl}$ can be written as $M_{\rm pl} \sim M^{1+n/2}V^{n/2}$, where M is the fundamental scale in higher dimensions. After this model, the warped extra dimension scenario has been proposed by Randall and Sundrum.[4,5] Their first model[4] is the small and warped extra dimension scenario for the gauge hierarchy problem. Their second model[5] gives us the new dimensional reduction procedure without compactification. Since this last model is simplest and non-trivial feature in the gravity aspects, cosmologists has paid attention to this model very much. In this review, we shall summarise the progress obtained so far in the second Randall-Sundrum model and describe remaining issues.

The rest of this review is organised as follows. In the section 2, we begin with the "Einstein" equation on the brane to see the whole view of the (self-gravitating)brane world. We will realise that the four dimensional gravity is recovered at low energy limit. In the section 3, we summarise the cosmology. The universe is described by the domain wall motion in anti-deSitter spacetime. In the section 4, the holographic aspects will be discussed. Finally we will give summary in the section 5.

2. Gravitational equation on the brane

The model proposed by Randall and Sundrum is simple structure:the bulk spacetime is five dimensional anti-deSitter(adS) spacetime and the brane tension is fine-tuned so that the brane geometry is four dimensional Minkowski spacetime. In this section we generalise the Randall-Sundrum toy model to cosmology.

We begin with the five dimensional Einstein equation

$$R_{ab} - \frac{1}{2} g_{ab} R = \kappa_5^2 T_{ab}. \tag{1}$$

Using the Gaussian normal coordinate around the brane, the five dimensional metric is

$$ds^2 = dy^2 + q_{\mu\nu}(y, x) dx^\mu dx^\nu, \tag{2}$$

where $\mu = 0, 1, 2, 3$ and it is supposed that the brane is located at $y = 0$. $q_{\mu\nu}(0, x)$ is the induced metric of the brane. The energy-momentum tensor T_{ab} is given by

$$T_{ab} = -\Lambda g_{ab} + \Big(-\lambda q_{ab} + \tau_{ab} \Big) \delta(y), \tag{3}$$

where Λ, λ and τ_{ab} are the bulk cosmological constant, the brane tension and the energy-momentum tensor on the brane, respectively ($\tau_{ab} n^b = 0$ for the unit normal vector of the brane, $n = \partial_y$).

Using the junction condition and Z_2-symmetry, the Gauss-Codazzi equations imply the "Einstein equation" on the brane[6,7,8]

$${}^{(4)}G_{\mu\nu} = -\Lambda_4 q_{\mu\nu} + 8\pi G_N \tau_{\mu\nu} + \kappa_5^4 \pi_{\mu\nu} - E_{\mu\nu} \tag{4}$$

and the conservation of the matter on the brane

$$\nabla_\mu \tau^\mu_\nu = 0 \tag{5}$$

where

$$\Lambda_4 = \frac{1}{2} \kappa_5^2 \left(\Lambda + \frac{1}{6} \kappa_5^2 \lambda^2 \right), \tag{6}$$

$$G_4 = \frac{\kappa_5^4 \lambda}{48\pi}, \tag{7}$$

$$\pi_{\mu\nu} = -\frac{1}{4} \tau_{\mu\alpha} \tau_\nu{}^\alpha + \frac{1}{12} \tau \tau_{\mu\nu} + \frac{1}{8} q_{\mu\nu} \tau_{\alpha\beta} \tau^{\alpha\beta} - \frac{1}{24} q_{\mu\nu} \tau^2, \tag{8}$$

and $E_{\mu\nu}$ is a part of the 5-dimensional Weyl tensor defined by $E_{\mu\nu} := {}^{(5)}C_{\mu a\nu b}n^a n^b$. Note that we did not use any approximations in the above derivations. In the Randall-Sundrum toy model, it is tuned so that $\Lambda_4 = 0$. Since the background spacetime is anti-deSitter, $E_{\mu\nu} = 0$ and the brane geometry is just four-dimensional Minkowski spacetime. We can recover the four dimensional Einstein gravity from the above equation if we can omit $E_{\mu\nu}$ and $\pi_{\mu\nu}$. It is easy to see that $\pi_{\mu\nu}$ is negligible at low energy scale. However, the problem is the evaluation of $E_{\mu\nu}$ because it is five dimensional quantity and the "Einstein" equation is not closed in four dimensions(as expected). To evaluate $E_{\mu\nu}$ on the brane, we have to solve the five dimensional equation for $E_{\mu\nu}$ with the appropriate boundary condition(See Ref. 7 for the linear perturbation). This gives us the serious problem to have the prediction in cosmology. But, we can do so for the linearised gravity. Indeed we can see $E_{\mu\nu}$ contains the contribution from the Kaluza-Klein graviton and it could be negligible at low energy scale.[9,7,10] The corrected Newton potential is given by

$$V(r) \sim -\frac{GM}{r}\left(1 + \frac{\ell^2}{r^2}\right), \tag{9}$$

where ℓ is the bulk curvature radius.

3. Braneworld Cosmology

3.1. *Homogeneous and isotropic universe*

Let see how the homogeneous and isotropic universe can be realised in the braneworld.[11] Fortunately, we have the exact solution. To do so we assume that the matter on the brane is perfect fluid:

$$\tau_{\mu\nu} = \rho t_\mu t_\nu + P(q_{\mu\nu} - t_\mu t_\nu), \tag{10}$$

where t^μ is the unit normal vector of the spacelike hypersurfaces. From the "Einstein" equation on the brane, we obtain the Friedmann equation

$$\left(\frac{\dot{a}}{a}\right)^2 = \frac{8\pi G_N}{3}\rho - \frac{K}{a^2} + \frac{\Lambda_4}{3} + \frac{\kappa_5^4}{36}\rho^2 + \frac{\mu}{a^4}. \tag{11}$$

The last term in the right-hand side comes from $E_{\mu\nu}t^\mu t^\nu$. Regardless of the bulk geometry, we can determine its scale factor dependence. From Eqs. (4) and (5), we see $\nabla_\mu E^\mu_\nu = 0$ on the homogeneous and isotropic brane. Since $E^0_0 = -(1/3)E^i_i$, we can see that E_{00} behaves like radiation. This is so-called dark radiation.

Although we did not discuss the bulk geometry, it turns out that the bulk is Schwarzschild-adS spacetimes due to the symmetry:

$$ds^2 = g_{\mu\nu}dx^\mu dx^\nu = -f(r)dT^2 + \frac{dr^2}{f(r)} + r^2 d\Sigma_K^2, \tag{12}$$

where $d\Sigma_K^2$ is a metric of a unit three-dimensional sphere, plane or hyperboloid for $K = +1, 0$ or -1, respectively, and

$$f(r) = K + \frac{r^2}{\ell^2} - \frac{\mu}{r^2}. \tag{13}$$

Here, ℓ (> 0) and μ (≥ 0 for $K = +1$ or $K = 0$, $\geq -l^2/4$ for $K = -1$) are constants. μ is the mass parameter of black hole.

If you are conservative, you might want to describe the following whole history of the universe. The quantum universe(brane) is created in the bulk or with bulk[12] like bubbles. After the creation, the induced geometry on the brane is deSitter spacetime. This is the inflation era. After the reheating on the brane, the radiation dominated universe will come out as usual. See Ref. 13 for the chaotic inflation on the brane. Note that the bulk scalar field can be inflaton for the brane.[14]

The cosmological perturbation has been formulated by many peoples.[13,15,16,17] It turns out that it is too difficult to evaluate the modification from the four dimensional Einstein theory, that is, the signal of the braneworld. But, I think we can do that near future(hopefully before MAP or PLANCK result). For the deSitter brane[12,13,16] or large scale[17], we can successfully compute the perturbations due to the symmetry.

3.2. *New fundamental process – brane collisions –*

Most new thing which the braneworld shows us is the brane collision process. There is no reason for thinking of only single brane model. We can or must consider other branes. Especially, if you are interested in the model for the gauge hierarchy problem, two branes exist at least. In such situation, the brane collision could be fundamental process. Through this process, the hot big-bang universe might begin.[18] Or our universe might be created via the collision two adS bubbles in the bulk.[20] Or the inflation will terminate when the bubble nucleated in the bulk hits the brane where we are.[19] You can write down such funny stories, but they might be real. If you are serious about such scenario, we should predict right cosmological fluctuations generated at the collision. It is, however, unlikely that the simple model does not work.[21] Skillful models is needed.

Since the brane collision is definitely key process, it is worthy to investigate that in the detail.[22]

4. Braneworld black-hole

4.1. *Solution?*

It is natural to ask how strong gravity is. Except for the homogeneous and isotropic universe, we could treat only the linearised theory(See Ref. 23 for the second order perturbation). The first purpose is simply to find the exact solution of black holes. We can naively expect that the gravity looks like five dimensional at smaller distance than the bulk curvature scale. At larger distance, the four dimensional gravity could be recovered as expected from the linearised theory. Chamblin et al[24] guessed the shape of the black hole through the Gregory-Laflamme instability of the black string solution[25,26](Although Gibbons and Hartnoll recently showed that the instability mode discussed in Ref. 26 is not normalizable in adS.). Unfortunately, except for the lower dimensional case, the solution has not been discovered so far. Since the three dimensional black holes does not exists, the solution in lower dimension[28] is not what we want. People is trying to find black hole solution in several ways. At this moment I think that we have to rely on the numerical procedure:*numerical braneworld!*

If one does not take care of the bulk behavior, we can freely assign the brane geometry satisfying the "Hamiltonian" constraint. For the vacuum case, the constraint becomes ${}^{(4)}R = 0$. So all of geometries satisfying this is candidate for the brane induced geometry. Indeed, the "Reissner-Nordstrom" solution is possible candidate[29] because the energy-momentum tensor of the Maxwell field is traceless. The metric is given by

$$ds^2 = -f(r)dt^2 + f(r)^{-1}dr^2 + r^2(d\theta^2 + \sin^2\theta d\phi^2), \tag{14}$$

where $f(r) = 1 - 2M/r - q^2/r^2$. Exactly say, this solution is not what we want. This is because the asymptotic behavior of the metric on the brane does not follow the linearised theory. In addition, the numerical analysis of the bulk behavior for the "Reissner-Nordstrom" solution indicates the bad signal.[30] Although this approach is not complete, it is meaningful as a first step.

The initial data has been constructed in numerical way.[31] The initial data describes a moment of the gravitational collapse. If we could follow the full time evolution, the final state could be the solution which we want. The relativistic star solution has been also numerically presented.[32] This is supposed to be static. The feature of the event horizon has been also investigated.[33]

Since the information inside of the black hole on the brane can affect the outside of that via bulk, the Birkhoff theorem on the brane does not hold. This means that the solution is not unique.

Last year Bruni et al[34] addressed the collapse of the homogeneous dust star on the brane[34] and they gave an implication on the non-existence of the *static* black hole. This conjecture might be supported by adS/CFT correspondence. We will briefly comment on this in the section 5.

4.2. *The uniqueness and non-uniqueness of the higher dimensional black-holes*

As stated before, small black hole is well approximated by the higher dimensional black hole. From the possibility of the black hole production in accelerators,[35] it is quite important to investigate the fundamental feature of the higher dimensional black holes. For example, the uniqueness of the stationary black hole. If we can prove that, we can safely use the unique solution to have some implications. Recently Emparan and Reall discovered the five dimensional stationary black ring solution.[36] On the other hand, we also have the higher dimensional Kerr solution.[37] Thus, the uniqueness theorem does not hold in *stationary* spacetimes. For the static and asymptotically flat spacetime, however, we can have the uniqueness theorem which has been recently proved[38](See also Ref. 39). This is not cheap extension of the four dimensional theorem.[40] In four dimensions one uses the four dimensional specialties, the Gauss-Bonett theorem etc, which we cannot use in higher dimensions. In higher dimensions than five, the asymptotically non-flat spacetime is also naturally permitted. In such case, the uniqueness theorem does not work.[38] Simplest example is the black string solution.

The issue of hoop conjecture[41] in higher dimensions is also important. This gives us the condition for the black hole formation and the cross section. People imagines that the black hole will be formed if the matter is confined small enough, say Schwarzschild radius. Therein the size is supposed to be measured by "hoop". In higher dimensions, the maximum size could be infinite because we have the black string solution as mentioned. So the "hoop" is not good measure. Thus we must reformulate the conjecture, say, using the maximum area in five dimensions.[42] The cosmic censorship in higher dimensions should be widely addressed.

5. Braneworld holography

We saw that a part of the deviation from the four dimensional gravity comes from the five dimensional Weyl tensor $E_{\mu\nu}$ and the quadratic term of the energy-momentum tensor $\pi_{\mu\nu}$ on the brane. They can be surprisingly regarded as a holography.[10,12,43,44,45,46] This is because the setup of the Randall-Sundrum model is resemble to that of the adS/CFT correspondence.[47] We might be able to use the adS/CFT correspondence in order to compute $E_{\mu\nu}$.

Remember that the dark radiation terms in the Friedmann equation comes from the black hole mass. We can see that the black hole entropy is identical to that of the dark radiation in a limit.[12] In general we can show that $E_{\mu\nu}$ is approximately identical to the energy-momentum tensor of the conformal field theory on the brane.[45] Furthermore, the trace of the quadratic term $\pi_{\mu\nu}$ is regarded as the trace anomaly of the conformal field theory on the brane.[45] Let see the detail below.

Following the adS/CFT correspondence, we can have the effective equation on the brane which is independently derived:[10,43,44]

$$G^{\mu\nu} = 8\pi G_4 T^{\mu\nu} + \frac{4}{\ell}\frac{1}{\sqrt{|g|}}\Big(\frac{\delta S_{\rm ct}^{(4)}}{\delta g_{\mu\nu}} + \frac{\delta\Gamma_{\rm CFT}}{\delta g_{\mu\nu}}\Big), \tag{15}$$

where $\Gamma_{\rm CFT}$ is the effective action of CFT living on the boundary and has the trace anomaly:[48,49,50]

$$\frac{\delta\Gamma_{\rm CFT}}{\delta g_{\mu\nu}} g^{\mu\nu} = \frac{\ell^3}{16}\sqrt{|g|}\Big(R_{\mu\nu}R^{\mu\nu} - \tfrac{1}{3}R^2\Big). \tag{16}$$

The quantity $S_{\rm ct}^{(4)}$ is R^2 terms of the counter-term which makes the action finite. For our purpose, we need not to write down the explicit form (See Ref. 49). What we will use is that $\delta S_{\rm ct}/\delta g_{\mu\nu}$ is traceless. Then, the trace part of Eq (15) is

$$R = -8\pi G_4 T - \frac{\ell^2}{4}\Big(R_{\mu\nu}R^{\mu\nu} - \tfrac{1}{3}R^2\Big) \tag{17}$$

On the other hand, we see from the "Einstein" equation in the section2 that the trace part can be written in terms of the four-dimensional quantities:

$$R = -8\pi G_4 T - \frac{\kappa_5^4}{4}\Big(T_{\mu\nu}T^{\mu\nu} - \frac{1}{3}T^2\Big) \tag{18}$$

Using the Einstein equation, $G_{\mu\nu} \simeq 8\pi G_4 T_{\mu\nu}$, approximately, we can check that Eq (18) is approximately identical with Eq (17). It is remarkable that ρ^2 terms appeared in the cosmological solution of the brane-world can be regarded as the non-linear contribution from CFT. In the linear order we easily obtain the relation between the energy-momentum tensor of CFT and a part of Weyl tensor:

$$E^{\mu\nu} \simeq -\frac{4}{\ell}\frac{1}{\sqrt{|g|}}\frac{\delta\Gamma_{\rm CFT}}{\delta g_{\mu\nu}}. \tag{19}$$

See Ref. 46 for more careful analysis.

Based on this holographic feature in the braneworld, Emparan et al gave a reason why we cannot discover the exact solution for the static black holes.[52] The idea is inspired by Tanaka[51] who discussed the Hawking radiation on the brane using the adS/CFT correspondence. Even if you think of the classical black hole solution in the bulk, the adS/CFT correspondence tells us that there is automatically a holographic quantum field theory on the brane, that is, the four dimensional black hole plus the quantum field on the brane. Natural guess is that the holographic field is regarded as the Hawking radiation. Thus the system on the brane will be dynamical. Emparan et al gave an interpretation for the exact static solution in lower dimensions. However, we cannot have the definite answer for this issue because it is conservative to think that the Bruni et al's work[34] simply suggests that the homogeneous dust collapse is impossible in the braneworld. To construct the static black hole, I think it is necessary to consider the inhomogeneous dust collapse. The absence of the Birkhoff theorem makes the situation difficult.

6. Summary

In this short review, we summarised the basics, the recent progress and remaining issues. In the braneworld the new items introduced is the extra dimensions and the brane. We saw that even a few introductions provides us the comprehensive structure, many implications and non-trivial issues:a geometrical solution to the gauge hierarchy problem, our universe as a domain wall, new creation scenario of the hot big-bang universe, holographic picture, black hole physics and so on.

The possibility of black hole production in experiments is impressive. If the fundamental scale is weak scale, the black hole formation is much easier than that in four dimensions. Indeed we have the chance to produce the black hole in experiments. The discovery of the black hole or so in experiments gives us the direct evidence for the extra dimensions or superstring theory.

Now we are in braneworld paradigm. Therein the technical and crucial problem is related to the treatment of the dynamics of the spacetime boundary under the physical boundary conditions in the bulk. In terms of "Einstein" equation, how can we evaluate $E_{\mu\nu}$. The method which we learned here is presumably useful beyond the braneworld paradigm, for example, for whole understanding of the spacetime in the superstring theory. The dynamics of the brane is essential.

Acknowledgments

TS is grateful to the organisers and secretaries of the CosPA conference for their generous hospitality. He would also like to thank A. Chamblin, G. W. Gibbons, D. Ida, K. Maeda, S. Mukohyama, H. S. Reall, M. Sasaki, M. Shibata, H. Shinkai and T. Torii for useful discussion and collaborations. This work is supported by Grant-in-Aid for Scientific Research from Ministry of Education, Science, Sports and Culture of Japan(No. 13135208, No.14740155 and No.14102004).

References

1. J. Polchinski, *String Theory* (Cambridge Univ. Press, Cambridge, 1998).
2. P. Horava and E. Witten, *Nucl. Phys.* **B460**, 506 (1996); *ibid* **B475**, 94 (1996)
3. N. Arkani-Hamed, S. Dimopoulos and G. Dvali, *Phys. Lett.* **B424**, 263 (1998); I. Antoniadis, N. Arkani-Hamed, S. Dimopoulos and G. Dvali, *Phys. Lett.* **B436**, 257 (1998); N. Arkani-Hamed, S. Dimopoulos and G. Dvali, *Phys. Rev.* **D59**, 086004 (1999); N. Arkani-Hamed, S. Dimopoulos, N. Kaloper, J. March-Russell, hep-ph/9903224.
4. L. Randall and R. Sundrum, *Phys. Rev. Lett.* **83**, 3370 (1999).
5. L. Randall and R. Sundrum, *Phys. Rev. Lett.* **83**, 4690 (1999).
6. T. Shiromizu, K. Maeda and M. Sasaki, *Phys. Rev.* **D62**, 024012 (2000).
7. M. Sasaki, T. Shiromizu and K. Maeda, *Phys. Rev.* **D62**, 024008 (2000).
8. R. Maartens, *Phys. Rev.* **D62**, 084023 (2000).
9. J. Garriga and T. Tanaka, *Phys. Rev. Lett.* **84**, 2778 (2000)
10. S. B. Giddings, E. Katz, and L. Randall, *JHEP* **0003**, 023 (2000).
11. A. Chamblin and H. S. Reall, *Nucl. Phys.* **B562**, 133 (1999); T. Nihei, *Phys. Lett.* **B465**, 81 (1999); N. Kaloper, *Phys. Rev.* **D60**, 123506 (1999); H. B. Kim and H. D. Kim, *Phys. Rev.* **D61**, 064003 (2000); P. Kraus, *JHEP* **9912**, 011 (1999); D. Ida, gr-qc/9912002; E. E. Flanagan, S.H.H. Tye and I. Wasserman, *Phys. Rev.* **D62**, 044039 (2000); A. Chamblin, A. Karch, A. Nayeri, hep-th/0007060; P. Bowcock, C. Charmousis and R. Gregory, hep-th/0007177; S. Mukohyama, *Phys. Lett.* **B473**, 241 (2000); S. Mukohyama, T. Shiromizu and K. Maeda, *Phys. Rev.* **D62**, 024028 (2000).
12. J. Garriga and M. Sasaki, *Phys. Rev.* **D62**, 043523 (2000).
13. R. Maartens, D. Wands, B. A. Bassett and I. P. Heard, *Phys. Rev.* **D62**, 041301 (2000).
14. J. Yokoyama and Y. Himemoto, *Phys. Rev.* **D64**, 083511 (2001).
15. S. Mukohyama, *Phys. Rev.* **D62**, 084015 (2000); H. Kodama, A. Ishibashi and O. Seto, *Phys. Rev.* **D62**, 064022 (2000); D. Langlois, *Phys. Rev.* **D62**, 126012 (2000); C. van de Bruck, M. Dorca and R. H. Brandenberger, *Phys. Rev.* **D62**, 123515 (2000); S. Mukohayma, *Class. Quant. Grav.* **17**, 4777 (2000); D. Langlois, *Phys. Rev. Lett.* **86**, 2212 (2001); H. A. Bridgman, K. A. Malik and D. Wands, *Phys. Rev.* **D63**, 084012 (2001); S. Mukohyama, *Phys. Rev.* **D64**, 064006 (2001); H. A. Bridgman, K. A. Malik and D. Wands, *Phys. Rev.* **D65**, 043502 (2002); K. Koyama and Jiro Soda, *Phys. Rev.* **D65**,

023514 (2002).

16. D. Langlois, R. Maartens and D. Wands, *Phys. Lett.* **B489**, 259 (2000).
17. D. Langlois, R. Maartens and M. Sasaki, *Phys. Rev.* **D63**, 084009 (2001).
18. J. Khoury, B. A. Ovrut and P. J. Steinhardt, *Phys. Rev.* **D64**, 123522 (2001); J. Khoury, B. A. Ovrut, N. Seiberg, P. J. Steinhardt and N. Turok, *Phys. Rev.* **D65**, 086007(2002); P. J. Steinhardt and N, Turok, hep-th/0111030; *Phys. Rev.* **D65**, 126003(2002); astro-ph/0204479; R. Kallosh, L. Kofman and A. D. Linde, *Phys. Rev.* **D64**, 123523 (2001); A. Linde, hep-th/0205259.
19. U. Gen, A. Ishibashi and T. Tanaka, *Phys. Rev.* **D66**, 023519 (2002).
20. M. Bucher, hep-th/0107148; J. J. Blanco-Pillado and M. Bucher, *Phys. Rev.* **D65**, 083517(2002).
21. D. H. Lyth, *Phys. Lett.* **B524**, 1 (2002); *Phys. Lett.* **B526**, 173 (2002); S. Tsujikawa, *Phys. Lett.* **B526**, 179 (2002); J. Hwang and H. Noh, *Phys. Rev.* **D65**, 124010 (2002); hep-th/0203193; F. Finelli and R. Brandenberger, *Phys. Rev.* **D65**, 103522 (2002); C. Gordon and N. Turok, hep-th/0206138 J. Martin, P. Peter, N. P. Neto and D. J. Schwarz, *Phys. Rev.* **D65**, 123513 (2002); J. Garriga and T. Tanaka, *Phys. Rev.* **D65**, 103506 (2002).
22. D. Langlois, K. Maeda and D. Wands, *Phys. Rev. Lett.* **88**, 181301 (2002).
23. H. Kudoh and T. Tanaka, *Phys. Rev.* **D64**, 084022 (2001).
24. A. Chamblin, S. W. Hawking and H. S. Reall, *Phys. Rev.* **D61**, 065007 (2000).
25. R. Gregory and R. Laflamme, *Phys. Rev. Lett.* **70**, 2837 (1993);
26. R. Gregory, *Class. Quantum Grav.* **17**, L125 (2000).
27. G. Gibbons and S. Hartnoll, hep-th/0206202.
28. R. Emparan, G. T. Horowitz and R. C. Myers, *JHEP* **0001**, 007 (2000).
29. N. Dadhich, R. Maartens, P. Papadopoulos and V. Rezania, *Phys. Lett.* **B487**, 1 (2000).
30. A. Chabmlin, H. S. Reall, H. Shinkai and T. Shiromizu, *Phys. Rev.* **D63**, 064015 (2001).
31. T. Shiromizu and M. Shibata, *Phys. Rev.* **D62**, 127502 (2000).
32. T. Wiseman, *Phys. Rev.* **D65**, 124007 (2002).
33. I. Giannakis and H. Ren, *Phys. Rev.* **D63**, 125017 (2001).
34. M. Bruni, C. Germani and R. Maartens, *Phys. Rev. Lett.* **87**, 231302 (2001).
35. P.C. Argyres, S. Dimopoulos and J. March-Russell, *Phys. Lett.* **B441**, 96 (1998); R. Emparan, G.T. Horowitz and R.C. Myers, *Phys. Rev. Lett.* **85**, 499 (2000); S. Dimupoulos and G. Landsberg, *Phys. Rev. Lett.* **87**, 161602 (2001); S.B. Giddings and S. Thomas, hep-th/0106219.
36. R. Emparan and H. S. Reall, *Phys. Rev. Lett.* **88**, 101101 (2002).
37. R. C. Myers and M.J. Perry, *Ann. Phys.* **172**, 304 (1986).
38. G. W. Gibbons, D. Ida and T. Shiromizu, gr-qc/0203004(to appear in the proceedings of YITP workshop "*Braneworld: Dynamics of Space-time Boundary*" Kyoto, Japan, 15-18 Jan 2002); *Phys. Rev. Lett.* **89**, 041101 (2002); hep-th/0206136(to appear in *Phys. Rev.* **D**); M. Rogatko, hep-th/0207187.
39. S. Hwang, Geom. Dedic. **71**, 5(1998).
40. W. Israel, *Phys. Rev.* **164**, 1776 (1967); W. Israel, *Commun. Math. Phys.* **8**, 245 (1968); P. Ruback, *Class. Quantum Grav.* **5**, L155 (1988); G.L. Bunting and A.K.M. Masood-ul-Alam, *Gen. Rel. Grav.* **19**, 147 (1987); A. K. M. Masood-ul-Alam, *Class. Quantum Grav.* **9**, L53 (1992); A. K. M. Masood-

ul-Alam, *Class. Quantum Grav.* **10**, 2649 (1993); M. Mars and W. Simon, gr-qc/0105023.
41. K. S. Thorne, in *Magic without Magic*, ed. J. Klauder(W.H. Freeman, San Francisco, 1972)
42. D. Ida and K. Nakao, gr-qc/0204082.
43. S. S. Gubser, *Phys. Rev.* **D63**, 084017 (2001).
44. S. Nojiri, S. D. Odintsov and S. Zerbini, *Phys. Rev.* **D62**, 64006 (2000); S. Nojiri and S. D. Odintsov, *Phys. Lett.* **B484**, 119 (2000); L. Anchordoqui, C. Nunez and K. Olsen, *JHEP* **0010**, 050 (2000); S. W. Hawking, T. Hertog and H. S. Reall, *Phys. Rev.* **D62**, 043501 (2000); K. Koyama and J. Soda, hep-th/0101164; T. Shiromizu, T. Torii and D. Ida, *JHEP* **0203**, 007 (2002); A. Padilla, *Phys. Lett.* **B528**, 274 (2002).
45. T. Shiromizu and D. Ida, *Phys. Rev.* **D64**, 044015 (2001).
46. S. Kanno and J. Soda, hep-th/0205188.
47. O. Aharony, S. S. Gubser, J. Maldacena, H. Ooguri and Y. Oz, *Phys. Rep.* **323**, 183 (2000).
48. S. de Haro, K. Skenderis and S. N. Solodukhin, hep-th/0002230.
49. S. de Haro, K. Skenderis and S. N. Solodukhin, hep-th/0011230.
50. M. Henningson and K. Skenderis, *JHEP* **9807**, 023 (1998); P. Kraus, F. Larsen and R. Siebelink, *Nucl. Phys.* **B563**, 259 (1999).
51. T. Tanaka, gr-qc/0203082, to appear in in the proceedings of YITP Workshop "*Braneworld: Dynamics of Space-time Boundary*" Kyoto, Japan, 15-18 Jan 2002
52. R. Emparan, A. Fabbri and N. Kaloper, hep-th/0206155.

Searching for Supersymmetric Dark Matter

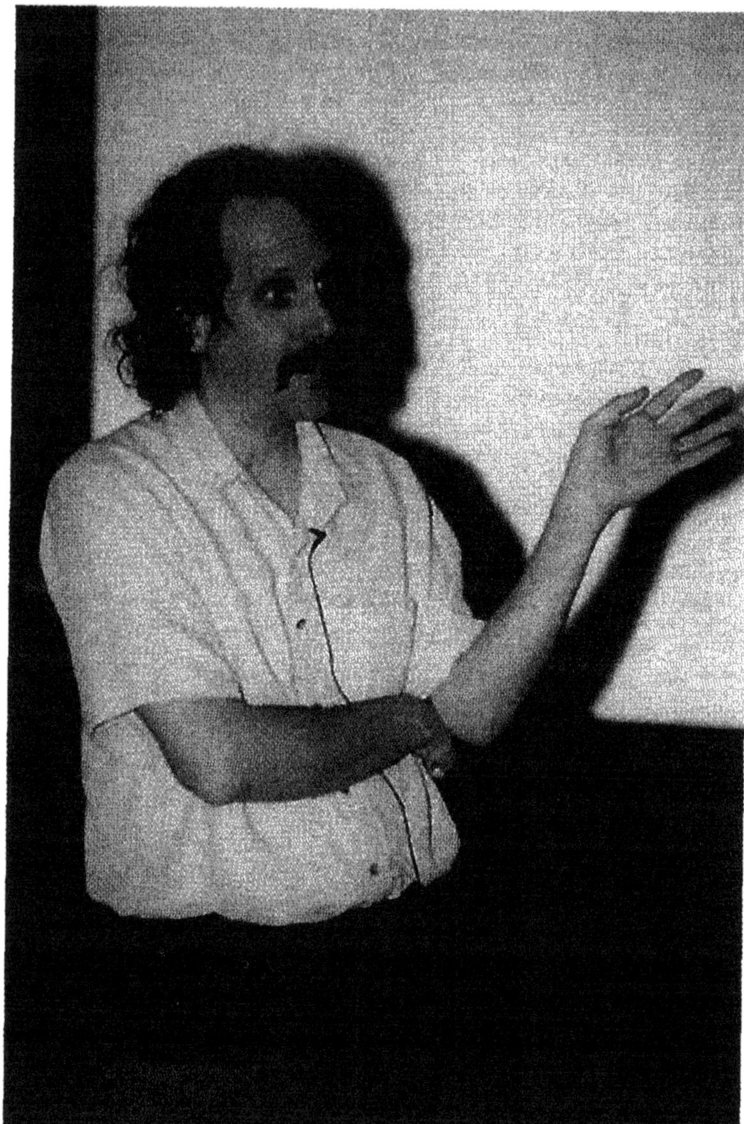

Keith Olive

SEARCHING FOR SUPERSYMMETRIC DARK MATTER

KEITH A. OLIVE

Theoretical Physics Institute, School of Physics and Astronomy, University of Minnesota, Minneapolis, MN 55455 USA

E-mail: olive@umn.edu

The supersymmetric extension to the Standard Model offers a promising cold dark matter candidate, the lightest neutralino. I will review the prospects for the detection of this candidate in both accelerator and direct detection searches.

1. Introduction

Although there are many reasons for considering supersymmetry as a candidate extension to the standard model of strong, weak and electromagnetic interactions,[1] one of the most compelling is its role in understanding the hierarchy problem[2] namely, why/how is $m_W \ll M_P$. One might think naively that it would be sufficient to set $m_W \ll M_P$ by hand. However, radiative corrections tend to destroy this hierarchy. For example, one-loop diagrams generate

$$\delta m_W^2 = \mathcal{O}\left(\frac{\alpha}{\pi}\right) \Lambda^2 \gg m_W^2 \tag{1}$$

where Λ is a cut-off representing the appearance of new physics, and the inequality in (1) applies if $\Lambda \sim 10^3$ TeV, and even more so if $\Lambda \sim m_{GUT} \sim 10^{16}$ GeV or $\sim M_P \sim 10^{19}$ GeV. If the radiative corrections to a physical quantity are much larger than its measured values, obtaining the latter requires strong cancellations, which in general require fine tuning of the bare input parameters. However, the necessary cancellations are natural in supersymmetry, where one has equal numbers of bosons B and fermions F with equal couplings, so that (1) is replaced by

$$\delta m_W^2 = \mathcal{O}\left(\frac{\alpha}{\pi}\right) |m_B^2 - m_F^2| \, . \tag{2}$$

The residual radiative correction is naturally small if $|m_B^2 - m_F^2| \lesssim 1$ TeV^2.

In order to justify the absence of interactions which can be responsible for extremely rapid proton decay, it is common in the minimal supersymmetric standard model (MSSM) to assume the conservation of R-parity. If

R-parity, which distinguishes between "normal" matter and the supersymmetric partners and can be defined in terms of baryon, lepton and spin as $R = (-1)^{3B+L+2S}$, is unbroken, there is at least one supersymmetric particle (the lightest supersymmetric particle or LSP) which must be stable. Thus, the minimal model contains the fewest number of new particles and interactions necessary to make a consistent theory.

There are very strong constraints, however, forbidding the existence of stable or long lived particles which are not color and electrically neutral.[3] Strong and electromagnetically interacting LSPs would become bound with normal matter forming anomalously heavy isotopes. Indeed, there are very strong upper limits on the abundances, relative to hydrogen, of nuclear isotopes[4], $n/n_H \lesssim 10^{-15}$ to 10^{-29} for 1 GeV $\lesssim m \lesssim$ 1 TeV. A strongly interacting stable relic is expected to have an abundance $n/n_H \lesssim 10^{-10}$ with a higher abundance for charged particles.

There are relatively few supersymmetric candidates which are not colored and are electrically neutral. The sneutrino[5] is one possibility, but in the MSSM, it has been excluded as a dark matter candidate by direct[6] and indirect[7] searches. In fact, one can set an accelerator based limit on the sneutrino mass from neutrino counting, $m_{\tilde{\nu}} \gtrsim 44.7$ GeV.[8] In this case, the direct relic searches in underground low-background experiments require $m_{\tilde{\nu}} \gtrsim 20$ TeV.[6] Another possibility is the gravitino which is probably the most difficult to exclude. I will concentrate on the remaining possibility in the MSSM, namely the neutralinos.

2. Parameters

The most general version of the MSSM, despite its minimality in particles and interactions contains well over a hundred new parameters. The study of such a model would be untenable were it not for some (well motivated) assumptions. These have to do with the parameters associated with supersymmetry breaking. It is often assumed that, at some unification scale, all of the gaugino masses receive a common mass, $m_{1/2}$. The gaugino masses at the weak scale are determined by running a set of renormalization group equations. Similarly, one often assumes that all scalars receive a common mass, m_0, at the GUT scale. These too are run down to the weak scale. The remaining parameters of importance involve the Higgs sector. There is the Higgs mixing mass parameter, μ, and since there are two Higgs doublets in the MSSM, there are two vacuum expectation values. One combination of these is related to the Z mass, and therefore is not a free parameter, while the other combination, the ratio of the two vevs, $\tan\beta$, is free.

If the supersymmetry breaking Higgs soft masses are also unified at the

GUT scale (and take the common value m_0), then μ and the physical Higgs masses at the weak scale are determined by electroweak vacuum conditions. This scenario is often referred to as the constrained MSSM or CMSSM. Once these parameters are set, the entire spectrum of sparticle masses at the weak scale can be calculated.

3. Neutralinos

There are four neutralinos, each of which is a linear combination of the $R = -1$ neutral fermions:[3] the wino $\tilde{W}^3$, the partner of the 3rd component of the $SU(2)_L$ gauge boson; the bino, $\tilde{B}$, the partner of the $U(1)_Y$ gauge boson; and the two neutral Higgsinos, $\tilde{H}_1$ and $\tilde{H}_2$. Assuming gaugino mass universality at the GUT scale, the identity and mass of the LSP are determined by the gaugino mass $m_{1/2}$, μ, and $\tan\beta$. In general, neutralinos can be expressed as a linear combination

$$\chi = \alpha\tilde{B} + \beta\tilde{W}^3 + \gamma\tilde{H}_1 + \delta\tilde{H}_2 \tag{3}$$

The solution for the coefficients α, β, γ and δ for neutralinos that make up the LSP can be found by diagonalizing the mass matrix

$$(\tilde{W}^3, \tilde{B}, \tilde{H}_1^0, \tilde{H}_2^0) \begin{pmatrix} M_2 & 0 & \frac{-g_2 v_1}{\sqrt{2}} & \frac{g_2 v_2}{\sqrt{2}} \\ 0 & M_1 & \frac{g_1 v_1}{\sqrt{2}} & \frac{-g_1 v_2}{\sqrt{2}} \\ \frac{-g_2 v_1}{\sqrt{2}} & \frac{g_1 v_1}{\sqrt{2}} & 0 & -\mu \\ \frac{g_2 v_2}{\sqrt{2}} & \frac{-g_1 v_2}{\sqrt{2}} & -\mu & 0 \end{pmatrix} \begin{pmatrix} \tilde{W}^3 \\ \tilde{B} \\ \tilde{H}_1^0 \\ \tilde{H}_2^0 \end{pmatrix} \tag{4}$$

where $M_1(M_2)$ is a soft supersymmetry breaking term giving mass to the U(1) (SU(2)) gaugino(s). In a unified theory $M_1 = M_2$ at the unification scale (at the weak scale, $M_1 \simeq \frac{5}{3}\frac{\alpha_1}{\alpha_2} M_2$). As one can see, the coefficients α, β, γ, and δ depend only on $m_{1/2}$, μ, and $\tan\beta$.

In Figure 1,[9] regions in the M_2, μ plane with $\tan\beta = 2$ are shown in which the LSP is one of several nearly pure states, the photino, $\tilde{\gamma}$, the bino, $\tilde{B}$, a symmetric combination of the Higgsinos, $\tilde{H}_{(12)}$, or the Higgsino, $\tilde{S} = \sin\beta\tilde{H}_1 + \cos\beta\tilde{H}_2$. The dashed lines show the LSP mass contours. The cross hatched regions correspond to parameters giving a chargino $(\tilde{W}^{\pm}, \tilde{H}^{\pm})$ state with mass $m_{\tilde{\chi}} \leq 45 GeV$ and as such are excluded by LEP.[10] This constraint has been extended by LEP[11] and is shown by the light shaded region and corresponds to regions where the chargino mass is $\lesssim 104$ GeV. The newer limit does not extend deep into the Higgsino region because of the degeneracy between the chargino and neutralino. Notice that the parameter space is dominated by the $\tilde{B}$ or $\tilde{H}_{12}$ pure states and that the photino only occupies a small fraction of the parameter space, as does the Higgsino combination $\tilde{S}$. Both of these light states are experimentally excluded.

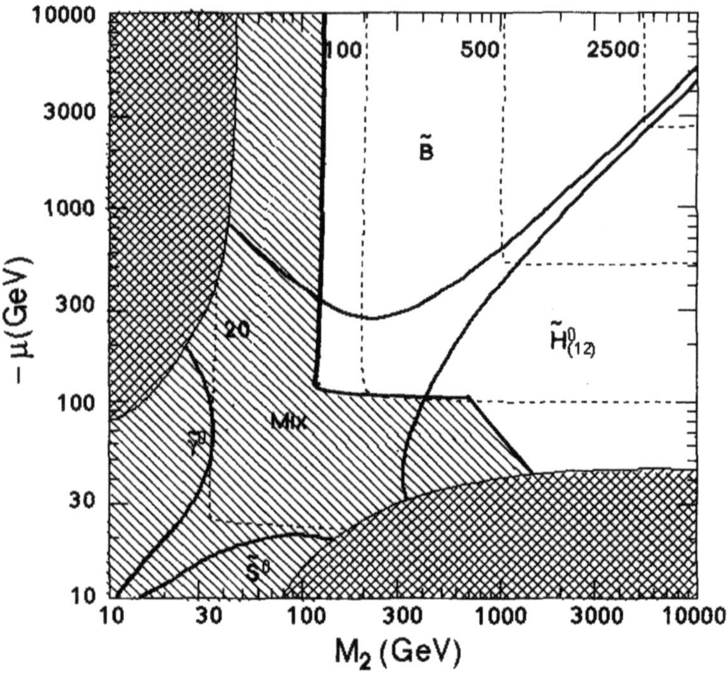

Figure 1. Mass contours and composition of nearly pure LSP states in the MSSM.[9]

4. The Relic Density

The relic abundance of LSP's is determined by solving the Boltzmann equation for the LSP number density in an expanding Universe. The technique[12] used is similar to that for computing the relic abundance of massive neutrinos.[13] The relic density depends on additional parameters in the MSSM beyond $m_{1/2}$, μ, and $\tan\beta$. These include the sfermion masses, $m_{\tilde{f}}$ and the Higgs pseudo-scalar mass, m_A,[a] derived from m_0 (and $m_{1/2}$). To determine the relic density it is necessary to obtain the general annihilation cross-section for neutralinos. In much of the parameter space of

[a]In general, the relic density depends on the supersymmetry-breaking tri-linear masses A (also assumed to be unified at the GUT scale) as well as two phases θ_μ and θ_A.

interest, the LSP is a bino and the annihilation proceeds mainly through sfermion exchange. Because of the p-wave suppression associated with Majorana fermions, the s-wave part of the annihilation cross-section is suppressed by the outgoing fermion masses. This means that it is necessary to expand the cross-section to include p-wave corrections which can be expressed as a term proportional to the temperature if neutralinos are in equilibrium. Unless the neutralino mass happens to lie near near a pole, such as $m_\chi \simeq m_Z/2$ or $m_h/2$, in which case there are large contributions to the annihilation through direct s-channel resonance exchange, the dominant contribution to the $\tilde{B}\tilde{B}$ annihilation cross section comes from crossed t-channel sfermion exchange.

Annihilations in the early Universe continue until the annihilation rate $\Gamma \simeq \sigma v n_\chi$ drops below the expansion rate given by the Hubble parameter, H. For particles which annihilate through approximate weak scale interactions, this occurs when $T \sim m_\chi/20$. Subsequently, the relic density of neutralinos is fixed relative to the number of relativistic particles. As noted above, the number density of neutralinos is tracked by a Boltzmann-like equation,

$$\frac{dn}{dt} = -3\frac{\dot{R}}{R}n - \langle\sigma v\rangle(n^2 - n_0^2) \tag{5}$$

where n_0 is the equilibrium number density of neutralinos. By defining the quantity $f = n/T^3$, we can rewrite this equation in terms of x, as

$$\frac{df}{dx} = m_\chi \left(\frac{1}{90}\pi^2\kappa^2 N\right)^{1/2} (f^2 - f_0^2) \tag{6}$$

The solution to this equation at late times (small x) yields a constant value of f, so that $n \propto T^3$. The final relic density expressed as a fraction of the critical energy density can be written as [3]

$$\Omega_\chi h^2 \simeq 1.9 \times 10^{-11} \left(\frac{T_\chi}{T_\gamma}\right)^3 N_f^{1/2} \left(\frac{\text{GeV}}{a x_f + \frac{1}{2} b x_f^2}\right) \tag{7}$$

where $(T_\chi/T_\gamma)^3$ accounts for the subsequent reheating of the photon temperature with respect to χ, due to the annihilations of particles with mass $m < x_f m_\chi$.[14] The subscript f refers to values at freeze-out, i.e., when annihilations cease. The coefficients a and b are related to the partial wave expansion of the cross-section, $\sigma v = a + bx + \ldots$. Eq. (7) results in a very good approximation to the relic density expect near s-channel annihilation poles, thresholds and in regions where the LSP is nearly degenerate with the next lightest supersymmetric particle.[15]

5. Phenomenological and Cosmological Constraints

For the cosmological limits on the relic density I will assume

$$0.1 \leq \Omega_\chi h^2 \leq 0.3. \tag{8}$$

The upper limit being a conservative bound based only on the lower limit to the age of the Universe of 12 Gyr. Indeed, most analyses indicate that $\Omega_{\rm matter} \lesssim 0.4 - 0.5$ and thus it is very likely that $\Omega_\chi h^2 < 0.2$. One should note that smaller values of $\Omega_\chi h^2$ are allowed, since it is quite possible that some of the cold dark matter might not consist of LSPs.

The calculated relic density is found to have a relevant cosmological density over a wide range of susy parameters. For all values of $\tan\beta$, there is a 'bulk' region with relatively low values of $m_{1/2}$ and m_0 where $0.1 < \Omega_\chi h^2 < 0.3$. However there are a number of regions at large values of $m_{1/2}$ and/or m_0 where the relic density is still compatible with the cosmological constraints. At large values of $m_{1/2}$, the lighter stau, becomes nearly degenerate with the neutralino and co-annihilations between these particles must be taken into account.[16] For non-zero values of A_0, there are new regions for which $\chi - \tilde{t}$ coannihilations are important.[17] At large $\tan\beta$, as one increases $m_{1/2}$, the pseudo-scalar mass, m_A begins to drop so that there is a wide funnel-like region (at all values of m_0) such that $2m_\chi \approx m_A$ and s-channel annihilations become important.[18,19] Finally, there is a region at very high m_0 where the value of μ begins to fall and the LSP becomes more Higgsino-like. This is known as the 'focus point' region.[20]

As an aid to the assessment of the prospects for detecting sparticles at different accelerators, benchmark sets of supersymmetric parameters have often been found useful, since they provide a focus for concentrated discussion. A set of proposed post-LEP benchmark scenarios[21] in the CMSSM are illustrated schematically in Fig. 2. Five of the chosen points are in the 'bulk' region at small $m_{1/2}$ and m_0, four are spread along the coannihilation 'tail' at larger $m_{1/2}$ for various values of $\tan\beta$. This tail runs along the shaded region in the lower right corner where the stau is the LSP and is therefore excluded by the constraints against charged dark matter. Two points are in rapid-annihilation 'funnels' at large $m_{1/2}$ and m_0. At large values of m_0, the focus-point region runs along the boundary where electroweak symmetry no longer occurs (shown in Fig. 2 as the shaded region in the upper left corner). Two points were chosen in the focus-point region at large m_0. The proposed points range over the allowed values of $\tan\beta$ between 5 and 50. The light shaded region corresponds to the portion of parameter space where the relic density $\Omega_\chi h^2$ is between 0.1 and 0.3.

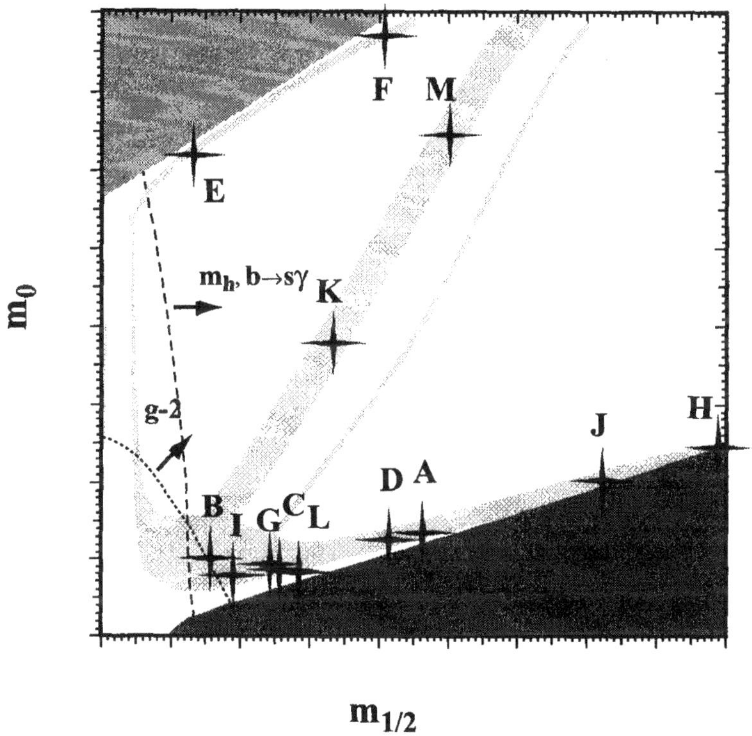

Figure 2. Schematic overview of the CMSSM benchmark points proposed in Ref. 21. The points are intended to illustrate the range of available possibilities. The labels correspond to the approximate positions of the benchmark points in the $(m_{1/2}, m_0)$ plane. They also span values of $\tan\beta$ from 5 to 50 and include points with $\mu < 0$.

The most important phenomenological constraints are also shown schematically in Figure 2. These include the constraint provided by the LEP lower limit on the Higgs mass: $m_H > 114.1$ GeV.[22] This holds in the Standard Model, for the lightest Higgs boson h in the general MSSM for $\tan\beta \lesssim 8$, and almost always in the CMSSM for all $\tan\beta$. Since m_h is sensitive to sparticle masses, particularly $m_{\tilde{t}}$, via loop corrections, the Higgs limit also imposes important constraints on the CMSSM parameters, principally $m_{1/2}$ as seen by the dashed curve in Fig. 2. The constraint imposed by measurements of $b \to s\gamma$ [23] also exclude small values of $m_{1/2}$. These measurements agree with the Standard Model, and therefore provide bounds on MSSM particles, such as the chargino and charged Higgs masses, in particular. Typically, the $b \to s\gamma$ constraint is more important for $\mu < 0$, but it is also relevant for $\mu > 0$, particularly when $\tan\beta$

is large. The BNL E821 experiment reported last year a new measurement of $a_\mu \equiv \frac{1}{2}(g_\mu - 2)$ which deviated by 2.6 standard deviations from the best Standard Model prediction available at that time.[24] The largest contribution to the errors in the comparison with theory was thought to be the statistical error of the experiment, which has been significantly reduced just recently.[25] However, it has recently been realized that the sign of the most important pseudoscalar-meson pole part of the light-by-light scattering contribution[26] to the Standard Model prediction should be reversed, which reduces the apparent experimental discrepancy to about 1.6 standard deviations ($\delta a_\mu \times 10^{10} = 26 \pm 16$). With the new data, the discrepancy with theory ranges from 1.6 to 2.6 σ, i.e., $\delta a_\mu \times 10^{10} = 26 \pm 10$ to 17 ± 11.[25] This constraint excludes very small values of $m_{1/2}$ and m_0. In Fig. 2, the $g-2$ constraint is shown schematically by the dotted line. It may also exclude very large values of the parameters as well as negative values of μ, if the discrepancy holds up.

Following a previous analysis,[27] in Figure 3 the $m_{1/2} - m_0$ parameter space is shown for $\tan\beta = 10$. The dark shaded region (in the lower right) corresponds to the parameters where the LSP is not a neutralino but rather a $\tilde{\tau}_R$. The cosmologically interesting region at the left of the figure is due to the appearance of pole effects. There, the LSP can annihilate through s-channel Z and h (the light Higgs) exchange, thereby allowing a very large value of m_0. However, this region is excluded by phenomenological constraints. Here one can see clearly the coannihilation tail which extends towards large values of $m_{1/2}$. In addition to the phenomenological constraints discussed above, Figure 3 also shows the current experimental constraints on the CMSSM parameter space due to the limit $m_{\chi^\pm} \gtrsim 103.5$ GeV provided by chargino searches at LEP.[11] LEP has also provided lower limits on slepton masses, of which the strongest is $m_{\tilde{e}} \gtrsim 99$ GeV.[28] This is shown by dot-dashed curve in the lower left corner of Fig. 3.

As one can see, one of the most important phenomenological constraint at this value of $\tan\beta$ is due to the Higgs mass (shown by the nearly vertical dot-dashed curve). The theoretical Higgs masses were evaluated using `FeynHiggs`,[29] which is estimated to have a residual uncertainty of a couple of GeV in m_h. The region excluded by the $b \to s\gamma$ constraint is the dark shaded (green) region to the left of the plot.

As many authors have pointed out,[30] a discrepancy between theory and the BNL experiment could well be explained by supersymmetry. As seen in Fig. 3, this is particularly easy if $\mu > 0$. The medium (pink) shaded region in the figure corresponds to the overall allowed region by the new experimental result: $-5 < \delta a_\mu \times 10^{10} < 46$. The two solid lines within the

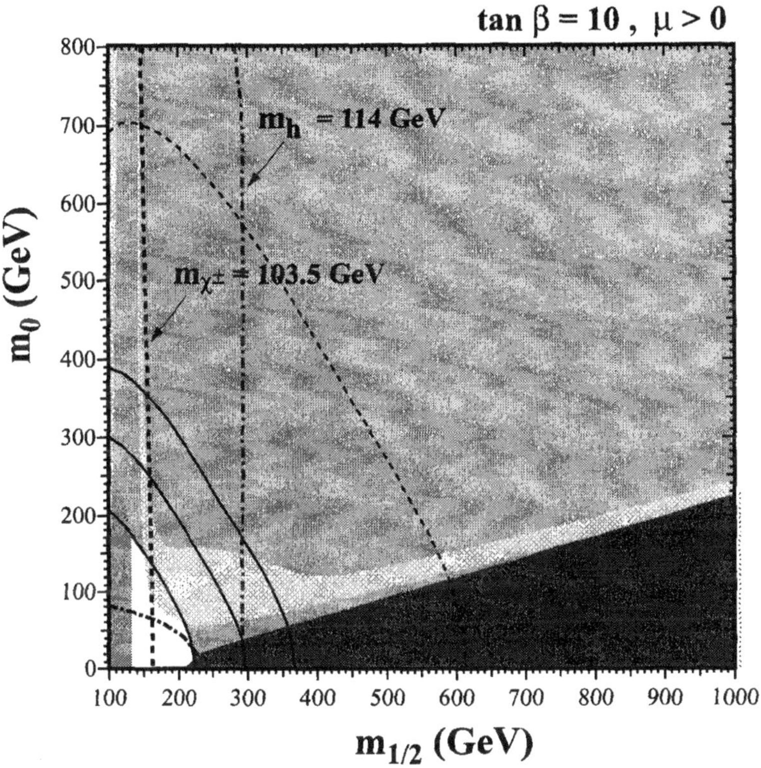

Figure 3. Compilation of phenomenological constraints on the CMSSM for $\tan\beta = 10, \mu > 0$, assuming $A_0 = 0, m_t = 175$ GeV and $m_b(m_b)^{\overline{MS}}_{SM} = 4.25$ GeV. The near-vertical lines are the LEP limits $m_{\chi^\pm} = 103.5$ GeV (dashed and black)[11], and $m_h = 114.1$ GeV (dotted and red)[22]. Also, in the lower left corner we show the $m_{\tilde{e}} = 99$ GeV contour.[28] In the dark (brick red) shaded regions, the LSP is the charged $\tilde{\tau}_1$, so this region is excluded. The light(turquoise) shaded areas are the cosmologically preferred regions with $0.1 \leq \Omega h^2 \leq 0.3$.[19] The medium (dark green) shaded regions are excluded by $b \to s\gamma$.[23] The shaded (pink) region in the upper right delineates the $2\,\sigma$ range of $g_\mu - 2$.

shaded region corresponds to the central values $\delta a_\mu \times 10^{10} = 17$ and 26 respectively. The optimistic 2σ lower bound of $\delta a_\mu \times 10^{10} = 6$ is shown as a dashed curve.

As discussed above, another mechanism for extending the allowed CMSSM region to large m_χ is rapid annihilation via a direct-channel pole when $m_\chi \sim \frac{1}{2} m_A$.[18,19] This may yield a 'funnel' extending to large $m_{1/2}$ and m_0 at large $\tan\beta$, as seen in Fig. 4.

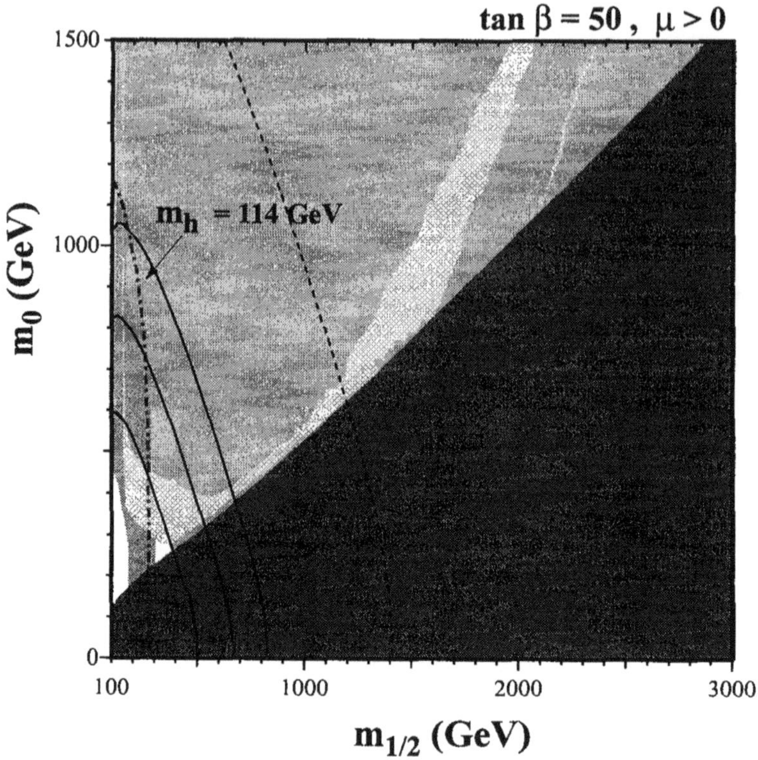

Figure 4. As in Fig. 3 for $\tan\beta = 50$.

In principle the true input parameters in the CMSSM are: μ, m_1, m_2, and B, where m_1 and m_2 are the Higgs soft masses (in the CMSSM $m_1 = m_2 = m_0$ and B is the susy breaking bilinear mass term). In this case, the electroweak symmetry breaking conditions lead to a prediction of $M_Z, \tan\beta$,and m_A. Since we are not really interested in predicting M_Z, it is more useful to assume instead the following CMSSM input parameters: M_Z, m_1, m_2, and $\tan\beta$ again with $m_1 = m_2 = m_0$. In this case, one predicts μ, B, and m_A. However, one can generalize the CMSSM case to include non-universal Higgs masses (NUHM), in which case the input parameters become:M_Z, μ, m_A, and $\tan\beta$ and one predicts m_1, m_2, and B.

The NUHM parameter space was recently analyzed[31] and a sample of the results found is shown in Fig. 5. While much of the cosmologically preferred area with $\mu < 0$ is excluded, there is a significant enhancement in the allowed parameter space for $\mu > 0$.

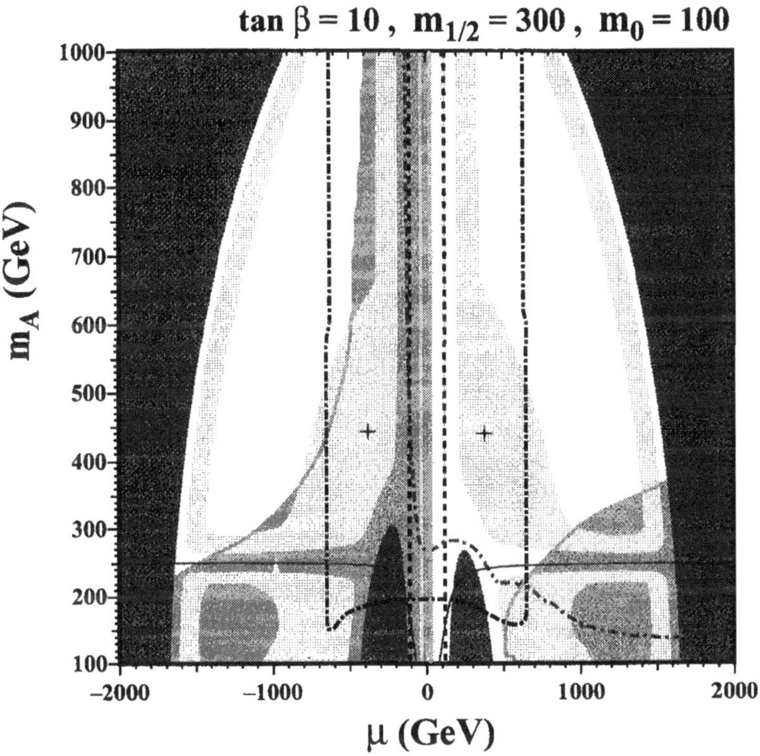

Figure 5. Compilations of phenomenological constraints on the MSSM with NUHM in the (μ, m_A) plane for $\tan\beta = 10$ and $m_0 = 100$ GeV, $m_{1/2} = 300$ GeV, assuming $A_0 = 0$, $m_t = 175$ GeV and $m_b(m_b)^{\overline{MS}}_{SM} = 4.25$ GeV. The shading is as described in Fig. 3. The (blue) solid line is the contour $m_\chi = m_A/2$, near which rapid direct-channel annihilation suppresses the relic density. The dark (black) dot-dashed line indicates when one or another Higgs mass-squared becomes negative at the GUT scale: only lower $|\mu|$ and larger m_A values are allowed. The crosses denote the values of μ and m_A found in the CMSSM.

5.1. *Detection*

Because the LSP as dark matter is present locally, there are many avenues for pursuing dark matter detection. Direct detection techniques rely on an ample neutralino-nucleon scattering cross-section. The effective four-fermion lagrangian can be written as

$$\begin{aligned}\mathcal{L} = & \bar{\chi}\gamma^\mu\gamma^5\chi\bar{q_i}\gamma_\mu(\alpha_{1i} + \alpha_{2i}\gamma^5)q_i \\ & + \alpha_{3i}\bar{\chi}\chi\bar{q_i}q_i + \alpha_{4i}\bar{\chi}\gamma^5\chi\bar{q_i}\gamma^5 q_i \\ & + \alpha_{5i}\bar{\chi}\chi\bar{q_i}\gamma^5 q_i + \alpha_{6i}\bar{\chi}\gamma^5\chi\bar{q_i}q_i \end{aligned} \quad (9)$$

However, the terms involving $\alpha_{1i}, \alpha_{4i}, \alpha_{5i}$, and α_{6i} lead to velocity dependent elastic cross sections. The remaining terms are: the spin dependent coefficient, α_{2i} and the scalar coefficient α_{3i}. Contributions to α_{2i} are predominantly through light squark exchange. This is the dominant channel for binos. Scattering also occurs through Z exchange but this channel requires a strong Higgsino component. Contributions to α_{3i} are also dominated by light squark exchange but Higgs exchange is non-negligible in most cases.

The results from a CMSSM and MSSM analysis[32,33] for $\tan\beta = 3$ and 10 are compared with the most recent CDMS[34] and Edelweiss[35] bounds in Fig. 6. These results have nearly entirely excluded the region purported by the DAMA[36] experiment. The CMSSM prediction[32] is shown by the dark shaded region, while the NUHM case[33] is shown by the larger lighter shaded region.

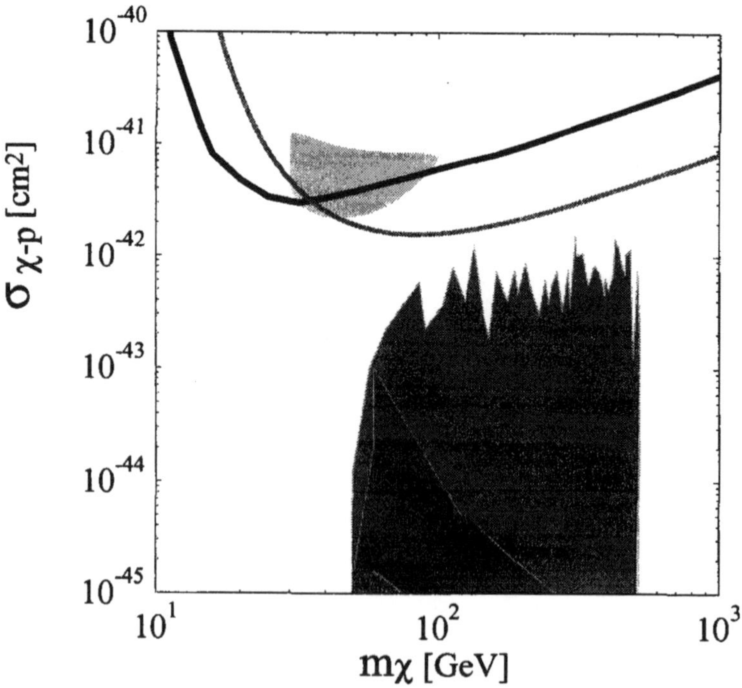

Figure 6. Limits from the CDMS[34] and Edelweiss[35] experiments on the neutralino-proton elastic scattering cross section as a function of the neutralino mass. The Edelweiss limit is stronger at higher m_χ. These results nearly exclude the shaded region observed by DAMA.[36] The theoretical predictions lie at lower values of the cross section.

I conclude by showing the prospects for direct detection for the benchmark points discussed above.[37] Fig. 7 shows rates for the elastic spin-independent scattering of supersymmetric relics, including the projected sensitivities for CDMS II [38] and CRESST[39] (solid) and GENIUS[40] (dashed). Also shown are the cross sections calculated in the proposed benchmark scenarios discussed in the previous section, which are considerably below the DAMA[36] range ($10^{-5}-10^{-6}$ pb). Indirect searches for supersymmetric dark matter via the products of annihilations in the galactic halo or inside the Sun also have prospects in some of the benchmark scenarios.[37]

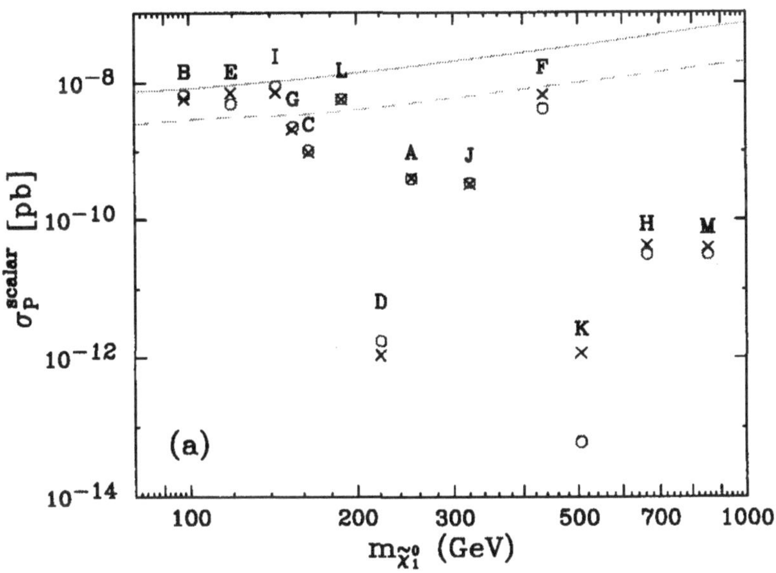

Figure 7. Elastic spin-independent scattering of supersymmetric relics on protons calculated in benchmark scenarios[37], compared with the projected sensitivities for CDMS II [38] and CRESST[39] (solid) and GENIUS[40] (dashed). The predictions of our code (blue crosses) and `Neutdriver`[41] (red circles) for neutralino-nucleon scattering are compared. The labels A, B, ...,L correspond to the benchmark points as shown in Fig. 2.

Acknowledgments

I would like to thank J. Ellis, T. Falk, A. Ferstl, G. Ganis, Y. Santoso, and M. Srednicki for enjoyable collaborations from which this work is culled. This work was supported in part by DOE grant DE-FG02-94ER40823 at Minnesota.

References

1. J. Wess and J. Bagger, *Supersymmetry and Supergravity*, (Princeton University Press, Princeton NJ, 1992);
   G.G. Ross, *Grand Unified Theories*, (Addison-Wesley, Redwood City CA, 1985);
   S. Martin, arXiv:hep-ph/9709356;
   J. Ellis, arXiv:hep-ph/9812235;
   K.A. Olive, arXiv:hep-ph/9911307.
2. L. Maiani, *Proceedings of the 1979 Gif-sur-Yvette Summer School On Particle Physics*, 1;
   G. 't Hooft, in *Recent Developments in Gauge Theories, Proceedings of the Nato Advanced Study Institute, Cargese, 1979*, eds. G. 't Hooft *et al.*, (Plenum Press, NY, 1980);
   E. Witten, *Phys. Lett.* **B105**, 267 (1981).
3. J. Ellis, J.S. Hagelin, D.V. Nanopoulos, K.A. Olive and M. Srednicki, *Nucl. Phys.* **B238**, 453 (1984);
   see also H. Goldberg, *Phys. Rev. Lett.* **50**, 1419 (1983).
4. J. Rich, M. Spiro and J. Lloyd-Owen, *Phys.Rep.* **151**, 239 (1987);
   P.F. Smith, *Contemp.Phys.* **29**, 159 (1998);
   T.K. Hemmick et al., *Phys. Rev.* **D41**, 2074 (1990).
5. L.E. Ibanez, *Phys. Lett.* **137B**, 160 (1984);
   J. Hagelin, G.L. Kane, and S. Raby, *Nucl., Phys.* **B241**, 638 (1984);
   T. Falk, K.A. Olive, and M. Srednicki, *Phys. Lett.* **B339**, 248 (1994).
6. S. Ahlen, et. al., *Phys. Lett.* **B195**, 603 (1987);
   D.D. Caldwell, et. al., *Phys. Rev. Lett.* **61**, 510 (1988);
   M. Beck et al., *Phys. Lett.* **B336** 141 (1994).
7. see e.g. K.A. Olive and M. Srednicki, *Phys. Lett.* **205B**, 553 (1988).
8. The LEP Collaborations, the LEP Electroweak Working Group, and the SLD Heavy Flavour and Electroweak Groups, CERN-EP-2000-016.
9. K.A. Olive and M. Srednicki, *Phys. Lett.* **B230**, 78 (1989); *Nucl. Phys.* **B355**, 208 (1991).
10. ALEPH collaboration, D. Decamp et al., *Phys. Rep.* **216**, 253 (1992);
    L3 collaboration, M. Acciarri et al., *Phys. Lett.* **B350**, 109 (1995);
    OPAL collaboration, G. Alexander et al., *Phys. Lett.* **B377**, 273 (1996).
11. Joint LEP 2 Supersymmetry Working Group, *Combined LEP Chargino Results, up to 208 GeV*,
    `http://lepsusy.web.cern.ch/lepsusy/www/inos_moriond01/charginos_pub.html`.
12. R. Watkins, M. Srednicki and K.A. Olive, *Nucl. Phys.* **B310**, 693 (1988).
13. P. Hut, *Phys. Lett.* **69B**, 85 (1977);
    B.W. Lee and S. Weinberg, *Phys. Rev. Lett.* **39**, 165 (1977).
14. G. Steigman, K. A. Olive and D. N. Schramm, *Phys. Rev. Lett.* **43**, 239 (1979);
    K. A. Olive, D. N. Schramm and G. Steigman, *Nucl. Phys.* **B180**, 497 (1981).
15. K. Griest and D. Seckel, *Phys. Rev.* **D43**, 3191 (1991).
16. J. Ellis, T. Falk, and K. Olive, *Phys. Lett.* **B444**, 367 (1998);
    J. Ellis, T. Falk, K. Olive, and M. Srednicki, *Astr. Part. Phys.* **13**, 181 (2000) [Erratum-ibid. **15**, 413 (2000)];

M. E. Gómez, G. Lazarides and C. Pallis, *Phys. Rev.* **D61**, 123512 (2000); *Phys. Lett.* **B487**, 313 (2000);
R. Arnowitt, B. Dutta and Y. Santoso, *Nucl. Phys.* **B606**, 59 (2001).
17. C. Boehm, A. Djouadi and M. Drees, *Phys. Rev.* **D62**, 035012 (2000); J. Ellis, K.A. Olive and Y. Santoso, arXiv:hep-ph/0112113.
18. M. Drees and M. M. Nojiri, *Phys. Rev.* **D47**, 376 (1993);
H. Baer and M. Brhlik, *Phys. Rev.* **D53**, 597 (1996); *Phys. Rev.* **D57**, 567 (1998);
H. Baer, M. Brhlik, M. A. Diaz, J. Ferrandis, P. Mercadante, P. Quintana and X. Tata, *Phys. Rev.* **D63**, 015007 (2001);
A. B. Lahanas, D. V. Nanopoulos and V. C. Spanos, *Mod. Phys. Lett.* **A16** 1229 (2001).
19. J. R. Ellis, T. Falk, G. Ganis, K. A. Olive and M. Srednicki, *Phys. Lett.* **B510**, 236 (2001).
20. J. L. Feng, K. T. Matchev and T. Moroi, *Phys. Rev. Lett.* **84**, 2322 (2000);
J. L. Feng, K. T. Matchev and T. Moroi, *Phys. Rev.***D61**, 075005 (2000);
J. L. Feng, K. T. Matchev and F. Wilczek, *Phys. Lett.* **B482**, 388 (2000).
21. M. Battaglia et al., *Eur. Phys. J.* **C22**, 535 (2001).
22. LEP Higgs Working Group for Higgs boson searches, OPAL Collaboration, ALEPH Collaboration, DELPHI Collaboration and L3 Collaboration, *Search for the Standard Model Higgs Boson at LEP*, ALEPH-2001-066, DELPHI-2001-113, CERN-L3-NOTE-2699, OPAL-PN-479, LHWG-NOTE-2001-03, CERN-EP/2001-055, arXiv:hep-ex/0107029; *Searches for the neutral Higgs bosons of the MSSM: Preliminary combined results using LEP data collected at energies up to 209 GeV*, LHWG-NOTE-2001-04, ALEPH-2001-057, DELPHI-2001-114, L3-NOTE-2700, OPAL-TN-699, arXiv:hep-ex/0107030.
23. M.S. Alam et al., [CLEO Collaboration], *Phys. Rev. Lett.* **74**, 2885 (1995); as updated in S. Ahmed et al., CLEO CONF 99-10; BELLE Collaboration, BELLE-CONF-0003, contribution to the 30th International conference on High-Energy Physics, Osaka, 2000;
See also K. Abe *et al.*, [Belle Collaboration], [arXiv:hep-ex/0107065];
L. Lista [BaBar Collaboration], [arXiv:hep-ex/0110010];
C. Degrassi, P. Gambino and G. F. Giudice, *JHEP* **0012**, 009 (2000);
M. Carena, D. Garcia, U. Nierste and C. E. Wagner, *Phys. Lett.* **B499**, 141 (2001);
P. Gambino and M. Misiak, *Nucl. Phys.* **B611**, 338 (2001);
D. A. Demir and K. A. Olive, *Phys. Rev.* **D65**, 034007 (2002).
24. H. N. Brown *et al.* [Muon g-2 Collaboration], *Phys. Rev. Lett.* **86**, 2227 (2001).
25. G.W. Bennet *et al.* [Muon g-2 Collaboration], arXiv:hep-ex/0208001.
26. M. Knecht and A. Nyffeler, *Phys. Rev.* **D65**, 073034 (2002);
M. Knecht, A. Nyffeler, M. Perrottet and E. De Rafael, *Phys. Rev. Lett.* **88**, 071802 (2002);
M. Hayakawa and T. Kinoshita, arXiv:hep-ph/0112102;
I. Blokland, A. Czarnecki and K. Melnikov, *Phys. Rev. Lett.* **88**, 071803 (2002);
J. Bijnens, E. Pallante and J. Prades, *Nucl. Phys.* **B626**, 410 (2002).

27. J. R. Ellis, K. A. Olive and Y. Santoso, *New Jour. Phys.* **4**, 32 (2002).
28. Joint LEP 2 Supersymmetry Working Group, *Combined LEP Selectron/Smuon/Stau Results, 183-208 GeV*, `http://alephwww.cern.ch/~ganis/SUSYWG/SLEP/sleptons_2k01.html`.
29. S. Heinemeyer, W. Hollik and G. Weiglein, *Comput. Phys. Commun.* **124**, 76 (2000);
S. Heinemeyer, W. Hollik and G. Weiglein, *Eur. Phys. J.* **C9**, 343 (1999).
30. L. L. Everett, G. L. Kane, S. Rigolin and L. Wang, *Phys. Rev. Lett.* **86**, 3484 (2001);
J. L. Feng and K. T. Matchev, *Phys. Rev. Lett.* **86**, 3480 (2001);
E. A. Baltz and P. Gondolo, *Phys. Rev. Lett.* **86**, 5004 (2001);
U. Chattopadhyay and P. Nath, *Phys. Rev. Lett.* **86**, 5854 (2001);
S. Komine, T. Moroi and M. Yamaguchi, *Phys. Lett.* **B506**, 93 (2001);
J. Ellis, D. V. Nanopoulos and K. A. Olive, *Phys. Lett.* **B508**, 65 (2001);
R. Arnowitt, B. Dutta, B. Hu and Y. Santoso, *Phys. Lett.* **B505**, 177 (2001);
S. P. Martin and J. D. Wells, *Phys. Rev.* **D64**, 035003 (2001);
H. Baer, C. Balazs, J. Ferrandis and X. Tata, *Phys. Rev.* **D64**, 035004 (2001).
31. J. Ellis, K. Olive and Y. Santoso, *Phys. Lett.* **B539**, 107 (2002).
32. J. R. Ellis, A. Ferstl and K. A. Olive, *Phys. Lett.* **B481**, 304 (2000); see also: *Phys. Lett.* **B532**, 318 (2002).
33. J. R. Ellis, A. Ferstl and K. A. Olive, *Phys. Rev.* **D63**, 065016 (2001).
34. D. Abrams *et al.* [CDMS Collaboration], arXiv:astro-ph/0203500.
35. R. Jakob, arXiv:hep-ph/0206271.
36. DAMA Collaboration, R. Bernabei *et al.*, *Phys. Lett.* **B436** (1998) 379.
37. J. Ellis, J. L. Feng, A. Ferstl, K. T. Matchev and K. A. Olive, *Eur. Phys. J.* **C24**, 311 (2002) [arXiv:astro-ph/0110225].
38. CDMS Collaboration, R. W. Schnee *et al.*, *Phys. Rept.* **307**, 283 (1998).
39. CRESST Collaboration, M. Bravin *et al.*, *Astropart. Phys.* **12**, 107 (1999).
40. H. V. Klapdor-Kleingrothaus, arXiv:hep-ph/0104028.
41. G. Jungman, M. Kamionkowski and K. Griest, *Phys. Rept.* **267**, 195 (1996); `http://t8web.lanl.gov/people/jungman/neut-package.html`.

Baryogenesis and Electric Dipole Moments in Minimal Supersymmetric Standard Model

Darwin Chang

BARYOGENESIS AND ELECTRIC DIPOLE MOMENTS IN MINIMAL SUPERSYMMETRIC STANDARD MODEL

DARWIN CHANG

NCTS and Physics Department, National Tsing-Hua University, Hsinchu 30043, Taiwan, R.O.C.

We make a brief review of the issues related to electroweak baryogenesis in minimal supersymmetric standard model(MSSM). The constraints from electric dipole moments of electron and neutron are emphasized.

The Standard Model(SM) of particle physics continues to pass a wide array of experimental tests except for an on-going modification associated with its neutrino sector. However, it also has a few longstanding weaknesses. One of these weaknesses is associated with its anomalous $B+L$ symmetry. Due to the instanton solution associated with weak $SU(2)_L$, the instanton can mediate $B+L$ breaking interaction by inducing a higher dimensional coupling between all the fermions in the SM which are $SU(2)$ doublet. This higher dimensional effective interaction is suppressed by very high powers of electroweak symmetry breaking scale M_2 and therefore harmless even though it mediates baryon violating interactions.

However, in the higher temperature environment (for $T \geq M_2$) such as in the early Universe, an effect closely related to instanton, called sphaleron, will induce large baryon violating interaction. If such interaction is in thermal equilibrium, it will wash out the baryon number generated at even earlier Universe. Such sphaleron mediated interaction is generically in thermal equilibrium at high temperature, therefore it makes any net baryon number generated in the earlier Universe disappears.

To generate baryon asymmetry of the Universe(BAU), there are a few proposals in the literature:

(a)To make electroweak phase transition strongly first order such that a net BAU can be generated at M_2 scale by the out-of-equilibrium sphaleron interaction at the domain wall during the transition. This is called electroweak baryogenesis.

(b)To generate net lepton number, L, through some nonequilibrium lepton

number violating processes in the earlier Universe, say at $T \sim M_{GUT}$, then, since sphaleron interaction preserves the anomaly free $B-L$ while washing away $B+L$, a net BAU survives after the sphaleron wash-out. This mechanism uses the sphaleron to convert lepton number into baryon number.
(c)One can also take advantage of the fact that sphaleron interaction is only effective on the fermions which are $SU(2)_L$ doublet. If there are $SU(2)_L$ singlet fermions which carries lepton or baryon number, such as the right-handed neutrinos, then they will not be washed out. Take right-handed neutrinos as example, in the early Universe, a CP violating process may generate a net left-handed lepton number L_L as well as a net right-handed one, L_R, while preserving the total lepton number. If the Universe is such that L_L and L_R cannot be in thermal equilibrium until the temperature drops below M_2, then the sphaleron will only wash out $B+L_L$ and convert L_L to baryon number while keeping the L_R untouched. At a lower temperature $T < M_2$, L_L becomes in thermal equilibrium with L_R and partly washes out each other.

One of the major problem facing the SM is that its natural mechanism for CP violation, the Kobayashi-Maskawa(KM) mechanism, cannot provide enough CP violation to generate the BAU. In addition, in order for the electroweak phase transition to be strongly first order, it is necessary to have additional strongly interacting light scalar boson which does not exist in the SM particle spectrum. It is of course still possible that BAU is a consequence of lepton number asymmetry generated in the early Universe. However, if one prefers to implement electroweak baryogenesis, it is necessary to extend the SM. The most popular extension of the next generation of unified field theory is supersymmetric. This supersymmetric theory in its simplest form, MSSM,[1] may help to solve many of the outstanding problems in the Standard Model. Two examples are the coupling-constant-unification problem and the generation of BAU. It is the latter problem that will be discussed here.

To extend the SM in order to generate large enough BAU[2], particles lighter in mass but stronger in coupling are needed to make the electroweak transition more first order. Additionally, a new CP violating phase is required to generate enough BAU. It is very appealing that MSSM naturally provides a solution to both requirements.[3] Recently, this issue has caught a lot of attention in the literature. We wish to briefly review the issue and make emphasis on the constraints from electric dipole moments(EDM's) derived recently.

Baryon Asymmetry of the Universe

Big Bang nucleosynthesis has been one of the most successful theory in cosmology. The theory requires a net access of baryon number (over the anti-baryon number) of about

$$2 \leq \eta_{10} \equiv (n_B - n_{\bar{B}})/n_\gamma \times 10^{10} \leq 3, \quad (1)$$

at least in our observable Universe. Different analyses quote different limits. Here I use the latest summary in Ref. 4.

First of all, it is very unlikely that baryons and antibaryons got separated in the early Universe such that we observe only baryons in our corner of the Universe. (Even though there are theories that try to create such effect). Therefore, we assume that the asymmetry is homogeneous.

Secondly, in the very early Universe, at very high temperature, there were large numbers of both baryons and antibaryons. As the Universe cooled, most of the antibaryons annihilated with the baryons and turned into photons. The number of antibaryons and baryons at that very early epoch should be around the same order as the number of photons in the current Universe. It is hard to imagine that the Universe would be fine tuned at the very beginning to have a slight (one in 10^{10}) additional baryons. For this reason, it is considered one of the important problem in cosmology to be able to explain this asymmetry in the process of the cosmic evolution.

Conditions for Baryogenesis

The conditions of baryogenesis was spelled out by Sakarov long time ago. There are three general conditions:

(a)Need for baryon violating interactions.

There are two main sources of baryon number violation in particle physics. The first one involves grand unified theories (GUT baryogenesis). The second one involves anomalous $B + L$ current. The $B + L$ symmetry is anomalous with respect to the $SU(2)_L$ gauge interaction. As a result, the $SU(2)$ instantons and the related sphalerons (at high temperature) mediate $B + L$ violating interactions. In this review we are mainly concerned with the second mechanism, called electroweak baryogenesis.

(b)Need for C and CP violating interactions.

C symmetry is maximally broken in SM. The leading mechanisms of CP violation in the literature include: Kobayashi-Maskawa(KM) mechanism of SM; the spontaneous CP violation in multi-Higgs models; soft CP violation; and others. In SUSY models which we are most concern with here, in addition to KM mechanism, there are additional sources of CP violation belonging to the soft breaking category.

(c)The thermal condition of the Universe has to be such that the responsible B and CP violating process is *out of* thermal equilibrium at the time of baryogenesis such that the interaction biased the direction toward the baryon production instead of reduction. It typically requires the interaction rate to be slower than the expansion rate.

Baryogenesis in Standard Model and Beyond

There are many obstructions to electroweak baryogenesis in SM. First of all, the lattice gauge theory tends to conclude that the electroweak phase transition is NOT strongly first order. The consensus is that for the transition to be strongly enough first order, the Higgs mass should be lighter than $75GeV$ which is ruled out experimentally. Secondly, the CP violating mechanism of KM naturally comes with various flavor suppression factors such that the resulting BAU would be too small even if the transition were first order.

Even worse, because the transition is typically of second order, the sphaleron process is in thermal equilibrium at and above the transition scale M_2. It serves the opposite purpose of washing out any baryon asymmetry which might have been generated at higher temperature in earlier Universe.

For this reason, there is a flourish of proposals for baryogenesis in the literature beyond SM. Here we shall content with making a partial list of some of these proposals just to illustrate the multitude of ideas being toyed with before we proceed to concentrate on the MSSM case.

(a)Using the CP violating phases in the soft SUSY breaking terms in MSSM or beyond.

(b)Lepto-baryogenesis – using the lepton number asymmetry generated at GUT or similar higher energy scale and then using the sphalerons to convert them into the baryon number at electroweak scale.

(c)doublet-baryogenesis – using the fact that sphaleron is only sensitive to $SU(2)$ doublet. The mechanism generates asymmetries of doublet and singlet fermions at GUT or similar higher energy scale and then using the sphalerons to convert doublet asymmetry into the baryon number at electroweak scale.[5]

(d)Affleck-Dine mechanism – using the flat direction in D terms in SUSY scalar potential.

(e)Using large extra dimensions with Randall-Sundrum compactification.

(f)Using CP violations beyond SM, such as left-right models or multi-Higgs-doublet models.

(g)Using stable or unstable topological defects to generate asymmetry.

(e)Using decaying black holes which are known not to respect the global symmetry such as the baryon number.

Supersymmetry to the Rescue

Supersymmetric extensions of SM naturally provide many opportunities to avoid the obstructions for electroweak baryogenesis in SM.
(a)One can take advantage of the existence of many SUSY particles in such models to help make the electroweak phase transition strongly first order. During the first order transition, many bubbles with domain walls are formed and baryon number are produced through collision of the particles with the moving domain walls.
(b)One can use the many additional CP violating phases in the soft SUSY breaking terms for CP breaking.
(It is not the purpose of this simple review to go through the formulation of supersymmetry. For this, we shall refer the reader to a popular recent review[1] whose notation we follow here.)

To make the transition more first order, one needs additional, relatively light particles that couple strongly to the electroweak symmetry breaking sector: Higgs bosons. The most favorable candidates for such particles are stops (either left $\tilde{t}_L$, or, right $\tilde{t}_R$) which couple strongly to the Higgs boson due to the large Yukawa coupling. In generic SUSY model, stops are typically lighter than the other squarks in low energy effective theory due to RG evolutionary effect.

The loop contribution of the stop particles to the Higgs potential can induce terms which are cubic in Higgs boson H, such as TH^3, needed for the first order transition. However, to make such terms large enough, one needs one of the soft SUSY breaking masses, such as $M_{\tilde{Q}}^2$ or $M_{\tilde{U}}^2$, to be small enough. In addition, large CP violating phases in the soft SUSY terms tend to give large electric dipole moments for electron, neutron or atoms. Therefore there are already enough experimental constraints on MSSM to make the task nontrivial.

Experimentally, in the MSSM, to make the lightest Higgs mass large enough one already requires large one-loop correction from the top quark sector to the Higgs potential. Due to SUSY nonrenormalization theorem, such correction can only come with SUSY breaking sources. One requires $M_{\tilde{Q}}^2$ or $M_{\tilde{U}}^2$ to be larger than the top quark mass in order to generate large enough correction to the lightest Higgs mass so that the LEP II limit $M_H > 106GeV$ is satisfied). In addition, since we already have very good agreement between the measured ρ parameter and that in SM, the data favors a light $\tilde{t}_R$($SU(2)$ singlet), instead of $\tilde{t}_L$(which is $SU(2)$ doublet). Detailed analysis showed that one needs to include two loop corrections in order to demonstrate that the transition can be strong enough first order. The constraints on masses of $\tilde{t}_L$ and $\tilde{t}_R$ are cited as [4]

$$120GeV < m_{\tilde{t}_R} \leq m_t, \tag{2}$$

$$m_{\tilde{t}_L} > (265GeV)exp(\frac{M_H - 95GeV}{9.2GeV}). \tag{3}$$

There are many additional CP violating phases in MSSM. They all come from the soft SUSY breaking terms. In addition to the KM phase and the strong CP θ phase, the new sources of phases are
(I)SUSY Higgs mass, the μ term: $\mu\hat{H}_1\hat{H}_2$ where the hat represents superfields. The phase of μ can be made real if not for the SUSY breaking terms.
(II)SUSY breaking terms:
(a)Trilinear scalar couplings, the A parameters: one for each Yukawa couplings (A_U, A_D, and A_E). The trilinear couplings are $A_f Y_f$ ($f = U, D, E$) where Y_f are Yukawa couplings. A_f in general can be matrices.
(b)Bilinear scalar masses, one for each species of sfermions ($m^2_{ff'}$).
(c)Bilinear Higgs mass, the B parameter: $B\mu H_1 H_2$.
(d)Gaugino (majorana) masses: M_3, M_2 and M_1.

For simplicity, one usually assumes "universality" at a high energy (SUSY breaking) scale: A_U, A_D, and A_E are all proportional to unit matrix and equal, $A_U = A_D = A_E = A$; $m^2_{ff'} = m^2\delta_{ff'}$ independent of f, and therefore CP conserving; all gaugino masses are equal. So the potential new CP phases are A_f, M_i, μ and B. Two of the phases are uphysical because they can be absorbed by using the two broken symmetries: $U(1)_{PQ}$ and $U(1)_R$. It is convenient to take the convention that $B\mu$ is real because in that case the Higgs potential is real and one can assume that the VEV's are all reals. In "universal" limit, there are only two new physical CP violating phases. We shall also take the convention that the gaugino masses are real.

There are two leading sources of CP violating phases in MSSM for purpose of baryogenesis:
(a)stop ($\tilde{t}_L - \tilde{t}_R$) mixing phase, which is the relative phase between A_t(A parameter associated with top Yukawa coupling) and μ.
(b)Chargino mixing phase, which is the relative phase between M_2 and μ.
It has been shown [3] that stop mixing typically cannot generate enough BAU. This is because collider data plus the role that stops have to play in producing strongly first order phase transition already impose severe constraint on the stop sector as quoted earlier. The limits from electric

dipole moments of electron and neutron also put strong constraints on the stop mixing CP violation.[6]

The process can be roughly described as follows. During the first order electroweak transition, bubbles are formed with fast moving domain walls. In the chargino mixing scenario, outside of the domain wall in the symmetric phase the eigenstates are gauginos ($\lambda^{\pm}$)and higgsinos ($\psi^{\pm} = (\tilde{h}_{1L}^{+}, \tilde{h}_{2L}^{+})$), while in the broken phase they are charginos $w_i^{\pm}(i = 1, 2)$. The gauginos($Y = 0$) and higgsinos ($Y = -1/2$ for $\tilde{h}_{1L}^{+}$, $Y = 1/2$ for $\tilde{h}_{2L}^{+}$) have different hypercharge quantum numbers. CP violating scatterings of these particles with the domain wall emit net hypercharge flux into the unbroken phase and produce separation of $\tilde{h}_{1L}^{+}$ and $\tilde{h}_{2L}^{+}$. The gauge interaction and top quark Yukawa interaction are strong enough to keep the charginos and stop, top quark in thermal equilibrium through interaction such as $\tilde{h} + \lambda \to t_L + t_R$ and others. Such interactions convert hypercharge density into chiral quark density. That is, asymmetries $n(t_L) - n(\bar{t_L})$ and $n(t_R) - n(\bar{t_R})$ are both nonzero, even though the net baryon number, $n(t_L) - n(\bar{t_L}) + n(t_R) - n(\bar{t_R})$, remains zero at this stage. As a result, the chemical potentials for left handed quarks are nonzero and provide the sources for the sphaleron conversion into net baryon number.

There are many versions of the detailed mathematical description of this complicated nonequilibrium process.[3] The emphasis, the uncertainty and the numerical results are quite different in different versions. We shall simply quote the summary of the latest analysis a bit later.[7] It turns out that in most parameter space of MSSM, a nearly maximal CP violating phase is needed to generate enough BAU. One immediate question is whether or not such a new source of CP violation is already severely constrained experimentally. It is not surprising that the most severe constraints are provided by the current experimental limits of the electric dipole moment(EDM) of the electron (d_e) and the neutron (d_n).

Electric Dipole Moment(EDM)

Fortunately the lowest order (one-loop) contributions to various EDM's through the chargino mixing can be easily suppressed by demanding that the first two generations of sfermions to be heavier than the third one.[8,9] For example, if one requires these sfermions to be heavier than 10 TeV, the one-loop induced EDM's will be safely small.[10] In fact, such a scenario can even be generated naturally in a more basic scheme referred to as the more minimal SUSY model.[11] However, despite the enlarged parameter space of MSSM, thanks to all the intricate limits provided by accumulated data from various collider experiments, there is only a small region of parameters left within MSSM for such baryogenesis to work.[3]

Recently, two papers[12] point out that even if the sfermions of the first two generations are assumed to be very heavy, there are important contributions to the EDM of the electron at the two-loop level via the chargino sector, as in Fig.1, that strongly constrain the chargino sector as the source for BAU in MSSM. Similar contributions to quark EDM also exist but the resulting constraint turns out to be relatively weaker. Two-loop contributions were also found to be more important than the one-loop ones before in EDM and other context before.[6,13,14,15,16] In particular, two loop contributions to EDM from stop mixing was found to give strong constraint on the CP violation in that sector as well.[6]

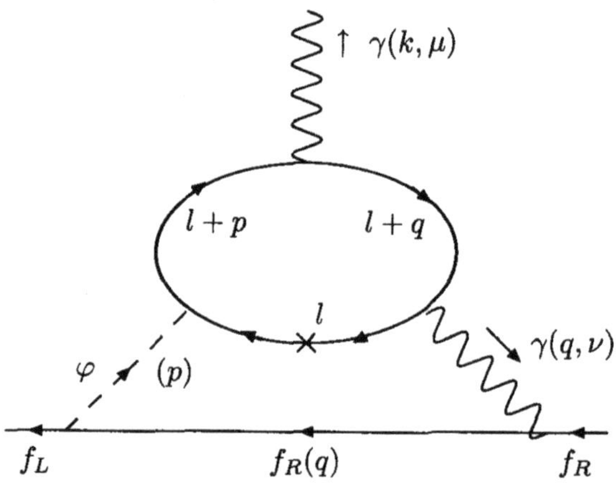

Fig. 1. A two-loop diagram of the EDM of the electron, or quarks. The chargino runs in the inner loop.

In the case of chargino contributions, the two-loop contribution is dominant because the one-loop contribution is suppressed when the sfermions are heavy. This aspect is similar to those in Refs. 6 and 16. In addition, the chargino case, the large CP violating phase in the chargino mixing and the light Higgs scalar, which is necessary to obtain a large baryon asymmetry, is also the same cause of a large EDM. Therefore, the resulting severe EDM constraint is very difficult to avoid in the mechanism of the chargino baryogensis by tuning parameters. We shall refer the reader to the literature for further details. It suffice to summarize the numerical result as follows.

Numerical analysis and baryogenesis

To our current knowledge, the experimental constraint on the electron EDM has become very restrictive:

$$|d_e| < 1.6 \times 10^{-27} e \text{ cm} \qquad (90\% \text{ C.L. Ref. } 17)\,. \tag{4}$$

In the many calculations of BAU in MSSM[3] the largest uncertainty seems to come from the calculation of the source term for the diffusion equations which couples to the left-handed quarks.[7,18] Using the latest summary of the situation in Ref. 4 as a reference point, large BAU requires $\tan\beta \lesssim 3$ with the wall velocity and the wall width close to their optimal values $v_w \simeq 0.02, l_w \simeq 6/T$, $\mu \simeq M_2$ and CP phase $\sin\phi$ close to one. Note that a smaller $\tan\beta$ gives a larger BAU, however, it tends to give a small lightest Higgs mass which violates the LEP II limit unless the left stop is much heavier than 1 TeV. Using the SUSY parameters in the above range, the numerical analysis in Ref. 12 indicates that the resulting EDM strongly constrains the allowed range of parameters involved.

On the the other hand, for the neutron EDM, our analysis indicates the current experimental limit in Eq.(12) gives only marginal constraint on MSSM parameters required for chargino BAU.

Conclusion

The baryogenesis in MSSM requires the lightest Higgs boson to be light in order to get a strong first order phase transition. It also requires the CP violating phase in chargino mixing to be large in order to get large enough BAU. As we have discussed, both requirements imply the predicted values of the EDM's of the electron and the neutron to be large. Note that even if electroweak baryogenesis is impossible for SM and MSSM, one can in general generate BAU through other processes such as lepto-baryogenesis or doublet-baryogenesis. Therefore, it does not rule out SM or MSSM as a low energy effective theory. However most alternatives require detailed higher energy physics. It is certainly exciting to entertain the possibility that BAU can be produced in the low energy framework through the existing low energy baryon violating mechanism: the sphaleron.

This work is in collaboration with Wai-Yee Keung of UIC and We-Fu Chang of TRIUMF and is supported by a grant from National Science Council(NSC) of Republic of China (Taiwan). We wish to thank H. Haber, H. Murayama, O. Kong, and K. Cheung for discussions and also thank theory groups at SLAC and LBL for hospitality during his visit.

References

1. S. P. Martin, hep-ph/9709356.
2. G. R. Farrar and M. E. Shaposhnikov, *Phys. Rev.* **D50**, 774 (1994) [arXiv:hep-ph/9305275].
3. M. Carena, M. Quiros and C. E. Wagner, *Phys. Lett.* **B380**, 81 (1996) [arXiv:hep-ph/9603420]; M. Carena, J. M. Moreno, M. Quiros, M. Seco and C. E. Wagner, *Nucl. Phys.* **B599**, 158 (2001) [arXiv:hep-ph/0011055]; M. Carena, M. Quiros, A. Riotto, I. Vilja and C. E. Wagner, *Nucl. Phys.* **B503**, 387 (1997) [arXiv:hep-ph/9702409]; J. M. Cline, M. Joyce and K. Kainulainen, *Phys. Lett.* **B417**, 79 (1998) [Erratum-ibid. **B448**, 321 (1998)] [arXiv:hep-ph/9708393]; J. M. Cline, M. Joyce and K. Kainulainen, *JHEP* **0007**, 018 (2000) [arXiv:hep-ph/0006119]. Erratum: arXiv:hep-ph/0110031; S. J. Huber, P. John and M. G. Schmidt, *Eur. Phys. J.* **C20**, 695 (2001) [arXiv:hep-ph/0101249]; P. Huet and A. E. Nelson, *Phys. Rev.* **D53**, 4578 (1996) [arXiv:hep-ph/9506477]; M. Quiros, *Nucl. Phys. Proc. Suppl.* **101**, 401 (2001) [arXiv:hep-ph/0101230]; M. Aoki, N. Oshimo and A. Sugamoto, *Prog. Theor. Phys.* **98**, 1179 (1997) [Archive:hep-ph/9612225]; *Prog. Theor. Phys.* **98**, 1325 (1997) [arXiv:hep-ph/9706287]; M. Carena, M. Quiros, M. Seco, C. E. M. Wagner, hep-ph/0208043.
4. J. M. Cline, hep-ph/0201286.
5. K. Dick, M. Lindner, M. Ratz and D. Wright, *Phys. Rev. Lett.* **84**, 4039 (2000) [arXiv:hep-ph/9907562]; H. Murayama and A. Pierce, hep-ph/0206177.
6. D. Chang, W.-Y. Keung and A. Pilaftsis, *Phys. Rev. Lett.* **82**, 900 (1999); *ibid.* **83**, 3972 (1999). The contribution of the chargino loop has been pointed out in this paper, but detailed analysis was only given to the stop-loop contribution. See also D. Chang, W.-F. Chang and W.-Y. Keung, *Phys. Lett.* **B478**, 239 (2000); A. Pilaftsis, *Phys. Lett.* **B471**, 174 (1999).
7. For a review see, J. M. Cline, *Pramana* **54**, 1 (2000) [*Pramana* **55**, 33 (2000)] [arXiv:hep-ph/0003029] and Ref. 18.
8. Y. Kizukuri and N. Oshimo, *Phys. Rev.* **D46**, 3025 (1992); J. Ellis, S. Ferrara and D. V. Nanopoulos, *Phys. Lett.* **B114**, 231 (1982); W. Buchmüller and D. Wyler, *Phys. Lett.* **B121**, 321 (1983); J. Polchinski and M. Wise, *Phys. Lett.* **B125**, 393 (1983); F. del Aguila, M. Gavela, J. Grifols and A. Mendez, *Phys. Lett.* **B126**, 71 (1983); D. V. Nanopoulos and M. Srednicki, *Phys. Lett.* **B128**, 61 (1983); M. Dugan, B. Grinstein and L. Hall, *Nucl. Phys.* **B255**, 413 (1985).
9. T. Ibrahim and P. Nath, *Phys. Rev.* **D57**, 478 (1998) [Erratum-ibid. **D58**, 019901 (1998)] [arXiv:hep-ph/9708456];
10. S. Abel, S. Khalil, and O. Lebedev, *Nucl. Phys.* **B606**, 151 (2001) [arXiv:hep-ph/0103320].
11. A. G. Cohen, D. B. Kaplan, and A. E. Nelson, *Phys. Lett.* **B388**, 588 (1996).
12. D. Chang, W.-F. Chang and W.-Y. Keung, hep-ph/0205084. A. Pilaftsis, hep-ph/0207277.
13. S. M. Barr and A. Zee, *Phys. Rev. Lett.* **65**, 21 (1990).
14. D. Chang, W. Y. Keung and T. C. Yuan, *Phys. Rev.* **D43**, 14 (1991). R. G. Leigh, S. Paban and R. M. Xu, *Nucl. Phys.* **B352**, 45 (1991). C. Kao and R.-M. Xu, *Phys. Lett.* **B296**, 435 (1992). D. Chang, W.-S Hou, W.-Y. Ke-

ung, *Phys. Rev.* **D48**, 217 (1993); D. Chang, W.-F. Chang, C.-H. Chou and W.-Y. Keung, *Phys. Rev.* **D63**, 091301 (2001).

15. D. Chang, W.-Y. Keung and T. C. Yuan, *Phys. Lett.* **B251**, 608 (1990); J. F. Gunion and D. Wyler, *Phys. Lett.* **B248**, 170 (1990); D. Chang, W.-F. Chang, M. Frank and W.-Y. Keung, *Phys. Rev.* **D62**, 095002 (2000).
16. T. Kadoyoshi, N. Oshimo *Phys. Rev.* **D55**, 1481 (1997) [arXiv:hep-ph/9607301]. This paper also used the same CP violating source from the chargino mixing, however, their two-loop amplitudes are not the dominant ones.
17. B.C. Regan et al., *Phys. Rev. Lett.* **88**, 071805 (2002). P. G. Harris et al., *Phys. Rev. Lett.* **82**, 904 (1999). K. F. Smith et al., *Phys. Lett.* **B234**, 191 (1990). S. K. Lamoreaux and R. Golub, *Phys. Rev.* **D61**, 051301(R) (2000).
18. H. Murayama and A. Pierce, hep-ph/0201261.

Detector Technologies for a New Generation of CMB Cosmology Experiments

Paul Richards

DETECTOR TECHNOLOGIES FOR A NEW GENERATION OF CMB COSMOLOGY EXPERIMENTS*

P. L. RICHARDS
Dept. of Physics
University of California
Berkeley, CA 94720, USA

Spectacular progress has been made in obtaining information about the early universe from the cosmic microwave background (CMB). Much more can be learned from observations with high sensitivity and precision. One important requirement for progress is improvements in receiver sensitivity. The types of detectors used for CMB research are approaching fundamental sensitivity limits. Future progress will come from the use of large format detector arrays. The detector types most suitable for large arrays are discussed. Several new experiments being developed by the Berkeley group are described which will make use of arrays of 300 to 1000 bolometric detectors.

1. Introduction

The Cosmic Microwave Background radiation (CMB) is the oldest electromagnetic radiation in the universe. Observations of the CMB give a detailed picture of the universe 300,000 years after the Big Bang, and they are one of the pillars of Big Bang cosmology. The smoothness of this radiation supports the idea of an inflationary expansion of the universe at an early epoch. The black body spectrum measured by COBE constrains energy release in the universe back to about two months after the Big Bang. Angular fluctuations in the CMB provide an ancient record of the interaction between matter and radiation in the early universe. COBE measured the primordial temperature fluctuations remaining after inflation. MAXIMA, BOOMERANG and DASI measurements of the small-scale temperature anisotropy have confirmed the 30 year old prediction that acoustic waves modify these fluctuations and play an important role in the formation of structure. The observed angular power spectrum of these fluctuations provides constraints on the contents of the universe, and shows that the

*This work is supported in part by NSF Grant FD97-31200, and by NASA/Ames Grant FDNAG2-1398.

universe is flat. These results are consistent with the picture that ordinary baryonic matter makes up only ~5% of the universe and that dark matter is ~25% and dark energy ~70%. MAP and Planck will test these results with exquisite precision, provide accurate values of most important cosmological parameters, and begin the exploration of the polarization anisotropy of the CMB.

Observations of the polarization characteristics of the CMB have the potential to provide a vision of the universe near the instant of its birth. Some mechanisms for polarization, such as scattering from density fluctuations, produce so-called E-mode polarization, which has no curl-like component. E-mode polarization will be measured over limited sky regions by ground based and balloon borne experiments, and it will be statistically characterized over the entire sky by MAP, and especially by Planck. Gravitational waves created in the earliest phase of an inflationary universe imprint a different and distinctive polarization pattern on the CMB radiation. The polarization field arising from tensor gravitational waves has a curl-like component and is called B-mode polarization. The detection of B-mode polarization would be a great triumph for inflationary cosmology, providing us a picture of physical processes in the universe 10^{-34} seconds after the Big Bang. A null result would still constrain the energy scale of inflation.

A deep probe of the polarization of the CMB will yield additional important results. Weak lensing by the intervening matter influences both the E and B-mode polarization fields. Lensing dominates the B-mode field for multipoles >100. Correlation of the E and B-mode fields can provide an accurate map of the mass distribution in the early universe out to multipoles of ~ 1000.

There is much current interest in measurements of the temperature anisotropy of the CMB on angular scales less than 10 arcmin. The CBI interferometer is producing information about the higher order acoustic peaks and the epoch of reionization of the universe. The AKBAR experiment will release data soon. Scattering of CMB photons by the hot electrons trapped in clusters of galaxies shifts the photons to higher energies, decreasing the brightness below the peak in the black body curve and increasing the brightness above the peak. Detailed studies of this Sunyaev-Zeldavich effect, can provide a wealth of information including a value of the Hubble constant and information about large scale flows from peculiar velocities. Because the temperature of the CMB increases with redshift, the SZ surface brightness does not decrease with the distance to the cluster. For this reason, the SZ effect is an excellent way to locate clusters back to the epoch of cluster formation. Large area SZ surveys, coupled with redshifts from optical follow-up, will test theories of structure development, and constrain the matter density of the universe and the equation of state for dark energy.

Measurements of the CMB require sensitive detectors at millimeter wavelengths. The sensitivity of such detectors has increased rapidly. For example, the 100 mK bolometers in the MAXIMA balloon anisotropy experiment measured ~25,000 pixels in three hours with comparable signal-to-noise ratio to that obtained by the COBE DMR experiment on ~5,000 pixels during four years of observation. Individual detectors are approaching fundamental sensitivity limits as will be described below. However, the sensitivity of receivers will continue to increase by the use of larger numbers of detectors. The largest current systems use 10-50 detectors. Plans are being made to increase these numbers to 10^3 or more detectors. This increase in the sensitivity of CMB receivers will lead to major new opportunities for future CMB experiments.

The next few sections of this paper describe the types of detector being used to study the CMB, and the improvements that are being made in receiver sensitivity. The final sections will describe new CMB experiments under development at Berkeley.

2. HEMT Amplifiers

One approach to CMB measurements at millimeter wavelength is to use low noise transistor amplifiers based on high electron mobility transistors (HEMT). A typical receiver will include a conical antenna, a HEMT amplifier, a band-pass filter and a diode detector. The HEMT amplifiers are cooled to ~20K to minimize their noise. Microwave integrated circuit (MIMIC) techniques and new materials are improving the manufacturability and the high frequency performance and reducing the noise in these amplifiers. Because linear phase-conserving amplification is involved, HEMT receivers are subject to quantum noise which corresponds to a fluctuation of $\pm$ 1 in the number of photons in the amplifier on the time scale determined by the inverse of the bandwidth. Reasonable projections of the development of HEMT receivers suggests that in the next few years they will have noise equal to ~3 times the quantum limit and bandwidths of 20-30% for frequencies up ~100GHz.

HEMT amplifiers have been used in many CMB temperature anisotropy and polarization anisotropy experiments. Modern systems use arrays of 10-15 receivers. In principle, much larger arrays of HEMT receivers could be used. However, issues of cost, complexity, power dissipation etc. must be addressed for close packed focal planes. Aperture synthesis interferometers such as DASI, CBI and AMIBA, which use HEMT amplifiers, are especially powerful for CMB anisotropy experiments. After amplification, the same signals are combined with different baselines to simultaneously measure many different spatial frequencies on the sky. This benefit is not significant for point source surveys where only long baselines are of interest.

3. Bolometers

Bolometers are thermal detectors which consist of an absorbing element, a resistive thermometer to measure the temperature rise and a weak thermal connection to a heat sink at some low temperature. As there is no phase conserving amplification involved in the detection process, bolometers do not produce quantum noise. For sufficiently low operating temperatures, bolometers can reach the fundamental noise limit set by the fluctuations in the rate at which photons are absorbed. Bolometric receivers use cooled baffles and filters to minimize the photons from sources other than the in-band signal from the CMB. Modern bolometers have optical efficiencies of $\sim$50%. With heat sink temperatures of $\sim$100 mK they can be photon noise limited on the signal from the CMB in 20-30% bandwidths.

A comparison between the sensitivities of HEMT and bolometer receivers depends on both the correlations in, and the rate of arrival, of the detected photons. If the photon occupation number is unity, as is the case when observing a black body in the Rayleigh-Jeans limit, then the correlated photon noise is exactly the same as the quantum noise. In practice, the sensitivities of optimized single polarization bolometers and HEMTS are essentially the same for CMB measurement at frequencies up to $\sim$60 GHz. At frequencies approaching the peak of the black body curve, however, the photon occupation number falls below unity, the photon correlations disappear and the performance of photon noise limited bolometric receivers becomes rapidly better than that of quantum noise limited HEMT receivers. Since the sensitivities of both HEMT and bolometric pixels are approaching fundamental limits, future improvements in receiver sensitivity will come from the use of larger arrays.

Well developed technologies exist for making arrays of a few hundred bolometers which operate at temperatures from 100 to 300 mK. These bolometers are supported on thin membranes of low-stress silicon oxy-nitride (LSN) and use metal film absorbers with an average sheet resistance of 377 Ohms per square followed by a reflecting backshort. The metalized membranes are usually patterned into a mesh or "spider web" to minimize heat capacity and cosmic ray cross section. These mesh are supported by thin legs of LSN which provide the thermal isolation from the heat sink. The thermistors are typically chips of neutron transmutation doped

(NTD) Ge which are biased with a constant current and read out through JFET Amplifiers which operate at ~100K to minimize their noise. An AC bias is used when low frequency stability is required. These devices have been perfected by the Caltech/JPL group and are used in a large number of important CMB experiments including BOOMERANG, MAXIMA, MAXIPOL, AKBAR, Planck/HFI etc. Polarization-sensitive bolometers are made by replacing the mesh with a one dimensional grid. Dual polarization bolometers have two closely spaced orthogonal grids. Each grid is attached to a separate thermistor and each is sensitive to a different linear polarization. Dual polarization bolometers from Caltech/JPL will be used on BOOMERANG, Planck/HFI, QUEST, BICEP, etc.

Bolometric receivers are designed to minimize the amount of background power from the instrument which reaches the detector. There is usually an array of cold horn antennas in the focal plane to define the detector field of view. Scalar (corrugated) feed horns preserve polarization and have a well-controlled antenna pattern that can view the ambient temperature optics directly. Smooth walled horns or Winston concentrators can be used when polarization is not an issue, but require a cooled aperture stop. Horn-coupled arrays produce a sparsely sampled pattern on the sky so that beams must be dithered or scanned to fill in a map. The horns are typically followed by a section of circular waveguide which acts as a high-pass filter, multi-layer metal mesh filters to define the frequency band being observed, the absorbing metal film of the bolometer and the reflecting backshort.

The current generation of bolometers gives excellent performance in many applications, but there are practical limits to the number of pixels that can be used. The JFET amplifiers limit the ability to produce arrays of more than a few hundred pixels. In addition to thermal and microphonics issues, the relatively high amplifier noise margin causes system problems. Because of these limitations, there is a need for revolutionary new long wavelength bolometers. Fortunately, very promising new approaches are under active development. The voltage-biased superconducting bolometer with transition edge sensor (TES) and SQUID readout amplifier is a negative feedback thermal detector with many favorable operating characteristics. It can be made entirely by thin film deposition and optical lithography. The feedback reduces the response time, improves the linearity, and isolates the bolometer responsivity from changes in infrared loading or heat sink temperature. There is also some suppression of Johnson noise. The SQUID amplifiers operate at bolometer temperatures, dissipate very little power and have significant noise margin. These bolometers are being developed at Berkeley and Caltech/JPL with appropriate architectures for horn-coupled arrays.

In addition, there is work on bolometers for CMB measurements that are coupled to the optics by planar lithographed antennas and superconducting microstrip transmission lines. The antennas are inherently polarization sensitive and the transmission lines can incorporate high performance microstrip bandpass filters. In essence, the low loss in superconductors is being used to extend MIMIC technology to higher frequencies. In this implementation, the absorbing element of the bolometer is a resistive termination to the line, which can have very small area. At low operating temperatures ≤ 100 mK, such small terminations can be combined with the TES and deposited directly on the Si substrate. The weakness of the coupling between electrons and phonons can provide adequate thermal isolation in these so called hot electron TES bolometers. In antenna-coupled bolometers, the submillimeter wave-signal propagates in a superconducting microstrip transmission line. These lines can branch to form diplexers so that one antenna can feed two or more bolometers which measure different frequency bands.

Antenna couple bolometers are attractive for measurements of the polarization anisotropy of the CMB. Requirements for a very high performance system include an antenna pattern narrow enough to couple efficiently to the telescope optics, simultaneous measurement of two orthogonal polarizations in each pixel, and simultaneous measurement in several frequency bands. Present designs meet two out of three of these requirements and work is in progress to meet all three. The Berkeley group is making crossed double-slot dipole antennas coupled to two bolometers. The difference between the outputs of the bolometers is sensitive to the polarization of the signal illuminating the pixel. One Si lens per pixel and a cold aperture stop are required to match the antenna pattern to the telescope beam. The Caltech/JPL group is studying an array of many slot antennas to give a narrow antenna pattern which couples efficiently to the telescope beam. It has a wide frequency bandwidth that can be divided into several photometric bands. Existing designs, however, do not couple simultaneously to two polarizations.

Large format arrays of TES bolometers require output multiplexing to avoid very large numbers of leads leaving the cryostat. Lines of 30-50 detectors can be multiplexed before amplification using superconducting thin film technology. The NIST group has developed a time-domain multiplexer which uses a SQUID for each bolometer to switch the outputs sequentially through a single SQUID amplifier. Groups in Berkeley and Helsinki are developing frequency-domain multiplexers which combine the signals from a row of bolometers, each of which is biased at a different frequency. The signals are then amplified by a single SQUID and recovered with ambient temperature lock-in amplifiers. The success of the TES array technology ultimately depends on the success of one or both of these multiplexers.

4. Berkeley Projects

The group at Berkeley started by P.L. Richards has a long history of pioneering mm-wave detector techniques that are in current use. They made the first liquid ^{4}He cooled mm-wave bolometers in 1960 and the first SIS mixers in 1979. Other developments included the introduction of bolometers with metal film absorbers and NTD thermistors, and the first astrophysical applications of bolometers at 300 and 100 mk. Since 1996 an effort in the group led by AT Lee has focused on the new TES bolometer technology.

Most of these detector developments were motivated by the needs of CMB observations done by the group. In the 1970's the Woody Richards experiment showed that the frequency spectrum of the CMB has a peak characteristic of a $\sim$ 3K black body. In the process, it explored many of the technologies later used in the FIRAS measurement of the CMB frequency spectrum. From late 1988 to 1996 the MAX experiment(with A.E. Lange, P.M. Lubin and G.F. Smoot) made the first observation of the temperature anisotropy of the CMB on degree scales and demonstrated the potential for bolometric balloon anisotropy experiments. Starting in the late 1990's the MAXIMA experiment (with A.T. Lee, S. Hannay and many others) has made major contributions to CMB anisotropy science and pioneered the use of bolometers cooled to 100 mK. Since 2000, MAXIMA has been converted to a polarization anisotropy experiment MAXIPOL. Leadership of MAXIPOL has passed to S. Hannay and his group at Minnesota. Berkeley participation includes A.T. Lee, P.L. Richards, and graduate students B. Rabii, C.D. Winant and J. Collins. MAXIPOL is ready for a first balloon flight in 9/02. Leadership of the Berkeley group has now shifted to Adrian T. Lee and the focus has shifted to three new CMB projects in addition to MAXIPOL. The list of participants involved either in detector development or in these new projects includes Senior Scientists: J.C. Clarke, W.L. Holzapfel, A.T. Lee, P.L. Richards and H. Spieler; Postdocs S. Cho, M. Dobbs, N. Halverson and H. Tran; Graduate Students J. Collins, T. Lanting, J. Mehl, R. O'Brient and D. Schwan.

5. Polar Bear

There are more than 20 experiments in progress or under development to measure the polarization anisotropy of the CMB, which has not yet been observed. A successful experiment must have adequate sensitivity to observe the small polarization signal and adequate calibration and control of systematic errors to know with confidence what is being observed. The sensitivity of different experiments can be compared in terms of a figure of merit called the effective integration time. This is the product of the expected integration time, the number of pixels and the square of the detector sensitivity.

The POLAR BEAR experiment proposed by Adrian T. Lee and the Berkeley group uses an array of 318 dual-polarization antenna coupled TES bolometers operated at 250 mK and configured for measurements at 150, 250 and 350 GHz on the 2 meter off-axis Viper telescope at the South Pole. With 3 years of observations, the effective integration time is essentially the same as for the Planck space mission. Two other capable bolometric experiments, BICEP and QUEST have estimated effective integration times a factor ~ 3 smaller than POLAR BEAR or Planck. All other experiments have significantly smaller values of this parameter. Some by a factor of 10^{-3}. The projections for the sensitivity of these proposed new ground based experiments assume that there will be no significant polarized sky noise, which is plausible, but not demonstrated. These experiments have no significant advantage over Planck that is not included in the effective integration time. They are able to make a deep observation of a relatively small region of the sky so as to minimize the noise on the polarization power spectrum. Planck is constrained to spread its observations over the whole sky. In terms of sensitivity and angular scale, POLAR BEAR is the most capable CMB polarization experiment yet proposed. Even so, the estimated sensitivity is just adequate to detect the contribution to B-mode polarization from GUT scale inflation. The power of POLAR BEAR comes from the use of the new TES bolometer technology. If the current technology developments are successful, the 318 bolometer POLAR BEAR focal plane could be expanded to 1000 or more multiplexed TES bolometers on the recently funded 8 meter off-axis South Pole telescope.

6. Photometers for SZ Surveys

Bolometric array technology has advanced to the point that it is attractive to build instruments to survey hundreds of square degrees of sky at 150 and 230 GHz to locate clusters of galaxies with the SZ effect. The

Caltech/JPK/Cardiff BOLOCAM project on the CSO is implementing 151 horn coupled NTD spider web bolometers. A survey for SZ sources is an important part of their program. The Berkeley group is building TES array receivers for SZ surveys on two new telescopes.

A European consortium including Onsala, ESO and the MPIfR at Bonn have purchased a 12 m on-axis ALMA prototype telescope. This Atacama Pathfinder Experiment (APEX) telescope will be installed at the ALMA site in the Chilean high desert in about 2 years. The Berkeley group with A.T. Lee as P.I. has been funded to do an SZ survey on this telescope at 150 and 230 GHz in collaboration with a group led by C. Menton at Bonn. A $\sim$ 300 pixel receiver will be built in Berkeley using horn-coupled TES spider web bolometers at 250 mK. The baseline design calls for one SQUID amplifier per pixel, but multiplexing will be used if it is available. In two seasons of observation it should be possible to survey 250 square degrees to a noise level of $100\mu K_{CMB}$ per 0.8' pixel at 150 GHz. Simulations show that all galaxy clusters larger than 4×10^{14} solar masses will be detected, regardless of redshift.

A consortium of many U.S. University groups headed by J. Carlstrom of Chicago has been funded to make an exhaustive SZ survey from the South Pole over the next 5 years. A new off-axis 8 meter telescope will be built for this project. The Berkeley part of this effort, with W.L. Holzapfel as P.I., will include the development, construction and deployment of a TES bolometer receiver. This receiver will be significantly more capable than the first generation APEX receiver. The plan is to have a larger telescope field of view which can illuminate $\sim$ 1000 multiplexed horn coupled bolometers.

7. Source of Additional Information

Far infrared to millimeter wave detectors, including those used for CMB measurements have undergone intensive development during the past decade. Hundreds of articles have been published describing these developments, most of which are either highly technical or out of date, or both. The most recent comprehensive review article[1] on bolometers was in 1994. In March 2002 NASA sponsored a Far-Infrared, Sub-mm and mm Detector Technology Workshop in Monterey, CA which has produced high quality up-to-date articles on many detector topics relevant to CMB measurements. Some of the manuscripts are now available on the Workshop Web Site at http://sofia-usra.arc.nasa.gov/det_workshop/papers/manuscript.html. Others will become available soon. It is expected that the workshop proceedings will be published[2] in late 2002.

In early 2002, NASA assembled a Detector Working Group led by E. Young, that included the author. The report of the Working Group "Detector Needs for Long Wavelength Astrophysics" will eventually be published by NASA. It is now available at http://sofia-usra.arc.nasa.gov/det_workshop/report/report.html.

The Berkeley experiments described above are mostly so new that web sites are only now being developed. There will soon be a linked set of web sites that can be located by searching topics such as MAXIMA, MAXIPOL, Berkeley Bolometer, POLAR BEAR, APEX Experiment and SOUTH POLE TELESCOPE. Unfortunately, the names now in use for the APEX and SOUTH POLE SZ surveys will probably soon change.

Acknowledgements

The author's lifetime interest in detectors of far-infrared to millimeter-wave radiation has been stimulated by a large number of students, collaborators and professional associates. The remarkable ongoing progress in this endeavor has been driven by many extremely capable investigators.

References

1. P.L. Richards, Bolometer for Infrared and Millimeter Waves, *J. Applied Phys.* **76**, 1 (1994).
2. Proceedings in FAR-IR, SUB-MM DETECTOR TECHNOLOGY WORKSHOP, Wolf J., Farhoomand J. and McCreight C.R. (eds) NASA/CP-211408, 2002 (in press).

Our Age of Precision Cosmology

George Smoot

OUR AGE OF PRECISION COSMOLOGY

GEORGE F. SMOOT
Department of Physics,
LBNL, & SSL,
University of California
Berkeley, CA, 94720, USA
E-mail: GFSmoot@lbl.gov

This talk addresses the current state of cosmology. I put forward two theses: (1) We have entered an era of precision cosmology. (2) New Physics is being sought as a result of this success. The combination of these two make both for as sense of excitement and crisis. The talks at this CosPA conference serve to illustrate and support these two theses and point a way toward the future.

1. Introduction

We have entered an era precision cosmology. Presently we have determined most of the parameters describing cosmology to the first decimal or about the 10% level. This development of a cosmological model and determining its primary parameters is a substantial intellectual achievement and of general interest.

Our current cosmological model describes observations well but insists that new physics must come into play.

New Physics is sought due to this success and the consequence that successful model demands it. Some of the new physics is sought because it is now possible to explore new avenues because we have a successful model and the data with which to probe new areas. This search for new physics is spread through out field which is rife with speculations of new physics and its signatures. This conference typifies this trend more than most meetings. Precise and crosslinking observations plus a successful baseline model makes these searches possible.

1.1. *Our Era of Precision and the Drive for New Physics*

It is a matter of pride and prestige that we have entered what we can call our age of precision cosmology. New and more accurate observations of the properties of the Universe seem to appear regularly. Our 'standard model of cosmology', the Inflationary Big Bang, is quite successful in accommodating and organizing these new observations. There is a minor 'industry' in adding in each new set of observations and producing a new set of best-fitted cosmological parameters.

But in the midst of this success and excitement, we find a strange disquiet, precisely because of the very great successes of both the standard model of particle physics and our new model of cosmology, both of which indicate whole new levels of physics are imminent. At this point speculations are rife. Most of the speculations are far from the main areas that must yield. However, this speculation means that a whole range of systematics have become important. There are large systematic uncertainties opened because the a new model is going to allow extra degrees of freedom. As we look for astrophysical and cosmological signatures of new physics, systematic effects we did not need consider previously can now be important.

1.2. *Evolution versus Revolution*

Scientists (Physicists) are conservative:

"If it is not broken, don't fix it!"

"If the data fit the model, one does not need nor should not add features/parameters." - c.f. A. Jaffe's talk.

On the other hand scientists revere "Revolutions" which are conceptual but link back and connect to the older theory in its domain of validity: e.g.

Relativity $\Longleftrightarrow$ Newtonian

Quantum Mechanics $\Longleftrightarrow$ Classical Mechanics

1.2.1. *How to manage and encourage creativity?*

There are techniques to manage creativity constructively. At this stage they require a careful understanding of Where we are? - current model and status What are current assumptions? What might be possible in the intermediate future?

Do we need a Cultural Revolution? Is the old model so flawed and out of date that it must be junked? How do we go about finding a new model?

(1) "Let a thousand flowers bloom!"

That is let everyone take their best guess and make up their theories. Why not let a thousand flowers bloom? We might get an unmanageable destructive chaos including training a whole new generation that is the appropriate and acceptable way one does science. It is not just in China, one has learned that this approach leads to a period of chaos and only later does some new order emerge after some calm and sanity return and selection effects and market forces (comparison to observations) have time and opportunity to act. There is a good reason that scientists are conservative and do not make up new theories at the drop of the hat. The most effective way to make progress is through evolution that matches to the environment.

However, one might come up with a way to harness this creative energy of random new theories springing up like wild flowers - a la Markov chain Monte Carlo. When one has too many degrees of freedom (dimensions of parameters) to be able to sum or integrate over numerically to find the optimum, then it is often more effective to randomly chose a point in the parameter space, evaluate the likelihood (function value) and change the probabilities for the choosing the next point to explore the volume of parameter space most effectively. Such a technique requires a metric for evaluating the "goodness" of the model and casting rule for generating the next model. Both of these are likely to be contrary to the spirits and personalities of those making up new models. It is, however, a technique that could work well in conjunction with the next approach which matches the disciplined and measured approach.

(2) Systematic Parameterization or Characterization of all allowable degrees of freedom. In this approach one sets up a frame work in which to develop a new physics theory that automatically will match to the older theory in its domain of validity. The goal is to set up the framework in a manner that is as close to exhaustive of straightforward potential new physics. Two general methodologies are:

(i) Perturbative: Take the current valid theory/model and allow perturbations or general parameterization, e.g. PPN formalism for gravity.

(ii) Correspondence Principle: Make a list of all possible additional degrees of freedom and determine how these could be incorporated in a larger class of theories which reduce (correspond) to the current model in relevant range.

A draw back to these approaches is that the first is likely to require a lot of work to parameterize the perturbations. For the second it is hard to be sufficiently imaginative and complete and, if one is, then the allowable degrees of freedom are likely to span a large number of dimensions parameter space with possible combinations exceeding our abilities to pursue.

The advantages of these approaches is that this can be a method that is fairly complete intellectually and one that gives good guidance to people in the field, particularly to the observers/experimentalists and to the data analysts.

2. Age of Precision Cosmology

We have new technologies and corresponding new observations. There are more precise and wider ranging observations of our Universe. From these we have developed a standard model of cosmology and we now have its parameters determined at the order 10% level. We anticipate achieving on order 1% level in a decade. These more precise data will allow precise tests of this standard model and more accurate determination of its parameters.

2.1. *Contents Sorted by Equation of State*

Content	Observed Density	Eqn of State $\mathbf{w} = \mathbf{P}/\rho$
Relativistic/Radiation:	$\Omega_R \leq 10^{-4}$	**w = 1/3**
Photons:	$\Omega_\gamma = 4 \times 10^{-5}$	CMB
Neutrinos:*	$0.003 < \Omega_\nu < 0.01$	
Gravity Waves	$\Omega_{GW} <?10^{-5}$	
Matter:	$\Omega_M = 0.33 \pm 0.04$	**w = 0**
Cold Dark Matter:	$\Omega_{CDM} = 0.29 \pm 0.04$	
Baryons:	$\Omega_B = 0.04 \pm 0.01$	large scale structure
Neutrinos:*	$0.003 < \Omega_\nu < 0.01$	
Curvature:	$\Omega_k = 0.03 \pm 0.07$	**w = -1/3**
Dark Energy:	$\Omega_{DE} = 0.7 \pm 0.1$	**-1 ≤ w ≤ -1/3**
Quintessence:	$\Omega_Q =?$	
Cosmological Constant:	$\Omega_\Lambda =?$	

2.2. *Current Conditions: (like initial conditions only later)*

Parameter	Current value
Current Temperature	$T_0 = 2.725 \pm 0.001$ K
Current Age:	$t_0 = 14 \pm 1$ Gyr
Current Expansion Rate:	$H_0 = 72 \pm 7$ km/s/Mpc
Primordial Perturbations:	$Q = 32 \pm 3\ \mu K$

2.3. *Resulting Science*

New Accurate Observations have lead to: More confidence in the 'standard model of cosmology' - Inflationary Big Bang. Some surprises but they seem to fit into the model. This model gives us a clear context and framework Francois Bouchet and Andrew Jaffe provided a discussion of how Large Scale Structure and CMB can be used for cosmology and more and more for astrophysics or conventional astronomy. Bruce McKellar shows how the model can accumulate neutrino oscillations in BBNS. But they raise larger questions (mysteries).

3. CosPA Interests

It is not an accident that this conference deals with these issues. The CosPA science was chosen to be cutting edge in these areas, I.e. Prof W-Y. Pauchy Hwang talk on CosPA and CPU outlines these. The organizers selected speakers to address these issues. The relevant results which came out in time for the CosPa meeting include the CBI results which predict greater signals for AMIBA and the new X-ray results also indicate that the cluster structure is complex and relevant for SZ science of AMiBA (plus X-rays).

Pauchy Hwang overview indicated how the SZ effect merges LSS, CMB, and with x-rays. Mark Birkinshaw reviewed and and made predictions on the SZ effect. Tzihong Chiueh gave a good discussion of AMiBA effort. AMiBA has good prospects but will have others hard on its heels for this science. Francois Bouchet discussed Planck's potential for SZ illustrating that CosPA has picked challenging science which will attract competing groups. CMB Anisotropies / Polarization with AMiBA review by Kin-Wang Ng - again an area of high interest and thus high competition.

4. Standard Model of Cosmology

The standard model of cosmology is in good shape, its predictions are well supported by observations. There are multiple cross-checks:

E.g., SZ effect independent of redshift, Agreement between LSS, CMB, Supernovae provides a concordant model of flat universe with dark energy, BBNS and CMB agreement on Ω_{baryon}.

Successfully survived challenges - thus far.

But what do we mean by a model? A model is the part of theory in which we describe the physical situation. That is our model involves an inventory of the contents of the Universe, i.e. Table 2.1, and an accounting of their condition, e.g. the current age, expansion rate, Temperature, perturbation level., i.e. many of the things in Table 2.2. Usually in physics we have a model listing the contents of the system to be discussed and an appropriate set of initial conditions to which we apply our known (dynamical) laws of physics and calculate how things evolve. Generally, this involves solving a set of second order differential equations so that we can evolve things backwards and forwards so that any complete set of conditions, e.g. positions and velocities is sufficient. In the case of the Universe, we need some information on a space-like hypersurface perpendicular to our time-like coordinate.

5. Standard Model of Cosmology Separable into Two Independent Components

The large scale isotropy and homogeneity of the Universe - "The Cosmological Principle" implies that the total metric is the Robertson-Walker background metric plus perturbations. For a flat cosmology this is simply (expressed in conformal time):

$$g_{\mu\nu} = a(t)^2 \left(\eta_{\mu\nu} + h_{\mu\nu}\right)$$

where $\eta_{\mu\nu} = diag[1, -1, -1, -1]$ is the Minkowski metric and $a(t)$ is the scale factor for the Universe. (Here I am jumping to the conclusion that the Universe is flat in order to demonstrate how simple the metric is.)

For a Cosmological Principle matching = Robertson-Walker Universe, then cosmology is reduced to finding scale factor $a(t)$ as function of time, i.e. the expansion history of the Universe. (Here I leave out determining the curvature of Universe and by using only a flat model.) This is the ground state to which to apply perturbation theory.

The perturbations $h_{\mu\nu}$ are separable into three classes:

1. Scalar - energy (mass) density fluctuations
2. Vector - vorticity and curl - *usually neglected*
3. Tensor - gravity waves - *often neglected*

5.1. *Gaps in Standard Model*

The standard model of cosmology depends on experimentally unproven pillars which are also theoretically softly defined.

1. Inflation
2. Baryogenesis/synthesis - cf Darwin Chang talk
3. Dark Matter
4. Dark Energy

Inspite or despite this, it is still possible to fit observations to the standard model and determine parameters and consistency. The standard approach is to use the simplest possible model or parameterization for these effect and calculate the consequences.

6. A Theory Of Knowledge And Ignorance

6.1. *Morsels of Knowledge, Banquets of Ignorance*

"A state of thoroughly conscious ignorance is the prelude to every real advance of knowledge." –James Clerk Maxwell

The title of this section is particularly appropriate given The superb selection of restaurants and meals by our hosts. insert figure showing concepts of knowledge, ignorance, and non-knowledge

6.2. *A Theory of Ignorance*

Ignorance is: What you know you don't know. Ignorance is generally on the increase, as one learns more (gains knowledge) and realizes what isn't known.

Being physicists we could approximate this process in the form of 'Assume a spherical cow...'. Assume the topology of our knowledge is roughly spherical, then one would expect that our ignorance just surrounds our knowledge like the skin on a grapefruit or orange. Then our ignorance I is proportional to our knowledge K to the 4/3 power: $I \propto K^{4/3}$. The more we know, the more ignorant we are. If our knowledge is expanding at a rate $\dot{K}$, then our ignorance is increasing at a rate of $\dot{I} \propto K^{1/3}\dot{K}$. The faster we learn, the more rapidly we become ignorant.

Of course, the topology of ignorance and knowledge is often more complicated than spherical: separate areas of knowledge may merge erasing their boundary of ignorance. So we can have breakthroughs.

Don't know the range over which a theory is valid until one knows where it is invalid.

I call the region outside of knowledge and ignorance the non-knowledge in that we don't know enough about it to be ignorant. We do not even know to pose the questions.

7. New Physics

There are many clear and evident motivations for new physics. Part of our cosmology model is based upon new physics, e.g. Inflation, Baryogenesis,

Dark Matter and Dark Energy. But we also depend upon having the basic dynamics correct. In the case of cosmology that is dominated by gravity.

Gravity in form of General Relativity has failure points:

1. Incompatibility with quantum mechanics
2. Holographic Principle - non-local
3. Singularities - black holes, the Universe?
4. Issues of large scale - e.g. topology - Proty Wu talk.

Can these be categorized and systematized? And then tested?

7.1. *Cosmo-archeology of New Physics*

1) Model independent comparison to observations is our goal.

E.g., model all possible modes excited and compare these to observations. Here we take a generic approaches limited by basic symmetries.

2) Find abundant relics and inspect carefully is another approach. E.g. measure/observer carefully the CMB, BBNS, LSS, other relics.

3) Treatment of effect via modeling and comparison to observations. Use model fit to find parameters and goodness of fit.

All of these approaches are subject to systematic biases - theoretical and experimental. Nevertheless, they are potentially among the best approaches.

7.2. *Cosmological Signatures of New Physics*

In this conference we heard of many of the possible cosmological signatures of new physics:

1) Relics: Baryogenesis, Dark Matter, Dark Energy - Supersymmetric particles ? - Keith Olive, Darwin Chang

2) Spectrum of Primordial Perturbations - A new high-energy scale - Trans-Planckian physics - Robert Brandenberger

3) Non-standard accelerating expansion - Special Axion, modified GR

4) Primordial Magnetic Fields - Extra dimensions

5) Topological Defects - Noncommuting geometries

6) Neutrino Sector surprises - A new high-energy scale

7.2.1. *Ultra-High Energy Cosmic Rays and especially Neutrinos as probes of new physics effects*

There was special emphasis of the highest energy cosmic rays and neutrinos as potential probes of new physical effects.

(1) Substantial interest in using high-energy neutrinos to probe new physics

- High-energy cross-sections
- Top-down scenarios, e.g. defects, X-particles
- Neutrino oscillations now accepted as given n_τ

(2) GZK crisis pointing towards new physics?
- Lorentz invariance violation
- Extragalactic magnetic fields
- Z-bursts, neutralinos, ... - T. Weiler
- Or error in experiments?

(3) Still no accepted mode of UHECR acceleration
- Alternate top-down and rotating MM - QH Peng

7.3. *Neutrino Oscillations*

Another topic of high interest these days and in this conference is neutrino flavor oscillations.

(1) Important Effect in Astrophysics
- Katsuhiko Sato discussed effect of neutrino oscillations on supernova flux
- Bruce McKellar discussed impact and use in BBNS

(1) Neutrino oscillations are now assumed for much work
- George Wei-Shu Hou's mountain neutrino telescope only works effectively for oscillations to n_τ
- Jin-Jun Tseng and Tsung-Wen Yeh calculations of galactic and atmospheric flux calculations also

7.4. *Extra Dimensions / BraneWorlds*

Many Compelling Ideas have come from the effort to reconcile gravity and quantum mechanics. Extra dimensions and brane worlds are prime examples. Henry Tye presented a talk that Brane Inflation natural outcome. Tetsuya Shiromizu presented work on Relativisitic Braneworld Cosmology. These ideas lead to many more potential degrees of freedom - that is more parameters and effects to study and understand.

7.5. *Unification of Dark Side*

Dark Matter plus Dark Energy plus Einstein Equations yield ansatz of standard model.

This immediately begets questions:
- Why roughly equal now? (dark matter and energy)
- What is the Dark Energy?

- Cosmological Constant too big embarrassment
- Quintessence current stalking horse - Wolung Lee

Extra Dimensions could change equations and lead to a possible unification of the dark - Je-An Gu.

BraneWorld generation of dark energy and matter? - Henry Tye

8. A little Controversy

"Cosmology is becoming an Experimental Science" - P. Huang

My response is that cosmology is already an observational science that depends upon keen observations, processing (filtering), and analysis.

If we made universes, then we would be *experimenters*.

If we were University President: e.g. Wei-Jao Chen or Frank Shu, then we would institute new University requirements: (University Education = universal education)

For Ph.D. thesis one must make a new universe that produces:

1. A great new culinary dish for mankind
2. New classics in art, literature, and music
3. New medical discovery
4. New business and management practices
5. New technology and insight into physics

9. New Technology Enables New Science

Paul Richards talked about the development of new technologies and the new measurements that they make possible.

He also argued for wide support of new technology programs esp. from theorists

Diana Worrall showed great new results from analysis of Chandra and XMM obs, demonstrating the power of good data and the continuing development of multiwavelength astronomy. Discussion of X-ray jets in radio-loud A.G.

It is both new (1) devices and (2) techniques and algorithms that are critical new technology.

10. Closing Comments

Cosmology is doing spectacularly well.

We anticipate even more progress shortly.

We will get more confidence in our model.

We will be more desperately seeking (susy?) new physics. (This is likely to take a decade or more.)

Hopefully there will be some surprises (crisis = opportunity) in the upcoming results.

Acknowledgments

This work supported in part by the Director, Office of Science, DOE under DE-AC03-76SF00098.

I would like to acknowledge and thank the speakers and participants, the organizers and staff, especially those doing the work e.g students gathering the viewgraphs and collecting the proceedings, the supporting organizations - CosPA. And especially want to express our appreciation for the visit to National Palace Museum and special dinners provided by our hosts.

Program

May 31

08:30	Registration
• *Chair: Xiao-Gang He (NTU)*	
09:00 - 09:30	– Opening – Dr. Wei-Jao Chen (President of NTU) W-Y. Pauchy Hwang (PI of CosPA) Cosmology: An Experimental Science for the New Century
09:30 - 10:15	Francois Bouchet (Paris, France) Perspectives on Large Scale Structures and Microwave Fluctuations*
10:15 - 10:35	Break
• *Chair: Goerge Wei-Shu Hou (NTU)*	
10:35 - 11:20	Andrew Jaffe (Imperial, UK) Cosmology and Astrophysics with the CMB in 2002
11:20 - 12:05	Mark Birkinshaw (Bristol, UK) The Sunyaev-Zel'dovich Effect: Surveys and Science
12:05 - 13:30	Lunch
• *Chair: Guey-Lin Lin (NCTU)*	
13:30 - 14:15	Tzihong Chiueh (NTU, Taiwan) Observation of SZ effect with AMiBA*
14:15 - 15:00	Kin-Wang Ng (AS, Taiwan) Observation of CMB Anisotropies with AMiBA*
15:00 - 15:45	J. H. Proty Wu (NTU, Taiwan) If the Universe is Finite
15:45 - 16:05	Break
• *Chair: Pei-Ming Ho (NTU)*	
16:05 - 16:50	Robert Brandenberger (Brown, USA) Trans-Planckian Physics and Inflationary Cosmology
16:50 - 17:15	Wolung Lee (AS, Taiwan) CMB Constraints on Quintessence and its Cosmological Implications*
17:15 - 17:40	Je-An Gu (NTU, Taiwan) A Way to the Dark Side of the Universe through Extra Dimensions
18:00 - 20:30	Reception

June 1

09:00 - 13:00 Excursion to the National Palace Museum

13:00 - 14:00 Lunch (on site)

- *Chair: Win-Fun Kao (NCTU)*

14:00 - 14:45 Diana Worrall (Bristol, UK)
X-ray Jets in Radio-loud Active Galaxies

14:45 - 15:30 Thomas J. Weiler (Vanderbilt, USA)
Neutrino Astrophysics at 10^{20} eV

15:30 - 15:50 Break

- *Chair: Wing-Huen Ip (NCU)*

15:50 - 16:35 George Wei-Shu Hou (NTU, Taiwan)
New Window for Observing Cosmic Neutrinos at 10^{15} to 10^{18} eV

16:35 - 17:00 Jie-Jun Tseng (NCTU, Taiwan)
Comparison between High-Energy Galactic and Atmospheric Tau Neutrino Flux*

17:00 - 17:25 Tsung-Wen Yeh (NCTU, Taiwan)
PQCD Analysis for High Energy Atmospheric Tau Neutrinos*

17:25 - 18:10 Qiu-He Peng (Nanking, China)
Origin of Cosmic Ray with Extreme High Energy*

19:00 - 21:00 Banquet

June 2

- *Chair: Tzihong Chiueh (NTU)*

09:00 - 09:45 Bruce McKellar (Melbourne, Australia)
Neutrinos in Nucleosynthesis*

09:45 - 10:30 Katsuhko Sato (Tokyo, Japan)
Supernova Neutrino Burst and Neutrino Oscillation*

10:30 - 10:50 Break

- *Chair: Miao Li (ITP and NTU)*

10:50 - 11:35 Henry Tye (Cornell, USA)
Brane Inflation: from Superstring to Cosmic Strings*

11:35 - 12:20 Tetsuya Shiromizu (Tokyo Inst. of Tech., Japan)
Relativistic Brane World Cosmology

12:20 - 14:30 Lunch, IAC Meeting

- *Chair: Wai-Yee Keung (UIC)*

14:30 - 15:15 Keith Olive (Minnesota, USA)
Searching for Supersymmetric Dark Matter

15:15 - 16:00 Darwin Chang (NTHU, Taiwan)
Baryogenesis and Electric Dipole Moments in MSSM

16:00 - 16:20 Break

- *Chair: W-Y. Pauchy Hwang (NTU)*

16:20 - 17:05 Paul Richards (Berkeley, USA)
Detector Technologies for a New Generation of CMB Cosmology*

17:05 - 17:50 George Smoot (Berkeley, USA)
Our Age of Precision Cosmology

18:30 - 21:00 Dinner

* title changed by author(s) in proceedings

List of Participants

Mark Birkinshaw	University of Bristol	Mark.Birkinshaw@Bristol.ac.uk
Francois Bouchet	Paris, France	bouchet@iap.fr
Robert Brandenberger	Brown University	rhb@het.brown.edu
Chuan-Tsung Chan	NCTU	ctchan@phys.ntu.edu.tw
Darwin Chang	NTHU	chang@phys.nthu.edu.tw
Feng-Yin Chang	NCTU	fychang.py90g@nctu.edu.tw
Chiang-Mei Chen	NTU	cmchen@phys.ntu.edu.tw
Tai-Chung Cheng	NTU	burningsky@yahoo.com.tw
Tzi-Hong Chiueh	NTU	chiuehth@phys.ntu.edu.tw
Han-Kuei Fu	NCKU	fjl2238@ms12.hinet.net
Dilip Ghosh	NTU	dghosh@phys.ntu.edu.tw
Je-An Gu	NTU	jagu@phys.ntu.edu.tw
Xiao-Gang He	NTU	hexg@phys.ntu.edu.tw
Pei-Ming Ho	NTU	pmho@phys.ntu.edu.tw
Wei-Shu Hou	NTU	wshou@phys.ntu.edu.tw
Yu-Kuo Hsiao	NTU	hexg4@phys.ntu.edu.tw
Athar Husain	NCTS	athar@phys.cts.nthu.edu.tw
W-Y. Pauchy Hwang	NTU	wyhwang@phys.ntu.edu.tw
Wing-Huen Ip	NCU	wingip@astro.ncu.edu.tw
Andrew Jaffe	Imperial College	a.jaffe@ic.ac.uk
Hung-Yu Jian	NTU	d8922203@ms89.ntu.edu.tw
Hsien-Chung Kao	TKU	hckao@mail.tku.edu.tw
Win- Fun Kao	NCTU	wfgore@cc.nctu.edu.tw
Wai-Yee Keung	NCTS Taiwan, and UIC, Illinois, USA	keung@uic.edu
Sun-Kun King	ASIAA	skking@asiaa.sinica.edu.tw
Otto Kong	NCU	otto@phy.ncu.edu.tw
Fei-Fain Lee	NCTU	u1717117.py87g@nctu.edu.tw
Wolung Lee	AS	leewl@phys.sinica.edu.tw
I-Hui Li	Affiliated Senior High School of NTNU	tornado@asiaa.sinica.edu.tw
Miao Li	NTU	mli@phys.ntu.edu.tw
Kwang-Chang Lai	NTU	kclai@phys.ntu.edu.tw
Wei-Chien Lai	NCKU	janet@phys.ncku.edu.tw
Feng-Li Lin	TKU	linfl@mail.tku.edu.tw
Guey-Lin Lin	NCTU	glin@cc.nctu.edu.tw
Kai-Yang Lin	NTU	d90222006@ms90.ntu.edu.tw
Lihwai Lin	NTU	d90222005@ms90.ntu.edu.tw
Shih-Yuin Lin	AS	sylin@phys.sinica.edu.tw
Wen-Long Lin	NTU	wenlin@phy03.phy.ntnu.edu.tw
William W. Liou	Western Michigan University	william.liou@wmich.edu
Guo-Chin Liu	ASIAA	gcliu@asiaa.sinica.edu.tw

Bruce McKellar	University of Melbourne	mckellar@physics.unimelb.edu.au
Kin-Wang Ng	AS	nkw@phys.sinica.edu.tw
Keith Olive	University of Minnesota	olive@mnhep.hep.umn.edu
Qiu-He Peng	Nankin, China	chou@phy.ncu.edu.tw
Paul Richards	University of California	richards@physics.berkeley.edu
Katsuhko Sato	University of Tokyo	sato@utap.phys.s.u-tokyo.ac.jp
Jian-Qing Shi	NTU	hexg1@phys.ntu.edu.tw
Tetsuya Shiromizu	Tokyo Inst. of Tech.	shiromizu@th.phys.titech.ac.jp
George Smoot	Lawrence Berkeley National Laboratory	smoot@smoot1.lbl.gov
Chopin Soo	NCKU	cpsoo@mail.ncku.edu.tw
Shang-Yuu Tsai	NTU	wyhwango@phys.ntu.edu.tw
Jie-Jun Tseng	NCTU	geny.py86g@nctu.edu.tw
Henry Tye	Cornell University	tye@mail.lns.cornell.edu
Keiichi UMETSU	ASIAA	keiichi@asiaa.sinica.edu.tw
John Wang	NTU	hllywd2@phys.ntu.edu.tw
I-Chin Wang	NCKU	wzboson@yahoo.com.tw
Kaiti Wang	NTU	ktwang@phys.ntu.edu.tw
Tom Weiler	Vanderbilt University	tom.weiler@vanderbilt.edu
Diana Worrall	University of Bristol	D.Worrall@Bristol.ac.uk
J. H. Proty Wu	NTU	jhpw@phys.ntu.edu.tw
Ronin Wu	NTU	r90222031@ms90.ntu.edu.tw
Hyun Seok Yang	NTU	hsyang@phys.ntu.edu.tw
Chen-Pin Yeh	NTHU	g893301@phys.nthu.edu.tw
Mao-Chuang Yeh	NTU	b5202023@ms.cc.ntu.edu.tw
Tsung-Wen Yeh	NCTU	twyeh@cc.nctu.edu.tw

Notes:

AS:	Academia Sinica
ASIAA:	Academia Sinica Institute of Astronomy and Astrophysics
NCKU:	National Cheng Kung University
NCTS:	National Center for Theoretical Sciences
NCTU:	National Chiao Tung University
NCU:	National Central University
NTHU:	National Tsing Hua University
NTNU:	National Taiwan Normal University
NTU:	National Taiwan University
TKU:	Tamkang University

www.ingramcontent.com/pod-product-compliance
Lightning Source LLC
LaVergne TN
LVHW071223011026
849280LV00007B/39
9789812382849